Current Topics in Microbiology and Immunology

Volume 359

Current Topics in Microbiology and Immunology

Previously published volumes

Further volumes can be found at www.springer.com

Vol. 328: **van Etten, (Ed.)**:
Lesser Known Large dsDNA Viruses. 2008.
ISBN 978-3-540-68617-0

Vol. 329: **Griffin, Diane E.**;
Oldstone, Michael B. A. (Eds.): Measles. 2009.
ISBN 978-3-540-70522-2

Vol. 330: **Griffin, Diane E.**;
Oldstone, Michael B. A. (Eds.):
Measles. 2009.
ISBN 978-3-540-70616-8

Vol. 331: **Villiers, E. M. de (Ed.)**:
TT Viruses. 2009.
ISBN 978-3-540-70917-8

Vol. 332: **Karasev A. (Ed.)**:
Plant produced Microbial Vaccines. 2009.
ISBN 978-3-540-70857-5

Vol. 333: **Compans, Richard W.**;
Orenstein, Walter A. (Eds.):
Vaccines for Pandemic Influenza. 2009.
ISBN 978-3-540-92164-6

Vol. 334: **McGavern, Dorian**; **Dustin, Micheal (Eds.)**:
Visualizing Immunity. 2009.
ISBN 978-3-540-93862-0

Vol. 335: **Levine, Beth**; **Yoshimori, Tamotsu**;
Deretic, Vojo (Eds.):
Autophagy in Infection and Immunity. 2009.
ISBN 978-3-642-00301-1

Vol. 336: **Kielian, Tammy (Ed.)**:
Toll-like Receptors: Roles in Infection and Neuropathology. 2009.
ISBN 978-3-642-00548-0

Vol. 337: **Sasakawa, Chihiro (Ed.)**:
Molecular Mechanisms of Bacterial Infection via the Gut. 2009.
ISBN 978-3-642-01845-9

Vol. 338: **Rothman, Alan L. (Ed.)**:
Dengue Virus. 2009.
ISBN 978-3-642-02214-2

Vol. 339: **Spearman, Paul**; **Freed, Eric O. (Eds.)**:
HIV Interactions with Host Cell Proteins. 2009.
ISBN 978-3-642-02174-9

Vol. 340: **Saito, Takashi**; **Batista, Facundo D. (Eds.)**:
Immunological Synapse. 2010.
ISBN 978-3-642-03857-0

Vol. 341: **Bruserud, Øystein (Ed.)**:
The Chemokine System in Clinical and Experimental Hematology. 2010.
ISBN 978-3-642-12638-3

Vol. 342: **Arvin, Ann M. (Ed.)**:
Varicella-zoster Virus. 2010.
ISBN 978-3-642-12727-4

Vol. 343: **Johnson, John E. (Ed.)**:
Cell Entry by Non-Enveloped Viruses. 2010.
ISBN 978-3-642-13331-2

Vol. 344: **Dranoff, Glenn (Ed.)**:
Cancer Immunology and Immunotherapy. 2011.
ISBN 978-3-642-14135-5

Vol. 345: **Simon, M. Celeste (Ed.)**:
Diverse Effects of Hypoxia on Tumor Progression. 2010.
ISBN 978-3-642-13328-2

Vol. 346: **Christian Rommel**; **Bart Vanhaesebroeck**;
Peter K. Vogt (Ed.):
Phosphoinositide 3-kinase in Health and Disease. 2010.
ISBN 978-3-642-13662-7

Vol. 347: **Christian Rommel**; **Bart Vanhaesebroeck**;
Peter K. Vogt (Ed.):
Phosphoinositide 3-kinase in Health and Disease. 2010.
ISBN 978-3-642-14815-6

Vol. 348: **Lyubomir Vassilev**; **David Fry (Eds.)**:
Small-Molecule Inhibitors of Protein-Protein Interactions. 2011.
ISBN 978-3-642-17082-9

Vol. 349: **Michael Kartin (Eds.)**:
NF-kB in Health and Disease. 2011.
ISBN 978-3-642-16017-2

Vol. 350: **Rafi Ahmed**; **Tasuku Honjo (Eds.)**:
Negative Co-receptors and Ligands. 2011.
ISBN 978-3-642-19545-7

Vol. 351: **Marcel, B. M. Teunissen (Ed.)**:
Intradermal Immunization. 2011.
ISBN 978-3-642-23689-1

Vol. 352: **Rudolf Valenta**; **Robert L. Coffman (Eds.)**
Vaccines against Allergies. 2011.
ISBN 978-3-642-20054-0

Vol. 353: **Charles E. Samuel (Ed.)**:
Adenosine Deaminases Acting on RNA (ADARs) and A-to-I Editing. 2011.
ISBN 978-3-642-22800-1

Vol. 354: **Pamela A. Kozlowski (Ed.)**:
Mucosal Vaccines. 2012.
ISBN 978-3-642-23692-1

Vol. 355: **Ingo K. Mellinghoff**; **Charles L. Sawyers (Eds.)**:
Therapeutic Kinase Inhibitors, 2012.
ISBN 978-3-642-28295-9

Vol. 356: **Cornelis Murre (Ed.)**:
Epigenetic Regulation of Lymphocyte Development. 2012.
ISBN 978-3-642-24102-4

Vol. 357: **Nicholas J. Mantis (Ed.)**:
Ricin and Shiga Toxins. 2012.
ISBN 978-3-642-27469-5

Benhur Lee · Paul A. Rota
Editors

Henipavirus

Ecology, Molecular Virology, and Pathogenesis

Responsible series editor: Michael B. A. Oldstone

Springer

Editors
Benhur Lee
Department of Microbiology, Immunology and Molecular Genetics, Department of Pathology and Laboratory Medicine
David Geffen School of Medicine at UCLA
Los Angeles, CA
USA

Paul A. Rota
Centers for Disease Control
Atlanta, GA
USA

ISSN 0070-217X
ISBN 978-3-642-29818-9 ISBN 978-3-642-29819-6 (eBook)
DOI 10.1007/978-3-642-29819-6
Springer Heidelberg New York Dordrecht London

Library of Congress Control Number: 2012945482

Printed on acid-free paper

Springer is part of Springer Science+Business Media (www.springer.com)

Preface

Henipaviruses form a new genus of emerging paramyxoviruses that are the deadliest human pathogens within the *Paramyxoviridae* family. Nipah and Hendra virus, prototype members of this *Henipavirus* genus, are zoonotic pathogens present in alarmingly high frequencies in its natural reservoir, the asian fruit bat. Henipavirus infections can cause respiratory and encephalitic illness in humans with mortality rates that exceed 70 %. Discovered around the turn of the century, the spillover of henipaviruses into animal and human populations dramatically illustrate the threat posed by emerging zoonoses when human and environmental factors force the unintended overlap of otherwise naïve ecological niches. Horses and pigs serve as amplifying reservoirs for Hendra and Nipah virus, respectively, but direct bat-to-human and human-to-human transmission, at least for Nipah virus, has been increasingly documented. Although Nipah and Hendra virus spillover events are currently limited to Southeast Asia and Australia, respectively, serological and molecular evidence for henipavirus-related pathogens in bat species has been documented in places as distant as West Africa, and most recently, in Central and South America. Indeed, as this book goes to press, an international consortium of virologists and viral ecologists led by Christian Drosten reports the identification of 19 novel viral species within the Henipavirus genus, which can now be classified into at least 23 distinct viral clades.[1] The phylogenetic diversity of henipaviruses is thus vastly greater than what was thought previously. Henipaviruses also exhibit a much broader species tropism than other paramyxoviruses, largely due to the highly conserved protein receptors that these viruses use. In the various taxonomic schemes proposed for the transitional dynamics of zoonotic pathogens,[2] all these features justifiably place henipavirus at or close to the penultimate stage for sustained transmission in human outbreaks.

---

[1] Drexler JF et al (2012) Bats host major mammalian paramyxoviruses. *Nat Commun* 3:796.

[2] Wolfe ND et al (2007) Origins of major human infectious diseases. *Nature* 447:279–283; Lloyd-Smith JO et al (2009) Epidemic dynamics at the human–animal interface. *Science* 326:1362–1367.

This timely volume deals with the many facets of henipavirus biology, and covers our current understanding regarding the ecology, molecular virology, and pathogenesis of henipavirus infections. It is an international effort written by a multidisciplinary panel of experts at the front lines of research into this lethal emerging group of paramyxoviruses. The first section introduces the epidemiology and ecology of Nipah and Hendra viruses in their respective endemic areas, including a first-hand account of the discovery of Nipah virus during its initial outbreak in Malaysia; the next section documents the molecular virology of henipaviruses, and the substantial advances made towards understanding the unique features of henipavirus entry and tropism; and this is followed by accounts of the clinical and pathologic features of henipavirus infections in their human and naturally infected animal hosts. The next sections on pathogenesis provide a comprehensive reference on how henipaviruses counteract the innate immune system, and the relevant pathogenic features in animal challenge models developed to test potential therapeutic strategies. The final sections describe our current and future capabilities for diagnosis and control, including an account of potentially effective immunization strategies that are currently being tested. This book will not only serve as a useful reference for the henipavirus field; it will be useful to basic and animal virologists, ecologists, epidemiologists, physicians, and others interested in emerging infectious viral diseases, as it showcases the multidisciplinary efforts required to understand the genesis, spread and hopefully, control, of a lethal emerging zoonotic pathogen.

Benhur Lee
Paul A. Rota

Contents

Contributors

Hector Aguilar-Carreno Department of Veterinary Microbiology and Pathology, Paul G. Allen School of Global Animal Health, College of Veterinary Medicine, Washington State University, Bustad 423, 100 Grimes Way, P.O. Box 617010, Pullman, WA 99164-7010, USA, e-mail: haguilar@vetmed.wsu.edu

Christopher F. Basler Department of Microbiology, Mount Sinai School of Medicine, 1 Gustave L. Levy Place, Box 1124, New York, NY 10029, USA, e-mail: chris.basler@mssm.edu

Katharine N. Bossart Department of Microbiology, Boston University School of Medicine, Boston, MA 02118, USA; National Emerging Infectious Diseases Laboratories Institute, Boston University School of Medicine, Boston, MA 02118, USA, e-mail: kbossart@bu.edu

Dr. Christopher C. Broder Department of Microbiology and Immunology, Emerging Infectious Diseases Graduate Program, Uniformed Services University, 4301 Jones Bridge Rd., B4106, Bethesda, MD 20814-4799, USA, e-mail: cbroder@usuhs.mil

Dr. K. B. Chua Temasek Lifesciences Laboratory, National University of Singapore, 1 Research Link, Singapore 117604, Republic of Singapore, e-mail: chuakb@tll.org.sg

Gary Crameri Australian Animal Health Laboratory, CSIRO Livestock Industries, Geelong, VIC 3220, Australia

Peter Daniels Australian Animal Health Laboratory, CSIRO Livestock Industries, 5 Portarlington Road, PO Bag 24, Geelong, VIC 3220, Australia

Heinz Feldman Division of Intramural Research, Laboratory of Virology, Rocky Mountain Laboratories, National Institute of Allergy and Infectious Diseases, National Institutes of Health, 903 South 4th Street, Hamilton, MT 59840, USA, e-mail: feldmannh@niaid.nih.gov

Hume Field Department of Employment, Economic Development and Innovation, Biosecurity Queensland, Brisbane, QLD, Australia, e-mail: hume.field@qld.gov.au

Thomas W. Geisbert Department of Microbiology and Immunology, Galveston National Laboratory, University of Texas Medical Branch, 301 University Blvd., L29059, Galveston, TX 77550-0610, USA, e-mail: tom.geisbert@ utmb.edu

Emily S. Gurley Centre for Communicable Diseases ICDDRB, GPO Box 128, Dhaka 1000, Bangladesh, e-mail: egurley@iccddrb.org

Ronald M. Iorio Department of Microbiology and Physiological Systems, University of Massachusetts Medical School, Worcester, MA 01655, USA

Nina Yu-Hsin Kung Department of Employment, Economic Development and Innovation, Biosecurity Queensland, Brisbane, QLD, Australia

Benhur Lee Department of Microbiology, Immunology and Molecular Genetics, David Geffen School of Medicine at UCLA, BSRB 251a, 615 Charles E. Young Drive East, Los Angeles, CA 90095-1489, USA, e-mail: bleebhl@ucla.edu

Michael K. Lo Centers for Disease Control and Prevention, MS-C-22, 1600 Clifton Road, Atlanta, GA 30333, USA

Steve P. Luby Department of Medicine, Stanford University, The Jerry Yang and Akiko Yamazaki Environment and Energy Building 473 Via Ortega, Room 185, MC 4020, Stanford, CA 94305, USA; Centre for Communicable Diseases ICDDRB, GPO Box 128, Dhaka 1000, Bangladesh, e-mail: sluby@stanford.edu

Deborah Middleton Australian Animal Health Laboratory, CSIRO Livestock Industries, 5 Portarlington Road, PO Bag 24, Geelong, VIC 3220, Australia, e-mail: Deborah.Middleton@csiro.au

Dimitar B. Nikolov Structural Biology Program, Memorial Sloan-Kettering Cancer Center, New York, NY 10065, USA, e-mail: nikolovd@mskcc.org

Jackie Pallister Australian Animal Health Laboratory, CSIRO Livestock Industries, 5 Portarlington Road, Geelong, VIC 3220, Australia, e-mail: Jackie.Pallister@csiro.au

Olivier Pernet Department of Microbiology, Immunology and Molecular Genetics, David Geffen School of Medicine at UCLA, BSRB 251a, 615 Charles E. Young Drive East, Los Angeles, CA 90095-1489, USA

Paul A. Rota Centers for Disease Control and Prevention, MS-C-22, 1600 Clifton Road, Atlanta, GA 30333, USA, e-mail: prota@cdc.gov

Chong Tin Tan Faculty of Medicine, Department of Medicine, University of Malaya, 50603 Kuala Lumpur, Malaysia

Yao E. Wang Department of Microbiology, Immunology and Molecular Genetics, David Geffen School of Medicine at UCLA, BSRB 251a, 615 Charles E. Young Drive East, Los Angeles, CA 90095-1489, USA

Lin-Fa Wang Australian Animal Health Laboratory, CSIRO Livestock Industries, 5 Portarlington Road, PO Bag 24, Geelong, VIC 3220, Australia, e-mail: Linfa.wang@csiro.au

H. M. Weingartl National Centre for Foreign Animal Disease, Canadian Food Inspection Agency, Winnipeg, Canada; Department of Medical Microbiology, University of Manitoba, Winnipeg, Canada, e-mail: hana.weingartl@inspection.gc.ca

Kum Thong Wong Faculty of Medicine, Department of Pathology, University of Malaya, 50603 Kuala Lumpur, Malaysia, e-mail: wongkt@ummc.edu.my

K. Xu Structural Biology Program, Memorial Sloan-Kettering Cancer, New York, NY 10065, USA

Introduction: Nipah Virus—Discovery and Origin

Kaw Bing Chua

Abstract Until the Nipah outbreak in Malaysia in 1999, knowledge of human infections with the henipaviruses was limited to the small number of cases associated with the emergence of Hendra virus in Australia in 1994. The Nipah outbreak in Malaysia alerted the global public health community to the severe pathogenic potential and widespread distribution of these unique paramyxoviruses. This chapter briefly describes the initial discovery of Nipah virus and the challenges encountered during the initial identification and characterisation of the aetiological agent responsible for the outbreak of febrile encephalitis. The initial attempts to isolate Nipah virus from the bat reservoir host are also described.

Contents

K. B. Chua (✉)
Temasek Lifesciences Laboratory, National University of Singapore,
1 Research Link, Singapore 117604, Singapore
e-mail: chuakb@tll.org.sg

Current Topics in Microbiology and Immunology (2012) 359: 1–9

DOI: 10.1007/82_2012_218

Published Online: 11 July 2012

1 Introduction

A major outbreak of febrile encephalitis affecting pigs and humans in peninsular Malaysia in 1998/1999 resulted in the death of 105 humans and the eventual culling of about 1.1 million pigs (Chua 2003a). The outbreak affecting pig farm workers was officially noted in September 1998 in a suburb of Ipoh, the capital city of Perak state, situated in the north-western part of peninsular Malaysia. From this epicentre, the outbreak subsequently spread to other pig farms within the state of Perak, and by December 1998, southward to pig farms in Sikamat which is situated in the state of Negeri Sembilan (Chua 2003a; Chua et al. 1999; CDC 1999; Mohd Nor et al. 2000; Chua et al. 2000). By February 1999, the outbreak reached the country's most concentrated pig-farming region, within the Bukit Pelanduk district inclusive of Sungai Nipah village in the state of Negeri Sembilan (Chua 2003a; Chua et al. 1999; CDC 1999; Mohd Nor et al. 2000). By March 1999, the outbreak was noted in Sepang and subsequently Sungai Buloh within the Selangor state (Chua 2003a; Chua et al. 1999, 2000; CDC 1999; Mohd Nor et al. 2000). The outbreak had also spread into Singapore through importation of live pigs from Malaysia (Paton et al. 1999; Chua 2010). The discovery of Nipah virus (NiV) as the aetiological agent of the outbreak in mid-March 1999 was pivotal and a significant turning point in instituting the effective control measures. This introductory chapter describes the sequence of events leading to the discovery of NiV and the first isolation of the virus from its reservoir host. It illustrates the importance of timely discovery of the aetiological agent responsible for the outbreak for institution of effective control measures, and also highlights challenges encountered in tracking the origin of NiV from its natural reservoir hosts.

2 Discovery of Nipah Virus

Malaysia is situated in Southeast Asia and has a tropical climate. It is separated into two geographical regions, peninsular Malaysia and Malaysian Borneo (Sarawak and Sabah). Peninsular Malaysia is bordered by Thailand in the north and Singapore in the south. Japanese Encephalitis (JE) is an important form of viral encephalitis in Asia which is also endemic in Malaysia causing sporadic outbreaks (Vaugh and Hoke 1992; Cardosa et al. 1995). Data from the Ministry of Health showed that the total number of reported JE cases from 1989 to 1998 was 529, ranging from 9 to 91 cases per year with an annual average of 53. The number of deaths recorded for the same period was 35, ranging from 6 to 10 per year with an average of 3.

2.1 Early Period and Suspicions of Japanese Encephalitis

As early as in 1997, clusters of an unusual deaths amongst pigs housed in some pigsties were noted by the pig farmers of two pig farms situated in Ampang district, suburb of Ipoh (personal communication). A similar event of unusual pig deaths recurred during mid-1998. Both these events were reported by the owners to veterinary officers of Veterinary Research Institute (VRI). The unusual swine mortality was initially assumed to be due to classical swine fever and no further investigation was subsequently undertaken. Oral history from the supervisor of one of the affected farms indicated that a private veterinarian working for one of the affected farms developed encephalitis and was treated in Ipoh General Hospital in 1997 (personal communication). This was substantiated by a later finding of six encephalitis patients who were treated at Ipoh General Hospital in 1997, and whose stored sera in the Institute for Medical Research (IMR) were found to have IgG to NiV (Ministry of Health Malaysia 2001).

In late September 1998, more cases of febrile encephalitis were noted to occur amongst pig farmers in the neighbouring districts. Based on the opinion of experts that the outbreak was due to JE virus, the control measures to eliminate the vector of JE virus which were already initiated by the Ministry of Health Malaysia were intensified (Chua 2004, 2010). Very intensive fogging with insecticides was undertaken in the outbreak areas and thousands of doses of JE vaccine were rushed into the country from Japan to vaccinate the farmers and those staying near the pig farms (Chua 2010). By end of 1998, serum from only four of the 28 patients tested positive for anti-JE IgM and postmortem brain tissues were sent to the WHO Collaborating Centre for Tropical Disease at the University of Nagasaki, Japan where no JE virus was isolated (Chua 2003a; Ministry of Health Malaysia 2001).

With continual devastating spread of the outbreak despite very intensive JE control measures, an urgent brainstorming session between the Ministries of Health and Agriculture convened in the Veterinary Research Institute, Ipoh on 24 February 1999. At the meeting, the unusual epidemiologic feature of the suspected JE outbreak in that mostly adults were affected was highlighted by government epidemiologists but authorities choose not to consider the possibility of etiologies other than JE (Anonymous 1999). At that time, there were some pig farmers who had already received two or even three doses of JE vaccine and still had succumbed to the illness. This phenomenon was interpreted as a failure of the inactivated JE vaccine and efforts were made to import live attenuated JE vaccine (SA14-14-2) from the Republic of Korea. At the meeting, a decision was reached to split future clinical samples into two portions for concurrent testing to be performed by the Department of Medical Microbiology, University of Malaya in addition to besides IMR.

2.2 Clinical Samples and Virus Isolation

At that time, the Department of Medical Microbiology was divided into three major units, Bacteriology, Virology, Mycology and Serology. The Virology Unit was physically and functionally further subdivided into Arbovirus (also served as the country's WHO Arbovirus Centre for Research and Reference), Hepatitis and

HIV, and the General Virology subunits. Samples were shared between the Arbovirus Unit to confirm the outbreak as JE and the General Virology Unit to attempt to isolate and characterise potentially novel strains of JEV. On 27 February 1999, the first patient with encephalitis from the outbreak area in Bukit Pelanduk was admitted to University of Malaya Medical Centre and clinical samples (serum and cerebrospinal fluid) from another five patients admitted to Seremban General Hospitals (SGH) with similar illness arrived at the Department of Medical Microbiology on 1 March 1999. Clinical samples were processed in the usual manner inside a biosafety class II cabinet and inoculated into a number of mammalian cell lines for virus isolation by the General Virology subunit.

2.3 Early Evidence for the Presence of a Novel Virus

On 5 March 1999, the cerebrospinal fluid sample of one of the patients admitted in SGH gave a syncytial type of cytopathic effect (CPE) in Vero cell culture, a cell-line derived from the kidney of African green monkey. The pattern of Vero cells dying was quite similar to the type of CPE seen in Hep2 cells after infection with respiratory syncytial virus. The CPE in Vero cells was immediately recognised as an unusual finding; however, several key individuals suggested that contamination was also a possibility. On 6 March 1999, two more original CSF samples from the five patients admitted to SGH showed the same pattern of CPE in Vero cells giving more support for the presence of a novel agent. On 7 March, extensive CPE was noted in the remaining Vero culture flasks that first showed the evidence of virus replication. The infected cells were carefully harvested and washed with sterile phosphate buffered saline by centrifugation. Aliquots of cell suspension were carefully transferred into wells of Teflon coated slides, air-dried within a biosafety class II cabinet. They were subsequently inactivated by UV-irradiation inside the cabinet and fixed in cold acetone to be used as antigens for identification of the virus using a panel of commercial monoclonal antibodies recognising common human viruses that were known to cause syncytial CPE in cultured cells. At the same time, an aliquot of clarified culture supernatant was carefully inoculated into a new flask of Vero cells in preparation for electron microscopy. Unfortunately, negative stained electron microscopy was not performed because of the danger of working with a potentially deadly virus, and because the only available electron microscope in the Faculty of Medicine was not functioning. The analysis would have had to be done at another institute located 2 km away. None of the panel of monoclonal antibodies for typing common human viruses gave any positive reactions in immunofluorescence. A couple of serum and CSF samples of patients from the outbreak areas were immediately processed for similar indirect immunofluorescent assay to verify that the samples contained antibodies that would bind the infected cells. With excitement and great anxiety, it was noted that the serum and CSF samples contained antibodies which reacted positively to the infected cells. Because the head of the department was concerned about the possibility of cross-reactive antibodies, a double-blinded test was performed the following morning. By late morning of 8 March 1999, the test results showed 100% accuracy in differentiating serum samples of patients with encephalitis in this outbreak from those who were previously infected with measles virus, mumps virus, parainfluenza virus and human herpesviruses 1 and 2.

On 9 March 1999, an expert panel (consisting of doctors, veterinarians and scientists) held a meeting in IMR which was coordinated by the Ministry of Health of Malaysia. At that time, a lot of discussion was focused on the outbreak of JE because the findings from the Arbovirus Unit of Department of Medical Microbiology, University of Malaya had already confirmed 40% of the cases as JE by either or both JE serological (IgM) or molecular (RT-PCR) tests. The group was stunned when the possibility of the outbreak being caused by a novel virus was finally announced towards the end of the meeting. On 10 March, a decision was made to source international assistance for confirmation and identification of the potential novel virus. On 11th March, the infected Vero cells following inactivation and 3 days of preparation were ready for examination under transmission electron microscope situated in the Institute of Higher Learning, University of Malaya. Unfortunately, because of poor maintenance of the scope, only some blurred images of potential envelope viral particles ranging in diameter between 100+ to 200+ nm and thickening of infected cell membrane were visible which was suggestive of paramyxovirus.

2.4 *Identification of a Novel Agent*

On 12 March 1999, materials consisting of inactivated Teflon coated antigen slides, ultra-thin sections of infected cells, virus isolates and some clinical samples were securely packed in accordance with biosafety requirement and transported initially to the Division of Vector Borne Diseases, Centers for Disease Control and Prevention (CDC), Fort Collins, Colorado, USA to test for novel arboviruses. In Fort Collins arbovirus was swiftly ruled out by indirect immunofluorescent assays using a panel of anti-sera against known arboviruses in all the three groups. The virus causing the outbreak was suggested to be a paramyxovirus based on the morphology of the virus particles and "herring-bone"-like structure of the viral nucleocapsids viewed as observed by electron microscopy. The electron micrograph of the virus was faxed to Malaysia on 15 March 1999 and that subsequently led to an immediate shift in the direction of control measures (Chua et al. 2007). Once the virus was suspected to be a paramyxovirus, all the materials were immediately transferred to the main campus of the CDC in Atlanta, Georgia, USA. Indirect immunofluorescence and sequence analysis rapidly confirmed the virus as being a novel paramyxovirus which was closely related to Hendra virus (HeV), another newly discovered paramyxovirus in Australia in 1994, by a team of scientists in CDC, Atlanta, USA (Ksiazek et al. 2011). The virus was called NiV, named after the residential village, Kampung Sungei Nipah, of the patient whom the virus was first isolated (Ksiazek et al. 2011).

3 Origin of Nipah Virus

As soon as the outbreak in pigs and humans came under control, surveillance of wildlife species for evidence of the origin of NiV was undertaken as an integral part of outbreak investigation. Knowledge of the similarities between NiV and Hendra virus was crucial in planning a more effective wildlife surveillance strategy for NiV. In contrast to the initial "wide-net" approach in the HeV wildlife investigations in Australia, targeted wildlife investigation with a priority focus on bats was undertaken. A team headed by Dr Hume Field from Queensland Department of Primary Industries and Dr Johara M Yob from VRI targeted the field investigation on wild-caught bats in the initial outbreak areas and fruit bats of the *Pteropid* species found in other parts of peninsular Malaysia. Blood and tissue samples from 324 bats from 14 species were collected and submitted for serological examination and virus isolation. Neutralising antibodies to NiV were found in 21 bats from five species with fruit bats of *Pteropid* species found to have the highest seroprevalence rate. However, no virus was isolated and all attempts to amplify NiV RNA on collected tissues were also negative (Mohd Yob et al. 2001).

3.1 Switching Focus From Laboratory to Field

Knowing that the diagnostic laboratory testing for patients and contacts was well taken care of by the CDC laboratory team, attention was redirected to the origin of NiV as soon as the work on laboratory confirmation of the presence of NiV in the respiratory secretions and urine of infected patients for strengthening precautionary measures to be adopted by health-care provider was completed. A number of field trips were made to the initial outbreak areas in the suburb of Ipoh. A few days were spent learning the techniques of handling bats from Hume Field and his team but most of the days were spent interacting with local pig farmers and hunters gathering information about any unusual situation that had happened in the previous 5 years prior to the outbreak and studying the surrounding environment of the outbreak pig farms.

3.2 New Approach for Sample Collection

A decision was made to undertake individual field investigation for isolation of NiV as soon as it was known that no virus was isolated from blood and tissues of bats collected by Field and his team. The conventional method of collecting clinical samples by trapping using mist net for smaller fruit bats and shooting for *Pteropid* species would not be possible due to lack of support and the work needed to be carried out single-handedly in most occasions. Based on the finding that NiV was isolated from human respiratory secretions and urine, similar types of specimens from bats were targeted. Thus, to overcome the challenge, an innovative unconventional approach was adopted to collect bat specimens using a plastic sheets placed under their roost without having to catch or kill them (Chua 2003b). In view of the fact that bats roost upside down, a hypothesis was made that bats should urinate and defecate the moment they were back to roost. The hypothesis was supported by the fact that bat guano accumulated in caves below bat roosts.

A preliminary work to collect bat urine on plastic sheets was carried out in mid-1999 from a colony of *Eonycteris spelaea* that was roosting in Gua Tempurong cave, situated about 25 km from NiV initial outbreak area. The work not only confirmed the hypothesis was sound, it also led to the isolation of a novel bacterium (Chua et al. 2005). On 17 August 1999, a field trip was made to collect urine samples from a colony of *Pteropus hypomelanus* roosted in Tioman Island situated along the eastern coast of peninsular Malaysia. Single-handedly, 50 pooled urine samples were collected and another novel paramyxovirus, Tioman virus, was isolated (Chua et al. 2001). On 23 August 1999, another trip was made to the same island with the assistance of a medical laboratory technologist. A total of 275 pooled urine samples were collected from the same colony of fruit bats but no virus was isolated.

On 11 June 2000, a third trip was made to the same colony of *P. hypomelanus* roosting in the same island. On arrival it was noted the colony of fruit bats had changed their usual roosting site consisting of five adjoining tall trees and three coconut palms. In the previous roosting site, the colony of bats appeared to have a social structural pattern of roosting with lactating females and young pups roosting in the canopy of coconut palms. Fully mature males and females (mainly in pairs) roost in the highest parts and the juveniles among the lower parts of tall trees. According to the residents of the islands, the colony of fruit bats had shifted about 10 days prior to the third visit due to the attack of sea eagles which preyed on them as the canopy of the roosting trees were damaged and therefore exposed them to being attacked. At that visit, most of the bats were noted to have crowded into a single nearby huge "sea-almond" tree and roost without the usual social structure of roosting. The whole colony of bats appeared to be in disharmony with odd behavioural observed. The collection of urine was made easier by their crowding, and 263 pooled urine samples collected. During the night, visits were also made to their feeding ground and more than 27 swabs of their freshly landed partially eaten fruits were collected for virus isolation. NiV was isolated from two pooled urine samples and a swab sample of partially eaten fruit (Chua et al. 2002). In the third trip, Tioman virus virus was re-isolated from five different pooled urine samples and a novel bat orthoreovirus was also isolated from 12 separate polled urine samples (Pritchard et al. 2006).

4 Challenges Encountered

There was not much of a challenge, except for biosafety concern, in the isolation of NiV from clinical materials as the virus replicates easily in a number of mammalian cell lines. In the first isolation of NiV from a human victim during the outbreak, the challenge lied in applying the "tactful art" to avoid conflict and overcome preconceived notions of the authority about the identity of the virus so as to gain access to clinical samples for carrying out the virus isolation work. However, for the first isolation of NiV from its natural host, the challenge lay in the ability to adopt unconventional scientific approach to collect clinical samples for virus isolation. The result of this unconventional approach has also shed light on the likely modes of transmission of NiV from its natural reservoir hosts to domestic pig population and also possibly into horses that were housed in the polo club within the initial outbreak area (Chua et al. 2002).

References

Anonymous (1999) JE claimed more adult victims than children, say don. *The Star*, February 25, Nation 10
Cardosa MJ, Hooi TP, Kaur P (1995) Japanese encephalitis virus is an important cause of encephalitis among children in Penang. Southeast Asia J Trop Med Public Health 26:272–275
CDC (1999) Outbreak of Hendra-like virus—Malaysia and Singapore, 1998–1999. Morb Mortal Wkly Rep (MMWR) 48:265–269
Chua KB (2003a) Outbreak of Nipah virus in Malaysia. J Clin Virol 26:265–275
Chua KB (2003b) A novel approach of collecting samples from fruit bats for isolation of infectious agents. Microbes Infect 5:487–490
Chua KB (2004) The discovery of Nipah virus: a person account. Neurol Asia 9:59–63
Chua KB (2010) Epidemiology, surveillance and control of Nipah virus infections in Malaysia. Malaysian J Pathol 32(2):69–73
Chua KB, Goh KJ, Wong KT, Kamarulzaman A, Tan PSK, Ksiazek TG, Paul G, Lam SK, Tan CT (1999) Fatal encephalitis due to a new paramyxovirus among pig-farmers in Malaysia. Lancet 354:1257–1259
Chua KB, Bellini WJ, Rota PA, Harcourt BH, Lam SK, Ksiazek TG, Rollin P, Zaki SR, Goldsmith CS, Roehig JT, Eaton B, Gould AR, Olson J, Field H, Daniel P, Ling AE, Anderson LJ, Mahy BW (2000) Nipah virus: a newly emergent deadly paramyxovirus. Science 288:1432–1435
Chua KB, Wang LF, Lam SK, Crameri G, Yu M, Wise T, Boyle D, Hyatt AD, Eaton BT (2001) Tioman virus, a novel paramyxovirus isolated from fruit bats in Malaysia. Virology 283(2):215–229
Chua KB, Koh CL, Hooi PS, Wee KF, Khong JH, Chua BH, Chan YP, Lim ME, Lam SK (2002a) Isolation of Nipah virus from Malaysian island flying-foxes. Microbes Infect 4:145–151
Chua KB, Chua BH, Wang CW (2002b) Anthopogenic deforestation, El Nino and the emergence of Nipah virus in Malaysia. Malays J Pathol 24:15–21
Chua PKB, Corkill JE, Hooi PS, Cheng SC, Winstanley C, Hart CA (2005) Isolation of *Waddlia malaysiensis*, a novel intracellular bacterium, from fruit bat (*Eonycteris spelaea*). Emerg Infect Dis 11:271–277
Chua KB, Wong EMH, Cropp BC, Hyatt AD (2007) Role of electron microscopy in Nipah virus outbreak investigation and control. Med J Malaysia 62:139–142
Ksiazek TG, Rota PA, Rollin PE (2011) A review of Nipah and Hendra viruses with an historical aside. Virus Res 162(1–2):173–183
Ministry of Health Malaysia (2001) Cabinet report: JE/Nipah outbreak in Malaysia, vol 1, pp 1–133
Mohd Nor MN, Gan CH, Ong BL (2000) Nipah virus infection of pigs in peninsular Malaysia. Rev Sci Tech Off Int Epiz 19(1):160–165
Mohd Yob J, Field HE, Rashdi AM, Morrissy C, van der Heide B, Rota P, Adzhar A, White J, Daniels P, Jamaluddin A, Ksiazek TG (2001) Nipah virus infection in bats (order chiroptera) in peninsular Malaysia. Emerg Infect Dis 7:439–441

Paton NI, Leo YS, Zaki SR, Auchus AP, Lee KE, Ling AE, Chew SK, Ang B, Rollin PE, Umapathi T, Sng I, Lee CC, Lim E, Ksiazek TG (1999) Outbreak of Nipah virus infection among abattoir workers in Singapore. Lancet 354:1253–1257

Pritchard LI, Chua KB, Cummins D, Hyatt A, Crameri G, Eaton BT, Wang LF (2006) Pulau virus: a new member of the genus orthoreovirus isolated from fruit bats in Malaysia. Arch Virol 151:229–239

Vaugh DW, Hoke CH (1992) The epidemiology of Japanese encephalitis: prospect for prevention. Epidemiol Rev 14:197–221

Ecological Aspects of Hendra Virus

Hume Field, Gary Crameri, Nina Yu-Hsin Kung and Lin-Fa Wang

Abstract Hendra virus, a novel and fatally zoonotic member of the family *Paramyxoviridae*, was first described in Australia in 1994. Periodic spillover from its natural host (fruit bats) results in catastrophic disease in horses and occasionally the subsequent infection of humans. Prior to 2011, 14 equine incidents involving seven human cases (four fatal) were recorded. The year 2011 saw a dramatic departure from the sporadic incidents of the previous 16 years, with a cluster of 18 incidents in a single 3-month period. The fundamental difference in 2011 was the total number of incidents, the geographic clustering, and the expanded geographic range. The 2011 cluster more than doubled the total number of incidents previously reported, and poses the possibility of a new HeV infection paradigm. Epidemiologic evidence suggests that compelling additional host and/or environmental factors were at play.

Contents

H. Field (✉) · N. Y.-H. Kung
Biosecurity Queensland, Department of Employment,
Economic Development and Innovation, Brisbane, QLD, Australia
e-mail: hume.field@qld.gov.au

G. Crameri · L.-F. Wang
Australian Animal Health Laboratory, CSRIO Livestock Industries, Geelong, VIC, Australia
e-mail: Linfa.wang@csiro.au

Current Topics in Microbiology and Immunology (2012) 359: 11–23
DOI: 10.1007/82_2012_214

Published Online: 3 April 2012

1 Introduction

Hendra virus (HeV) was first identified at the stables of laconic racehorse trainer Vic Rail who a few years earlier had captured the imagination of the Australian public with his champion sprinter Vo Rogue, his unorthodox training techniques, and a larger than life personality. On September 23, 1994, the Brisbane suburb of Hendra (where Rail had his training stables) was transformed; by the end of the outbreak the trainer and 13 of his horses were dead, and one of his stable-hands was recovering in hospital. The syndrome in horses was typically characterized by severe respiratory signs and high mortality. Clinical signs included fever, facial swelling, severe respiratory distress, ataxia, and terminally, copious frothy blood-tinged nasal discharge. The trainer and stable hand, both of whom were directly involved in nursing the primary case Drama Series, became ill with a severe influenza-like illness within a week. The trainer was hospitalized and subsequently died after respiratory and renal failure (Selvey et al. 1995). HeV had emerged. But the causal agent was initially unknown, and the incident sparked a major disease outbreak investigation. Within a week, scientists in Queensland and at the CSIRO Australian Animal Health Laboratory in Geelong, Victoria had detected a novel virus, subsequently shown to be a previously undescribed virus of the family *Paramyxoviridae* (Eaton et al. 2007; Murray et al. 1995). The virus was initially named equine morbillivirus (EMV), but was later renamed HeV after the Brisbane suburb where the outbreak occurred (Murray et al. 1998; Wang et al. 1998). Hypotheses as to the origin of the virus included contaminated biological products, illegal performance-enhancing substances and malicious intent. When investigations failed to support any of these scenarios, consideration was given to the possibility that the virus had emerged from a wildlife reservoir. Within 18 months, neutralizing antibodies to HeV were discovered in fruit bats of the genus *Pteropus* (commonly known as flying-foxes, Fig. 1), heralding the beginning of an enduring integrated program of research into the ecology of HeV (Field et al. 2001; Halpin et al. 2000, 2011; Young et al. 1996). There is currently no effective prevention or treatment modality for HeV infection in horses or humans, although trials of a monoclonal antibody therapeutic for human use and of a subunit vaccine for horses are currently underway (Pallister et al. 2011a, b; Bossart et al. 2009, 2011; Promed 2011).

Fig. 1 Flying-foxes are bats of the genus *Pteropus* (Order Chiroptera) that forage nocturnally on fruit and floral resources, and roost diurnally in colonies sometimes numbering in tens or hundreds of thousands. Individuals may weigh over 1 Kg and have a wingspan greater than a meter

2 Hendra Virus Incidents 1994–2010

2.1 Horses and Humans

The outbreak of HeV in 1994 remains the largest single incident; subsequent incidents have involved smaller numbers of horses. While the virus is highly lethal, it has subsequently been shown to have low transmissibility, and it is believed that the large number of cases in the 1994 outbreak reflected inadvertent human-assisted transmission associated with the trainer's treatment of the sick and healthy horses. It is possible that without this single large event HeV may have remained unknown in Australia until more recent times. In the 10 years following the first HeV outbreak, a further four separate HeV incidents were recorded, all in the state of Queensland, at Mackay (1994), Cairns (1999, 2004), and Townsville (2004) (Fig. 2; Table 1). In the five years from 2006 to 2010, there was a marked increase in the frequency and number of incidents, with two incidents occurring in every year with the exception of the single incident in 2010 (Fig. 2; Table 1). These incidents were spread over a 1,500 km distance from Cairns in the far north of Queensland to Murwillumbah in the adjoining state of New South Wales. The incidents ranged from single horse events on rural properties to an outbreak in a veterinary practice involving five horses and two staff. There have been seven

Fig. 2 Locations and years of hendra virus outbreaks from 1994 to 2010

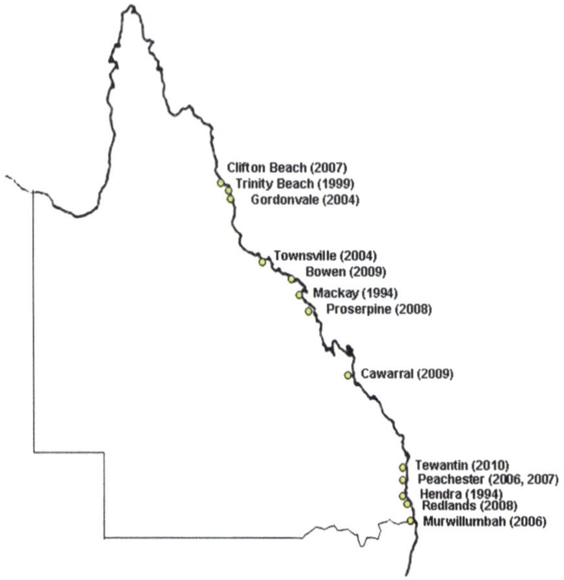

humans infected with the virus, four of whom have succumbed fatally. All human cases can be traced back to infected horses; there has been no flying-fox to human transmission, no evident transmission from flying-foxes to other species, and no human to human transmission.

2.2 Ecological Studies

Initial disease ecology studies sought to scope the 'history' of HeV infection in Australian flying-fox populations using serological surveillance. Field et al. (2001) found neutralizing anti-HeV antibodies in all four mainland flying-fox species (*Pteropus alecto*, *P. conspicillatus*, *P. poliocephalus*, *P. scapulatus*) across their geographic range (Fig. 3), and in archived samples that predated the first reported infection in horses. More recent disease ecology studies have focused on elaborating the spatial and temporal 'footprint' of virus excretion in flying-foxes, and on flying-fox-horse transmission scenarios (Field et al. 2001). HeV genome was detected by quantitative PCR in urine samples collected under roosting flying-foxes, and infectious virus isolated from a number of these samples. Although experimental studies had not demonstrated virus excretion in flying-fox urine and consequently not demonstrated experimental flying-fox to horse transmission (Williamson et al. 1998), the repeated isolation of virus from under flying-fox roosts suggests that the experimental model was flawed, and indicates that the screening of urine and other samples under roosting and foraging flying-foxes is a robust mechanism for monitoring natural infection and excretion. As part of a 3 year

Table 1 Hendra virus incidents 1994–2010

Date	Location	Horses cases (fatal cases)	Humans cases (fatal cases)
1994 August	Mackay, QLD	2(2)	1(1)
1994 September	Hendra, QLD	20(13)	2(1)
1999 January	Trinity Beach (Cairns), QLD	1(1)	0(0)
2004 October	Gordonvale (Cairns), QLD	1(1)	1(0)
2004 December	Townsville, QLD	1(1)	0(0)
2006 June	Peachester, QLD	1(1)	0(0)
2006 October	Murwillumbah, NSW	1(1)	0(0)
2007 June	Peachester, QLD	1(1)	0(0)
2007 July	Clifton Beach (Cairns), QLD	1(1)	0(0)
2008 July	Redlands, QLD	5(4)	2(1)
2008 July	Proserpine, QLD	3(2)	0/0
2009 July	Cawarral, QLD	4(3)	1(0)
2009 September	Bowen, QLD	2	0(0)
2010 May	Tewantin, QLD	1	0(0)
Total	14 events	45(31)	7(4)

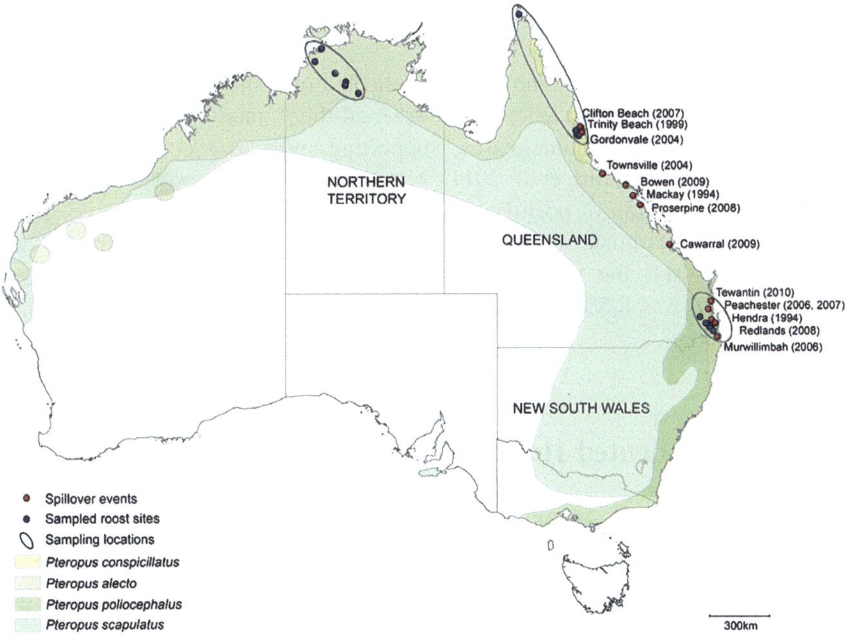

Fig. 3 The geographic range of the four mainland flying-fox species in Australia, and hendra virus antibody prevalence in the various Australian states

longitudinal study, pooled urine samples were collected from plastic sheets placed under flying-fox colonies in multiple locations in Queensland and in the Northern Territory (Field et al. 2011). Sampling locations were influenced by the location of

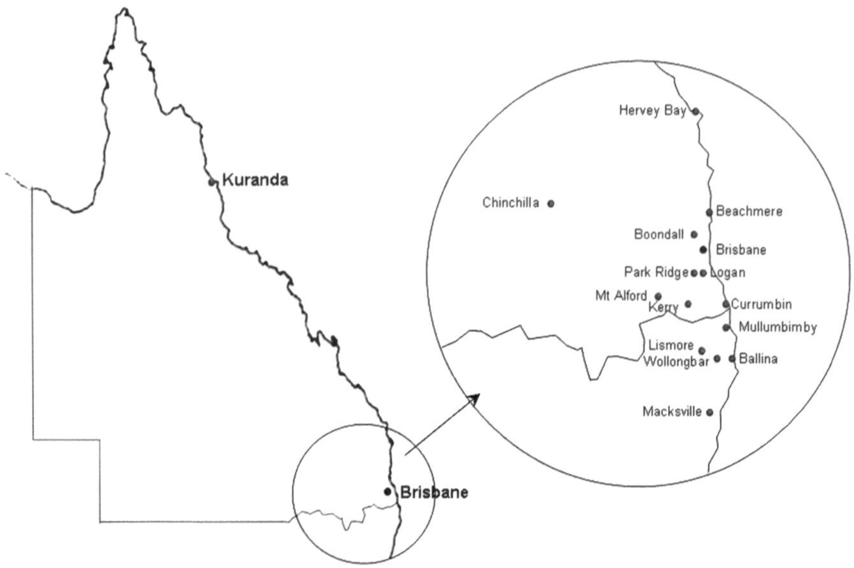

Fig. 4 Locations of hendra virus outbreaks in 2011

previous equine cases, and samples were collected on a monthly basis. Sampling locations were later expanded to include 'real-time' monitoring of flying-fox colonies proximate to equine cases, supporting molecular epidemiology and phylogenetic studies (Smith et al. 2011). Notwithstanding the pooled nature of the samples, the proportion of positive samples and the proportion of positive sheets provided a robust estimate of infection and excretion in the colony. In the 3 years prior to June 2011, the median proportion of positive samples from positive sampling events was 8.5%, in contrast to findings in the 3 months subsequent to June 2011 (below).

3 An Unprecedented Hendra Year 2011

3.1 More Incidents, More Locations

In 2011, in graphic contrast to the sporadic incidents and limited case numbers of the previous 17 years, a supercluster of 18 incidents (23 equine cases) occurred in a 12-week period in eastern Australia. Starting in late June, incidents were consecutively or simultaneously reported over a 1,800 km distance from Kuranda in far north Queensland, to Macksville in central New South Wales (Fig. 4; Table 2). Not only was the number of incidents unprecedented, but the location of the incidents extended the known southern occurrence of equine cases 300 km further south, and dramatically increased the number and proportion of incidents occurring in

Table 2 Hendra virus incidents in 2011

Date	Location	Horses cases (fatal cases)	Human cases (fatal cases)
June	Kerry, QLD	1(1)	0(0)
	Wollongbar, NSW	2(2)	0(0)
	Macksville, NSW	1(1)	0(0)
	Mt Alford, QLD	3(3)	0(0)
July	Park Ridge, QLD	1(1)	0(0)
	Kuranda, QLD	1(1)	0(0)
	Hervey Bay, QLD	1(1)	0(0)
	Boondall, QLD	1(1)	0(0)
	Lismore, NSW	1(1)	0(0)
	Logan, QLD	1(1)	0(0)
	Chinchilla, QLD	1(1)	0(0)
	Mullumbimby, NSW	1(1)	0(0)
August	Ballina, NSW	1(1)	0(0)
	South Ballina, NSW	2(2)	0(0)
	Mullumbimby, NSW	1(1)	0(0)
	Currumbin, QLD	1(1)	0(0)
	North Ballina, NSW	1(1)	0(0)
September	Beachmere, QLD	2(1)	0(0)
Total	18 events	23(22)	0(0)

New South Wales. In addition, one Queensland incident occurred 250 km inland and on the western side of the Great Dividing Range, extending the known inland occurrence of incidents by over 200 km. In seeking to understand why 2011 was so dramatically different from the previous years (during which 9 years had two reported incidents, 2 years had one reported incident, and 9 years had no reported incidents), it is useful to examine the historic data to establish how it was different. First, incidents have previously clustered seasonally, with 9/14 (64%) incidents prior to 2011 occurring from June to September (winter and early spring in Australia). Second, incidents have previous occurred near-simultaneously in disparate geographic locations [Mackay and Brisbane (1994); Peachester and Clifton Beach (2007); Redlands and Proserpine (2008); Cawarral and Bowen (2009)]. Third, equine infection has previously been reported in NSW (1/14 pre-2011 incidents occurred in Murwillumbah 2006).

3.2 What Was Different?

What was predominantly different in 2011 was the number of incidents, the geographic clustering in southeast QLD and northern NSW, and the unprecedented number and proportion of incidents (8/18) in NSW. We propose that multiple host and environment factors drove this situation. To elaborate further, all but one of the 18 incidents occurred within a 350 km radius of the first reported 2011 incident at Kerry, 75 km south of Brisbane. It is possible that the outlying Kuranda incident simply reflected the 'normal' background frequency of HeV incidents (1–2 per year), and that the drivers for the cluster of cases in south east Queensland and northern New South Wales, reflect relatively local factors. In addition, viral genome sequence from cases in horses from multiple incidents was phylogenetically consistent with that seen in previous years, strongly suggesting the 2011 cluster of incidents was not due to a new more virulent or transmissible strain of the virus, but due to nonviral factors. Finally, preliminary analysis of the pooled urine data from flying fox colonies proximate to equine cases in the 2011 cluster suggests that excretion levels were substantially higher than in the previous 3 years. The cause and relevance of this to the magnitude of the 2011 cluster is yet to be determined, as is the phylogenetic relatedness of viral genome sequence from horses and flying-foxes.

The 2011 cluster of incidents added another dimension to the history of HeV in Australia, with neutralizing antibodies to HeV detected in a dog associated with one of the equine incidents in south-east Queensland. The dog (a 2-year-old Kelpie) showed no evident clinical signs of disease, but serological testing demonstrated neutralizing anti-HeV antibodies, indicating exposure to HeV infection. The dog, from a property that had three equine cases, was subsequently euthanized. While earlier limited experimental studies had suggested that dogs could become subclinically infected, this was the first evidence of natural infection in a dog. No evidence of virus either by isolation or genetic testing was identified from the dog, precluding estimation of exposure, incubation or other epidemiological aspects. Although there is laboratory evidence of HeV infection of other species including pigs, cats, hamsters, and ferrets (Guillaume et al. 2009; Li et al. 2010; Williamson et al. 1998), and field evidence of Nipah virus infection of pigs, dogs, and horses (Chua et al. 2000), this is the first evidence of HeV infection of a species other than equine in the field. There has been no evidence of clinical signs in any other species on infected properties during current and previous incidents, and laboratory testing of 'in-contact' known susceptible species has not previously identified evidence of HeV infection.

The absence of human cases was a notable feature of the 2011 cluster. In the 14 reported equine incidents prior to 2011, there were 7 human cases, 4 with fatal outcomes; a horse trainer (1994), a farmer assisting a veterinarian perform a necropsy (1994), and two veterinarians (2008, 2009). Veterinarians and those assisting veterinarians represent five of the total seven human cases. In 2011, we would have mathematically expected nine human cases in the 18 equine incidents, assuming the same rate of infectivity and exposure. Genetic studies identify no change to the virus that might account for decreased infectivity, suggesting that altered exposure may be responsible for the absence of human cases. Notwithstanding the pathogenicity of HeV, horse to human transmission has historically been inefficient, with all human cases having direct and substantial contact with infectious body fluids of case horses. The absence of human cases in 2011 suggests that direct and substantial contact was not occurring, or not occurring with the same frequency as previously. Plausible explanations for this include (individually or combined) increased veterinarian and horse owner awareness of HeV (fostered by a sustained campaign by animal and public health authorities since 2008), early consideration of HeV as a differential diagnosis (which likely increased as the number of cases and associated publicity grew), the implementation of risk-minimizing protocols and appropriate personal protective equipment in the management of suspect cases, timely submission of appropriate samples to laboratories and rapid diagnostic turnaround, and effective quarantine and movement controls on case properties.

3.3 An Effective Vaccine?

An effective vaccine for horses offers reduced risk of infection in both horses and humans. Two vaccines for horses are currently being developed. The first is a canary pox virus-based recombinant vaccine that carries the glycoprotein (G) gene of Nipah virus. Previous trials showed that this vaccine construct generates high levels of neutralizing antibody in pigs, and delivers protection against Nipah virus disease without shedding of infectious virus (Weingartl et al. 2006). An HeV version of this vaccine is currently undergoing trials to establish efficacy in generating neutralizing antibodies; it has yet to be trialled for protection efficacy in horses. The second vaccine is based on recombinantly expressed soluble versions of the HeV G glycoprotein (sG) (Pallister et al. 2011b), formulated with adjuvant as an inactivated vaccine. Various forms of this preparation have been used under laboratory conditions to induce neutralizing antibody in cats and ferrets, and to protect from disease following challenge by Hendra and Nipah viruses (McEachern et al. 2008; Mungall et al. 2006). The sG Hendra vaccine has now been formulated with a proprietary adjuvant for use in trials in horses. The preliminary data show seroconversion of vaccinated horses, prevention of disease following exposure to an otherwise lethal HeV challenge, elimination of viral shedding, and prevention of viral replication in tissues (D. Middleton and L.-F.Wang, unpublished results). Trials into longevity of antibody level and protection are currently underway with the target date for registration for use by the horse-owning community in 2013.

4 Fruit Bats and Hendra Virus

An insight into the natural history and ecology of HeV infection in flying-foxes is useful at this point. Serologic evidence of infection occurs in all four species, and across their geographic range in Australia (Field et al. 2001). Infection is readily detected in the urine of roosting flying-foxes, is not evident in colonies continuously, but has been detected in all months of the years across colonies (Field et al. 2001). PCR-positive urine samples and flying-fox food debris have been identified beneath flowering/fruiting trees under which horses rest, suggesting a plausible transmission pathway (C. Smith, C. de Jong, N. Kung, and H. Field, unpublished results). There is some evidence for a positive association between pregnancy and HeV status (Breed et al. 2011; Plowright et al. 2008), between ecological stress and HeV status (Plowright et al. 2008), between incident location and proximity to flying fox roost (C. Smith, C. de Jong, N. Kung, and H. Field, unpublished results), and between lower rainfall periods and HeV incidents (C. Smith et al., unpublished results).

Thus, there is support for a complex causality for HeV spillover from flying-foxes to horses. That equine incidents are seasonally clustered when virus excretion is detected year-round supports this hypothesis; although the previous occurrence of incidents outside this June to September period (October to December, January and May) underlines that spillover can occur at any time, it is more likely to occur at sometimes than others because the presence of additional factors make transmission (excretion/exposure/infection) more likely or more efficient. In epidemiologic terms, this scenario is explained by the concept of 'necessary and sufficient cause'; the virus is a necessary component (infection cannot occur without it), but the likelihood of infection is moderated by the presence or absence of multiple host and environmental factors, sometimes in complex interaction, and providing 'sufficient' cause for spillover to occur. Thus, in seeking to explain the events of 2011, we have hypothesized host factors such as physiological and immunological status, and environmental factors such as food resource availability and climatic variables as plausibly supporting an increased level of infection in flying-foxes, as well as the extended excretion of virus by flying-foxes, the prolonged survival of virus in the environment and the increased likelihood of equine exposure and infection. From our monitoring of flying-fox colonies in south-east QLD and northern NSW during this cluster of incidents, we know that multiple colonies were infected throughout the region at the time, that the prevalence of virus excretion in these colonies was higher than the previous 3-year median, and that virus excretion continued for longer than previously observed (Field et al. 2011). Simplistically, the probability of flying-fox to horse transmission depends on the presence of infection in a proximate colony at that time, the proportion of susceptible flying-foxes, and the size of the colony size (i.e. the potential viral load), plus the number and density of horses, and the management of the horses (i.e. effective contact). During the epidemiological investigation of the 2011 HeV outbreaks, it was observed that additional factors such as pasture state and horse behavior may also play a role in the probability of flying-fox to horse transmission (N. Kung et al., unpublished results).

HeV has a number of epidemiological features in common with the related Nipah virus. Nipah virus, a co-member of the genus henipavirus, was first identified in Malaysia in 1999 as the cause of an outbreak of respiratory and/or neurological disease in intensively farmed pigs and associated pig workers (Chua et al. 2000). As with HeV in Australia, species of flying fox were identified as the natural host, with Nipah virus spilling from bats to pigs, and from pigs to people (versus bats-horses-people with HeV) (Field et al. 2007; Yob et al. 2001).

Since 2001, Nipah virus has been identified as causing clusters of typically fatal human disease in India and Bangladesh (Epstein et al. 2006). In contrast to HeV in Australia, human cases in Bangladesh are not typically associated with an intermediate livestock host, but rather with indirect bat-human transmission (Gurley et al. 2007). While the ecology of Hendra and Nipah virus infection remains to be fully elaborated, the multiple host and environmental factors discussed above in relation to HeV are likely equally likely to drive Nipah virus infection and spillover dynamics. Both viruses exhibit marked spatial and temporal ‘patterns’, reinforcing the likely role of physiological, biological, ecological, and anthropogenic drivers. Nipah virus is discussed in detail in the companion chapters of this book.

5 Epilogue

The 2011 cluster of incidents has more than doubled the total number of incidents previously reported, and poses the possibility of a new HeV infection paradigm. This new paradigm does not relate to the horse-to-horse or horse-to-human infectivity of the virus (which thankfully remains low), but to the occurrence and frequency of spillover events, previously regarded as sporadic. While it is probable that the increased frequency of identification of equine hendra cases since 2004 may reflect (in part) increased awareness of horse owners and veterinarians, the epidemiologic evidence suggests that compelling additional host and/or environmental factors were at play in 2011. These factors remain unclear, and are the focus of an accelerated research program funded by the state governments of Queensland and New South Wales and the Australian federal government. Time will tell whether the supercluster of 2011 was an anomaly, or if it indeed heralded a paradigm-shift.

References

Bossart KN, Geisbert TW, Feldmann H, Zhu Z, Feldmann F, Geisbert JB, Yan L, Feng YR, Brining D, Scott D, Wang Y, Dimitrov AS, Callison J, Chan YP, Hickey AC, Dimitrov DS, Broder CC, Rockx B (2011) A neutralizing human monoclonal antibody protects african green monkeys from hendra virus challenge. Sci Transl Med 5:e1000642

Bossart KN, Geisbert TW, Feldmann H, Zhu Z, Feldmann F, Geisbert JB, Yan L, Feng YR, Brining D, Scott D, Wang Y, Dimitrov AS, Callison J, Chan YP, Hickey AC, Dimitrov DS, Broder CC, Rockx B (2011) A neutralizing human monoclonal antibody protects african green monkeys from hendra virus challenge. Sci Transl Med 3:105–103

Breed AC, Breed MF, Meers J, Field HE (2011) Evidence of endemic hendra virus infection in flying-foxes (*Pteropus conspicillatus*)—implications for disease risk management. PLoS ONE 6:e28816

Chua KB, Bellini WJ, Rota PA, Harcourt BH, Tamin A, Lam SK, Ksiazek TG, Rollin PE, Zaki SR, Shieh W, Goldsmith CS, Gubler DJ, Roehrig JT, Eaton B, Gould AR, Olson J, Field H, Daniels P, Ling AE, Peters CJ, Anderson LJ, Mahy BW (2000) Nipah virus: a recently emergent deadly paramyxovirus. Science 288:1432–1435

Eaton BT, Mackenzie JS, Wang L-F (2007) Henipaviruses. In: Knipe DM, Griffin DE, Lamb RA et al. (eds) Fields Virology. Lippincott Williams & Wilkins, Philadelphia, pp 1587–1600
Epstein JH, Field HE, Luby S, Pulliam JR, Daszak P (2006) Nipah virus: impact, origins, and causes of emergence. Curr Infect Dis Rep 8:59–65
Field H, Young P, Yob JM, Mills J, Hall L, Mackenzie J (2001) The natural history of Hendra and Nipah viruses. Microbes Infect 3:307–314
Field HE, Mackenzie JS, Daszak P (2007) Henipaviruses: emerging paramyxoviruses associated with fruit bats. Curr Top Microbiol Immunol 315:133–159
Field H, de Jong C, Melville D, Smith C, Smith I, Broos A, Kung YH, McLaughlin A, Zeddeman A (2011) Hendra virus infection dynamics in Australian fruit bats. PLoS ONE 6:e28678
Guillaume V, Wong KT, Looi RY, Georges-Courbot MC, Barrot L, Buckland R, Wild TF, Horvat B (2009) Acute Hendra virus infection: Analysis of the pathogenesis and passive antibody protection in the hamster model. Virology 387:459–465
Gurley ES, Montgomery JM, Hossain MJ, Bell M, Azad AK, Islam MR, Molla MA, Carroll DS, Ksiazek TG, Rota PA, Lowe L, Comer JA, Rollin P, Czub M,Grolla A, Feldmann H, Luby SP, Woodward JL, Breiman RF (2007) Person-to-person transmission of Nipah virus in a Bangladeshi community. Emerg Infect Dis 13:1031–1037
Halpin K, Young PL, Field HE, Mackenzie JS (2000) Isolation of Hendra virus from pteropid bats: a natural reservoir of Hendra virus. J Gen Virol 81:1927–1932
Halpin K, Hyatt AD, Fogarty R, Middleton D, Bingham J, Epstein JH, Rahman SA, Hughes T, Smith C, Field HE, Daszak P, The H (2011) Pteropid bats are confirmed as the reservoir hosts of henipaviruses: a comprehensive experimental study of virus transmission. Am J Trop Med Hyg 85:946–951
Li M, Embury-Hyatt C, Weingartl HM (2010) Experimental inoculation study indicates swine as a potential host for Hendra virus. Vet Res 41:33
McEachern JA, Bingham J, Crameri G, Green DJ, Hancock TJ, Middleton D, Feng YR, Broder CC, Wang LF, Bossart KN (2008) A recombinant subunit vaccine formulation protects against lethal Nipah virus challenge in cats. Vaccine 26:3842–3852
Mungall BA, Middleton D, Crameri G, Bingham J, Halpin K, Russell G, Green D, McEachern J, Pritchard LI, Eaton BT, Wang LF, Bossart KN, Broder CC (2006) Feline model of acute Nipah virus infection and protection with a soluble glycoprotein-based subunit vaccine. J Virol 80:12293–12302
Murray K, Selleck P, Hooper P, Hyatt A, Gould A, Gleeson L, Westbury H, Hiley L, Selvey L, Rodwell B, et al. (1995) A morbillivirus that caused fatal disease in horses and humans. Science 268:94–97
Murray K, Eaton B, Hooper P, Wang L, Williamson M, Young P (1998) Flying foxes, horses, and humans: a zoonosis caused be a new member of the *Paramyxoviridae*. In: Scheld WM, Armstrong D, Hughes JM (eds) Emerging Infections. ASM Press, Washington, pp 43–58
Pallister J, Middleton D, Broder CC, Wang L-F (2011a) Henipavirus vaccine development. J Bioterrorism and Biodefense:S1:005
Pallister J, Middleton D, Wang LF, Klein R, Haining J, Robinson R, Yamada M, White J, Payne J, Feng YR, Chan YP, Broder CC (2011b) A recombinant Hendra virus G glycoprotein-based subunit vaccine protects ferrets from lethal Hendra virus challenge. Vaccine 29:5623–5630
Plowright RK, Field HE, Smith C, Divljan A, Palmer C, Tabor G, Daszak P, Foley JE (2008) Reproduction and nutritional stress are risk factors for Hendra virus infection in little red flying foxes (Pteropus scapulatus). Proc Biol Sci 275:861–869
Promed (2011) 20110525.1589 (2011) Hendra virus—Australia. Vaccine 06
Selvey LA, Wells RM, McCormack JG, Ansford AJ, Murray K, Rogers RJ, Lavercombe PS, Selleck P, Sheridan JW (1995) Infection of humans and horses by a newly described morbillivirus [see comments]. Med J Aust 162:642–645
Smith I, Broos A, de Jong C, Zeddeman A, Smith C, Smith G, Moore F, Barr J, Crameri G, Marsh G, Tachedjian M, Yu M, Kung YH, Wang LF, Field H (2011) Identifying Hendra virus diversity in pteropid bats. PLoS ONE 6:e25275

Wang LF, Michalski WP, Yu M, Pritchard LI, Crameri G, Shiell B, Eaton BT (1998) A novel P/V/C gene in a new member of the Paramyxoviridae family, which causes lethal infection in humans, horses, and other animals. J Virol 72:1482–1490

Weingartl HM, Berhane Y, Caswell JL, Loosmore S, Audonnet JC, Roth JA, Czub M (2006) Recombinant Nipah virus vaccines protect pigs against challenge. J Virol 80:7929–7938

Williamson MM, Hooper PT, Selleck PW, Gleeson LJ, Daniels PW, Westbury HA, Murray PK (1998) Transmission studies of Hendra virus (equine morbillivirus) in fruit bats, horses and cats. Aust Vet J 76:813–818

Yob JM, Field H, Rashdi AM, Morrissy C, van der Heide B, Rota P, bin Adzhar A, White J, Daniels P, Jamaluddin A, Ksiazek T (2001) Nipah virus infection in bats (order Chiroptera) in peninsular Malaysia. Emerg Infect Dis 7:439–441

Young PL, Halpin K, Selleck PW, Field H, Gravel JL, Kelly MA, Mackenzie JS (1996) Serologic evidence for the presence in Pteropus bats of a paramyxovirus related to equine morbillivirus. Emerg Infect Dis 2:239–240

Epidemiology of Henipavirus Disease in Humans

Stephen P. Luby and Emily S. Gurley

Abstract All seven recognized human cases of Hendra virus (HeV) infection have occurred in Queensland, Australia. Recognized human infections have all resulted from a HeV infected horse that was unusually efficient in transmitting the virus and a person with a high exposure to infectious secretions. In the large outbreak in Malaysia where Nipah virus (NiV) was first identified, most human infections resulted from close contact with NiV infected pigs. Outbreak investigations in Bangladesh have identified drinking raw date palm sap as the most common pathway of NiV transmission from *Pteropus* bats to people, but person-to-person transmission of NiV has been repeatedly identified in Bangladesh and India. Although henipaviruses are not easily transmitted to people, these newly recognized, high mortality agents warrant continued scientific attention.

Contents

S. P. Luby (✉)
Department of Medicine, Stanford University, The Jerry Yang and Akiko Yamazaki Environment and Energy Building 473 Via Ortega, Room 185, MC 4020, Stanford, CA 94305, USA
e-mail: sluby@stanford.edu

S. P. Luby · E. S. Gurley
Centre for Communicable Diseases icddr,b, GPO Box 128, Dhaka 1000, Bangladesh
e-mail: egurley@icddrb.org

Current Topics in Microbiology and Immunology (2012) 359: 25–40
DOI: 10.1007/82_2012_207

Published Online: 1 July 2012

1 Hendra Outbreaks

Hendra virus (HeV), previously referred to as equine morbillivirus, was first identified in an outbreak in September 1994 in Hendra, a suburb of Brisbane, Queensland Australia in September 1994 (Selvey and Sheridan 1995, Selvey et al. 1995). The first recognized infection occurred in a pregnant mare that was staying in an open paddock when noted to be ill. The mare was moved into a stable in Hendra and died within 2 days. Between 8 and 11 days after the mare's death 18 other horses residing in or near the stable became ill. Affected horses had depression, loss of appetite, fever, ataxia, tachycardia, tachypnea, dyspnea, and a copious frothy nasal discharge. Among 18 horses with clinical illness, 14 died, 12 horses from the Hendra stable, one horse staying in the paddock adjoining the stable, and one horse living on a neighboring property that had very close contact with horses from the Hendra stable. Autopsy findings from the horses were notable on gross pathology for heavy edematous lungs with hemorrhage and froth in the airway. Histopathological investigations identified interstitial pneumonia with focal necrotizing alveolitis and syncytial giant cells within the vascular endothelium (Murray et al. 1995).

Two employees at the stable, a 40-years-old male stable hand and a 49-years-old male horse trainer had particularly close contact with the index mare during the final stages of her fatal illness. The horse trainer, whose hands and arms had abrasions, attempted to force feed the mare by placing his bare hands with food into the sick mare's mouth. Both the stable hand and the horse trainer became ill in 5–6 days after the death of the mare with fever, myalgia, headaches, lethargy, and vertigo. The stable hand remained lethargic for several weeks but eventually recovered. The horse trainer developed progressive respiratory failure and died. His autopsy findings were consistent with interstitial pneumonia, with focal alveolitis and syncytial formation (Selvey et al. 1995). An identical virus, which was ultimately named Hendra virus (HeV) was grown from samples from both the affected horses and the people (Murray et al. 1995).

Since its identification and the first two recognized human infections, five additional human infections with HeV, have been recognized all in Queensland, Australia. The third person with recognized Hendra infection was a 35–years-old male who lived in a horse stud farm (O'Sullivan et al. 1997). He had cared for two sick horses, one with acute respiratory distress and the other with a rapid onset of neurological symptoms. Both horses died. He assisted a veterinary surgeon during the necropsy of the two horses. Throughout caring for the horses and the necropsy, the assistant never wore gloves, mask, or protective eyewear. A few days after assisting with the autopsies he became ill with headache, drowsiness and stiff neck, and sought medical attention. Results of his cerebrospinal fluid were negative for bacteria and viruses and he was provisionally diagnosed as partially treated meningitis and made a full recovery. Thirteen months later he presented with irritable mood and seizures. His illness progressed to dense right hemiplegia, depressed consciousness, and death. Brain specimens collected at autopsy stained positive with rabbit antiserum against HeV and PCR evaluation of a serum sample collected during his acute illness 13 months earlier amplified a 500 nucleotide sequence of the matrix gene of HeV.

McCormack and colleagues evaluated people who had contact with HeV infected humans and horses during these first two recognized outbreaks of HeV. They collected serum samples from 159 people who had contact with HeV infected human patients, 16 who participated in necropsies on HeV infected horses, 6 who had other close contact with HeV infected horses and 113 who had less intense contact with HeV infected horses (McCormack et al. 1999). None of the tested study subjects had neutralizing antibody to HeV.

The fourth recognized human infection with HeV occurred in a recent veterinary graduate who conducted a limited autopsy on a 10-years-old horse that died of a rapidly progressive respiratory illness with large amounts of blood stained frothy nasal secretions (Hanna et al. 2006). Although she initially wore gloves, she soon removed them, because they were not appropriately designed and had become contaminated inside. She did not use any other personal protective equipment. She reached deep into the carcass to examine internal organs and became heavily contaminated with the horse's body fluids. After completing the autopsy, the veterinarian returned home and showered. Seven days later she developed a dry cough, sore throat, fever, body aches, and fatigue. She recovered after 8 days. Serial serological samples demonstrated seroconversion of IgM and IgG antibodies against HeV. The two autopsy assistants and an adult family member who held the dying animal's head and were exposed to frothy bloody nasal secretions did not develop clinical illness and were seronegative for HeV infection (Hanna et al. 2006).

The fifth and sixth recognized human infections with HeV were a 33–years-old male veterinarian and a 21-years-old female veterinary nurse who worked at a veterinary hospital in Thornlands, Queensland during an outbreak of HeV that affected five horses in the practice (Playford et al. 2010). Both the veterinarian and the nurse performed nasal cavity lavage to a horse during the 3 days before the horse developed symptoms of what was ultimately confirmed to be a HeV infection. Both the veterinarian and nurse became ill and had HeV RNA detected by reverse transcription PCR from both serum and nasopharyngeal aspirate specimens. The outbreak investigation identified 83 other people who had contact with the sick horses. Sixteen reported mild symptoms, but none developed a clinical illness. None had HeV RNA or HeV antibodies. Among the 28 persons who reported contact with potentially infected equine body fluids only the two cases developed HeV infection. One veterinary worker who had a percutaneous blood exposure from an infected horse also had no evidence of infection.

The seventh recognized human infection with HeV was a veterinarian who examined a horse that died the next day. A pony and a horse on the same property died of confirmed HeV infection in the subsequent 11 days (Perkins 2009).

These seven people confirmed infected with HeV were infected through contact with only five HeV infected horses. Most infected horses do not transmit HeV to people. Indeed, of the 67 recognized equine HeV infections through January 2012, only five have resulted in human infection (Hess et al. 2011; Queensland Horse Council Inc 2012). In the originally identified outbreak in the Hendra stable, all of the infected horses developed illness within one incubation period (8–11 days after the death of the index mare). This suggests that the mare was a superspreader and the absence of a successive wave of infection among horses, and the low attack rate of HeV among persons who had contact with HeV infected horses suggest that such superspreaders are exceptional. Each of the seven recognized human cases of HeV had intimate contact with a HeV infected horse, usually with heavy exposure to respiratory secretions and without wearing personal protective equipment. Other people with close contact of these same horses did not develop HeV infection. These observations suggest that HeV is not easily transmitted from horse to human. It apparently requires a horse that is an unusually efficient transmitter and a person with a high exposure to infectious secretions.

All humans confirmed with HeV infections had contact with HeV infected horses. The absence of human cases among healthcare workers and among family members suggests that HeV is not easily transmissible from person-to-person. Selvey and colleagues identified 128 people who cared for Pteropid bats, the wildlife reservoir of HeV (Selvey et al. 1996). The bat carers included volunteers who cared for injured or orphaned bats and professionals who cared for captive bats. Bat carers had a median 48 months of bat contact; 74 % reported daily contact with flying foxes. Moreover, 74 % reported having been bitten, 88 % scratched, and 60 % reported exposure to flying fox blood. None of the bat carers tested positive for antibodies to HeV. While direct transmission of HeV from flying foxes to humans remains theoretically possible, the study suggested that, if such transmission does occur, it occurs rarely.

2 Nipah Virus Outbreaks

2.1 Malaysia/Singapore

Human Nipah virus (NiV) infection was first recognized in a large outbreak in peninsular Malaysia from September 1998 to May 1999 (Chua 2003; Chua et al. 2000; Paton et al. 1999). The initial human cases were identified among pig farmers who lived near the city of Ipoh within the state of Perak in northwestern peninsular Malaysia in late September 1998. Patients presented with fever and headache. Many developed a reduced level of consciousness (Goh et al. 2000). Among 28 early cases, four had IgM antibodies against Japanese encephalitis. The government declared the outbreak was due to Japanese encephalitis and initiated measures to reduce mosquito populations. By December 1998, larger clusters of similar cases were reported within the Port Dickson District of Negri Sembilan, 300 km south of Ipoh (CDC 1999). In March 1999, a novel paramyxovirus was isolated from the cerebrospinal fluid of a patient from Sungai Nipah village (Chua et al. 1999) that was confirmed to be the cause of the outbreak (Chua et al. 2000). Ultimately, the Malaysian Ministry of Health reported 283 cases with 109 (39 %) fatalities (Chua 2003).

Parashar and colleagues conducted a case–control study to explore the risk factors for human illness with NiV during the outbreak (Parashar et al. 2000). They enrolled 110 NiV antibody confirmed cases from Port Dickson and two sets of controls, 147 community farm controls from among persons who either lived or worked on pig farms with no reported human encephalitis cases, and 107 case farm controls who were selected from among NiV antibody negative persons who lived on farms where there was a known case of human NiV infection. Case patients were more likely than community farm controls to report increased numbers of sick or dying pigs, dogs and chickens on their farms. Case patients were more likely than case farm controls to have direct contact with pigs that appeared sick and to have close contact with pigs through feeding pigs, processing baby pigs, assisting in breeding of pigs, assisting in birth of pigs, injecting or medicating pigs, and handling dead pigs.

In contrast to the severe illness manifested by HeV infected horses, most pigs infected with NiV had mild illness. Forty-one percent of human NiV infected cases who worked on pig farms reported no increase in sick or dying pigs on their farm (Parashar et al. 2000). Case fatality among adult infected pigs was low, ranging from <1 to 5 % (Mohd Nor et al. 2000). Among three pigs infected with NiV through experimental oral inoculation or sharing a cage with an inoculated pig, all developed asymptomatic infections (Middleton et al. 2002). A subset of NiV infected pigs was severely affected and developed fever, agitation, trembling, and twitching accompanied by rapid labored respirations, increased drooling, and a nonproductive loud barking cough (Mohd Nor et al. 2000). Pathological examination of severely affected pigs demonstrated extensive involvement of the lungs with a giant cell pneumonia with multi-nucleated syncytial cells containing NiV antigen in the lungs and epithelial cells lining the upper airways (Chua et al. 2000). NiV was recovered from respiratory secretions of infected pigs, and NiV antigen was detected in renal tubular epithelial cells (Chua et al. 2000; Middleton et al. 2002).

Between March 10 and 19 in 1999, 11 workers in one of Singapore's abattoirs developed NiV associated with encephalitis or pneumonia (Paton et al. 1999). One patient died. Compared to controls who were also abattoir workers, cases were more likely to be exposed to pig urine or feces from pigs that had been imported from Malaysia during the Malaysian NiV outbreak. NiV RNA recovered from autopsy specimens from the one worker who died, had a nucleotide sequence that was identical to the sequences of NiV isolates from humans and from pigs in Malaysia (Paton et al. 1999).

The isolation of NiV from pigs' lungs and respiratory secretions combined with the observation that human cases of NiV infection had closer contact with pigs and so more contact with pigs' secretions and excretions than controls suggests that NiV was transmitted from infected pigs to humans through contaminated saliva and possibly urine. The human outbreak of NiV infection ceased after widespread deployment of personal protective equipment to people contacting sick pigs, restriction on livestock movements, and culling over 900,000 pigs (Uppal 2000). Since the outbreak ended in May 1999 through December 2011 no human or porcine NiV infections have been reported from Malaysia.

Mathematical modeling suggests that multiple spillovers to the pig population were necessary to create a dynamic population with sufficient newly susceptible pigs to sustain NiV transmission within pigs for months (Pulliam et al. 2011). All human NiV infections, in the outbreak in Malaysia/Singapore in 1998–1999, may have been linked to a single NiV transmission from an infected bat to an immunologically primed pig population, leading to a sustained porcine epidemic which, in turn, led to a human epidemic. The genomic sequences from Malaysian NiV isolates from pigs and people were nearly identical (AbuBakar et al. 2004; Chua et al. 2000).

Not all people infected with NiV in Malaysia had contact with pigs. In the Port Dickson case–control study, two human NiV infected cases reported no contact with pigs (Parashar et al. 2000). KS Tan provided details on two additional NiV patients who had no direct contact with pigs (Tan KS Tan and Goh 1999). One NiV patient who did not enter or go near a pig farms prior to his illness, worked repairing pig cages. His illness suggests that pig secretions/excretions remain infectious at least for hours and perhaps for days. The Port Dickson case–control study noted an increased risk of dying dogs on farms where NiV cases were confirmed (Parashar et al. 2000). Serological studies in dogs in Malaysia demonstrated that they were commonly infected (Field et al. 2001; Mills et al. 2009). One NiV patient who had no pig exposure worked as a cabinet maker and lived near a pig farm. His two pet dogs became seriously ill and died before the patient became ill with NiV infection (Tan KS 1999).

There was limited evidence of person-to-person transmission of NiV in Malaysia. Multiple cases in families may have resulted from shared exposures. A large cohort study enrolled healthcare workers from the three hospitals that admitted over 80 % of patients with suspected NiV encephalitis (Mounts et al. 2001). The study enrolled 363 health care workers who provided direct patient care to encephalitis patients. More than 60 % reported contact with encephalitis patient before infection and control measures were instituted on March 19, 1999. Many reported episodes of high-risk exposure including skin exposure to body fluids of NiV infected patients ($n = 89$) splash of patient body fluids to mucosal membranes ($n = 39$) or needle stick injuries ($n = 12$). None reported any serious illness, encephalitis, or hospital admission. None of the first serum samples was positive by EIA for NiV IgG or IgM antibody. In the second round of antibody

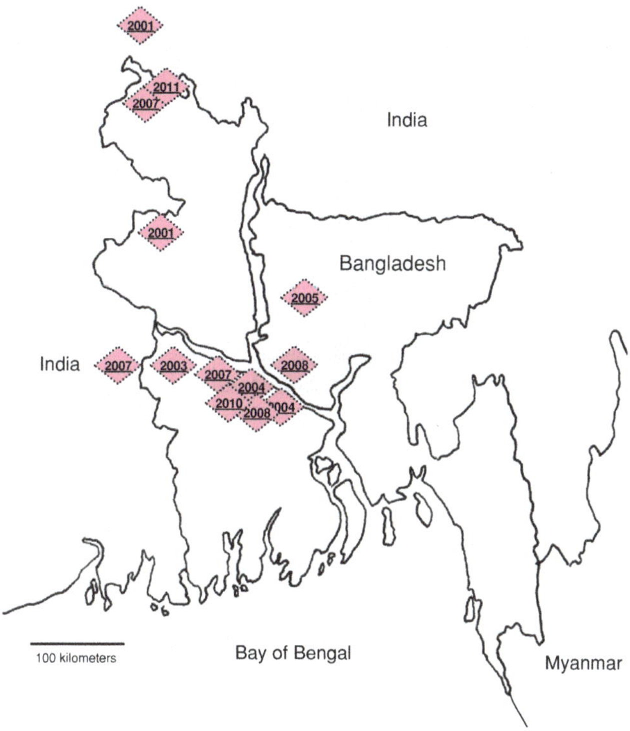

Fig. 1 NiV outbreaks Bangladesh and India 2001–2011

testing conducted 30 days later, 3 of 293 serum samples (1 %) from exposed health care workers were positive for NiV IgG antibody, although none had detectable IgM and all three were negative for anti-NiV neutralizing antibodies. All three were nurses who cared for outbreak-related encephalitis patients for more than 30 days compared with a mean of 10 days in nurses with negative IgG antibodies (Mounts et al. 2001). One of the nurses with NiV IgG antibody reported a febrile illness before the first serum sample was obtained, and the second reported a febrile illness between the two serum samples. One of the nurses reported a mucosal splash exposure. In a separately reported investigation, a nurse who cared for hospitalized NiV infected patients and had antibody against NiV but was asymptomatic, had MRI findings characteristic of NiV infection (Tan et al. 2000).

2.2 NiV Epidemiology Bangladesh/India

The epidemiology of NiV in Bangladesh/India has been quite different than in Malaysia. Since 2001, recurrent outbreaks have been identified almost annually in

Fig. 2 Month of NiV illness onset, Bangladesh 2001–2011

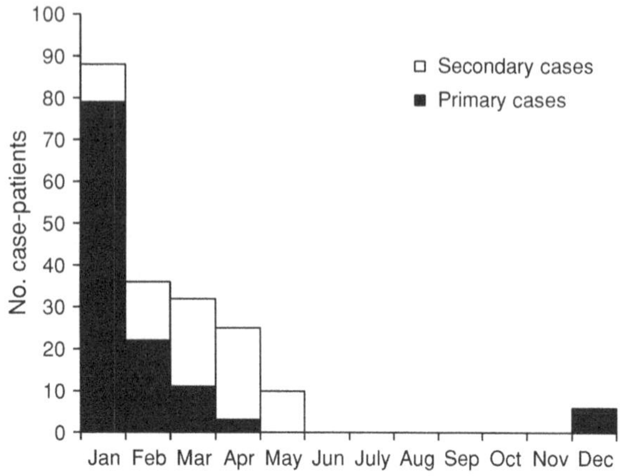

a particular region of western/northwestern Bangladesh and over the border in West Bengal, India (Fig. 1).

3 NiV Transmission Through Date Palm Sap

Outbreak investigations in Bangladesh have identified drinking raw date palm sap as the most common pathway of NiV transmission from *Pteropus* bats to people. In the 2005 outbreak investigation in Tangail, Bangladesh NiV cases were 7.9 times more likely to report drinking raw date palm sap in the 10 days before they developed illness than neighborhood matched controls (Luby et al. 2006). Similarly, in the 2008 outbreak in Manikgonj and Rajbari districts in Bangladesh cases were 25 times more likely than controls to report drinking raw date palm sap (Rahman et al. 2012). In outbreaks in Faridpur, Bangladesh in 2010, and in Lalmonirhat in 2011 cases were again significantly more likely than controls to report drinking raw date palm sap in the 2 weeks prior to the onset of illness (Chakraborty 2011; Sazzad 2010). The outbreaks of human NIV infection in Bangladesh and India coincide with the cool dry date palm sap harvesting season (Luby et al. 2009a) (Fig. 2).

In Bangladesh, date palm sap harvesters collect sap beginning in December with the first cold night and continue collecting most regularly from January till early February, although some harvesters continue to collect in at least a few trees from March till early April. At the beginning of the season, the bark is shaved off of one side of the tree (*Phoenix sylvestris*) near the top in a V shape and a small hollow bamboo tap is placed at the base of the V (Nahar et al. 2010). In the late afternoon, the date palm sap harvester climbs the tree, scrapes the area where the bark is denuded so the sap can flow freely, and ties a 2–4 liter clay pot underneath the tap. During the night as the sap rises to the top of the tree, some sap oozes out from where the bark is denuded, flows through the tap and drips into the clay pot. Palm sap collectors climb up the trees at daybreak to gather the clay pots.

Most date palm sap in Bangladesh is cooked and made into molasses that is a popular sweetener for cakes and other desserts (Halim et al. 2008; Nahar et al. 2010). A smaller amount of date palm sap is sold fresh for immediate consumption. Indeed, after a few hours, presumably due to fermentation, the date palm sap is less sweet and sap sellers have to lower the price. Sap harvesters will often share fresh sap as a treat with family members and walk house-to-house near where the sap was collected and offer to sell it to neighbors.

Sap harvesters and villagers report that bats and other animals sometimes visit the trees during sap collection. Sap harvesters commonly find bat excrement outside of the clay pot or floating in the sap and occasionally find drowned dead bats floating in the pots (Luby et al. 2006; Nahar et al. 2010). Infrared wildlife photography confirms that *Pteropus* bats, the presumed reservoir of NiV in Bangladesh, commonly visit date palm trees during collection and lick the sap stream (Khan et al. 2010). Infrared cameras placed in the seven trees that were the source of fresh date palm sap drunk by the human NiV cases in the Manikgonj/Rajbari outbreak in 2008, identified an average of four *Pteropus* bat visits per tree where the bat licked the sap stream, per night of observation (Rahman et al. 2012).

Date palm sap is a plausible vehicle for transmission of NiV from *Pteropus* bats to people. *Pteropus* bats occasionally shed NiV in their saliva (Middleton et al. 2007; Reynes et al. 2005; Wacharapluesadee et al. 2005). The infrared camera studies confirm that *Pteropus* bats directly lick raw date palm sap and urinate near the sap collection pot (Khan et al. 2010). NiV placed in mango flesh, mango juice, pawpaw juice, and lychee juice for up to 3 days is recoverable at high concentrations (Fogarty et al. 2008). Although survival in date palm sap has not been specifically tested, survival of NiV in these other juices was primarily dependent upon pH, and the pH of date palm sap at 7.2 (Aidoo et al. 2006) suggests that survival would be similar. To date, in outbreak investigations NiV has not been isolated directly from date palm sap (Rahman et al. 2011). This is not surprising, because *Pteropus* shedding of NiV is intermittent (Wacharapluesadee et al. 2010), and with the median 10 day-incubation period from exposure to date palm sap to illness (Rahman et al. 2011), and the time required to recognize an outbreak of NiV, outbreak investigation teams have only been able to collect sap samples from trees weeks after the likely transmission event.

Other direct pathways of NiV transmission from *Pteropus* to people have not been confirmed. In the 2004 outbreak in Rajbari District, Bangladesh, cases were more likely to climb trees than controls (83 versus 51 %, $p = 0.025$) (Montgomery et al. 2008). It is possible that children climbing trees had direct contact with NiV contaminated bat urine or bat saliva that subsequently infected their respiratory or gastrointestinal tract and lead to infection; however, this pathway of transmission has not been identified in any of the subsequent outbreak investigations through 2011. Moreover, 91 % of cases in the 2004 Rajbari outbreak reported drinking raw date palm sap (Montgomery et al. 2008). The father of two of the cases was a date palm sap harvester and case patients were clustered in families and groups of friends, suggesting a common source of exposure (Emily Gurley personal communication). Although there was insufficient statistical power to implicate date palm sap in the Rajbari outbreak investigation (91 versus 72 %, $p = 0.33$), the subsequent repeated implication of date palm sap as the vehicle of transmission in other outbreaks, and the high level of exposure among cases (91 %) suggests that date palm sap was the likely vehicle of NIV infection in this outbreak.

4 NiV Transmission from Domestic Animals

A second route of transmission for NiV from bats to people in Bangladesh is via domestic animals. Fruit bats commonly drop partially eaten saliva-laden fruit. Domestic animals in Bangladesh forage for such food. Date palm sap that is contaminated with bat feces and so is unfit for human consumption is also occasionally fed to domestic animals (Nahar et al. 2010). Animal husbandry practices in Bangladesh are quite different than in Malaysia. In Malaysia, thousands of pigs were raised together on large factory farms. By contrast, in Bangladesh many rural families keep just a few domestic animals. If a domestic animal in Bangladesh contracts NiV, there are few susceptible mammals close enough to transmit NiV to, hence, rather than sustained transmission as was observed in the Malaysian outbreak, in Bangladesh the chain of transmission would be short.

NiV has never been isolated from a domestic animal in Bangladesh, but there have been human NiV cases linked to apparent domestic animal infections. The index case in the Meherpur District 2001 outbreak developed illness on April 20, the latest post winter onset of any confirmed NIV outbreak in Bangladesh (Fig. 2), past the end of the date palm sap season in most communities. NiV cases in Meherpur were eight times more likely to report contact with a sick cow than controls (Hsu et al. 2004). In the Naogaon outbreak in 2003, NiV cases were six times more likely than controls to report contact with a pig herd that visited the community 2 weeks before the human outbreak (ICDDRB 2003). In 2004, a child developed NiV infection 2 weeks after playing with two goats that developed an illness that began with fever, and progressed to difficulty walking, frothing at the mouth and death (Luby et al. 2009b).

5 NiV Person-To-Person Transmission

In contrast to limited evidence of person-to-person transmission of NiV in Malaysia, person-to-person transmission of NiV has been repeatedly identified in Bangladesh/India. The first NiV outbreak recognized in the Indian subcontinent was a large outbreak affecting 66 people in Siliguri, India in 2001. The outbreak apparently originated from an unidentified patient admitted to Siliguri District Hospital who transmitted infection to 11 additional patients, all of whom were transferred to other facilities. In two of the facilities, subsequent transmission infected 25 staff and 8 visitors (Chadha et al. 2006).

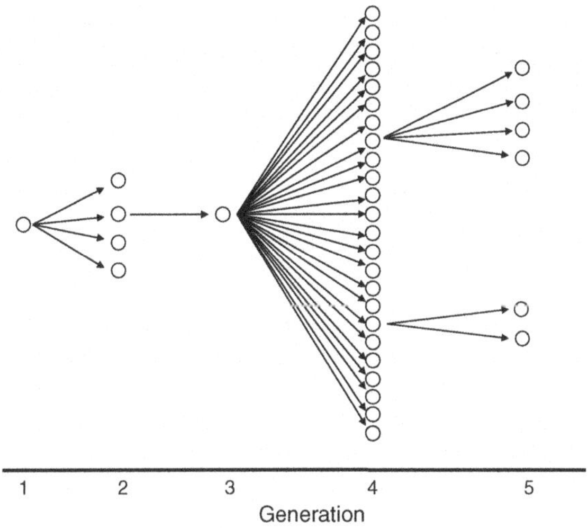

Fig. 3 Chain of person-to-person transmission in NiV outbreak in Faridpur, Bangladesh, 2004. (From (Luby et al. 2009b))

The longest sustained chain of person-to-person transmission of NiV so far identified in Bangladesh occurred in an outbreak in Faridpur in 2004. Friends and family members who provided direct care to NiV infected patients, or helped to carry them or transport them to health facilities when they were near death, sustained a chain of transmission through five generations (Gurley et al. 2007) (Fig. 3). One NiV patient was a popular religious leader who was visited by many of his family members and followers when he became ill. Twenty-two of these visitors developed NiV infection.

While the outbreaks in Siliguri in 2001 and Faridpur in 2004 were the largest examples, person-to-person transmission has been identified in other outbreaks in Bangladesh (Homaira et al. 2010; Sazzad 2010). In a review of the 122 NiV cases identified in Bangladesh from 2001 to 2007, 62 (51 %) developed illness 5–15 days after close contact with another NiV patient (Luby et al. 2009a).

Outbreak investigations in Bangladesh suggest that respiratory secretions are the primary vehicle of person-to-person transmission of NiV. In the 2004 Faridpur outbreak, cases were more likely than controls to report touching an NiV infected patient who later died (OR 5.5, 95 % CI 2.1, 16) (Gurley et al. 2007). Similarly, in Thakurgaon in 2007 six family members and friends who cared for an NiV infected patient developed NiV infection. NiV infected cases in Thakurgaon were more likely than uninfected controls to have been in the same room when the index case was coughing (100 vs. 0 %, $p = 0.04$) (Homaira et al. 2010). Across all recognized outbreaks in Bangladesh from 2001 to 2007, NiV patients who had difficulty breathing during their illness were more likely to transmit NiV than NiV patients who did not have difficulty breathing (12 vs. 0 %, $p = 0.03$) (Luby et al. 2009a).

NiV RNA has been frequently identified in the saliva of NiV patients (Chua et al. 2001; Harcourt et al. 2005). In Bangladesh, family members and friends without health care or infection control training provide nearly all the hands on care to ill patients both at home and in the hospital (Hadley et al. 2007). During the Faridpur 2004 outbreak, care providers shared eating utensils, ate leftovers of food offered to NiV patients, commonly slept in the same bed with a sick, coughing NiV patient, and often fed and hugged the dying patient (Blum et al. 2009).

6 Other Plausible Pathways of NiV Transmission

There are a number of plausible pathways of NiV transmission from *Pteropus* bats to people that have been explored in outbreak investigations in Bangladesh, but have not been implicated as pathways of transmission. One of these pathways is living underneath a bat roost. *Pteropus* bats intermittently shed NiV in their urine (Wacharapluesadee et al. 2010). Although some homes are located quite close to bat roosts, in outbreak investigations living near a bat roost has not been found more commonly among NiV cases than controls (Luby et al. 2009b).

Another plausible pathway of transmission is consumption of bat bitten fruit. Both birds and *Pteropus* bats often drop fruit after taking a single bite. In Bangladesh, where child malnutrition is widespread (NIPORT 2007), ripe tasty dropped fruit is commonly picked up and consumed by rural Bangladeshi residents. In each of the outbreak investigations in Bangladesh consumption of dropped fruit has been evaluated as a potential exposure, but in none of the outbreaks have cases been reported to have eaten dropped fruit significantly more commonly than controls.

7 Open Questions in the Epidemiology of Henipaviruses

Both known viral species of henipavirus, HeV, and NiV, are widely distributed among *Pteropus* bats, but human infection occurs in a much more restricted region. Apparently, the frequency of a specific human behavior that is uncommon across the human population but more common in these areas provides an opportunity for henipavirus transmission to humans. In Queensland, the popularity of horse racing leads to many horses sharing the natural environment with *Pteropus* bats, and people come in close contact with ill horses. In Bangladesh, *Pteropus* bats are present across the entire country, and presumably shed virus throughout the year (Wacharapluesadee et al. 2010). We hypothesize that people living in the outbreak infected regions in Bangladesh are more likely to drink fresh date palm sap or have other activities that put them in more contact with bat secretions compared with people living in other regions with *Pteropus* bats, but without recognized human NiV cases.

Outbreak investigations have confirmed several pathways for transmission of NiV to people. However, there are many cases, both during outbreaks as well as sporadic cases who have no apparent exposure to date palm sap, sick animals, or sick people. Further exercises in hypothesis generation, thorough outbreak investigations, and comprehensive analysis of all outbreaks investigated to date may identify other important exposures. We may also better understand pathways of transmission by deeper investigation of circumstances where henipaviruses are not transmitted. Why aren't people who have direct close contact with *Pteropus* bats including bat carers in Australia, bat hunters in Bangladesh and persons who live under *Pteropus* roosts at increased risk of henipavirus infection? Do HeV infected horses that efficiently transmit the virus have a different clinical presentation, pathology, or viral load than HeV infected horses that do not transmit?

The etiologic agents responsible for encephalitis and other severe infections are frequently unrecognized. Improving diagnostic capacity for henipaviruses throughout the range of *Pteropus* bats could broaden our understanding of the frequency, geographic distribution, and epidemiology of human infections.

A broad open question is how much strain differences in henipavirus are responsible for observed epidemiological differences. Are some strains of henipavirus more likely to infect humans, to cause respiratory disease, or to cause person-to-person transmission than other strains? If strain differences confer different infectious competencies, what are the genetic characteristics that underlie these competencies?

To prevent the devastating consequences of human infections with henipaviruses productive areas for future research include developing and evaluating interventions to prevent transmission from bats to humans as well as development of low cost antiviral therapies and vaccines with cost effective strategies for their deployment.

8 Conclusion

Careful investigation over the last 18 years has clarified the basic transmission pathways of human henipavirus infection. These organisms are not easily transmitted to people. When humans do become infected, only occasional superspreaders infected with NiV transmit illness. To date, transmission has not been sufficiently efficient to maintain person-to-person transmission. However, these agents are newly recognized, and there are a number of scientific questions with important public health implications that are worthy of ongoing scientific investigation.

Acknowledgments The authors thank Dr. Jahangir Hossain who coordinated Nipah outbreak investigations in Bangladesh from 2001 to 2011, and Sonia Hegde who analyzed the data and drafted Fig. 2.

References

AbuBakar S, Chang LY, Ali AR, Sharifah SH, Yusoff K, Zamrod Z (2004) Isolation and molecular identification of Nipah virus from pigs. Emerg Infect Dis 10:2228–2230
Aidoo KE, Nout MJR, Sarkar PK (2006) Occurrence and function of yeasts in Asian indigenous fermented foods. FEMS Yeast Res 6:30–39
Blum LS, Khan R, Nahar N, Breiman RF (2009) In-depth assessment of an outbreak of Nipah encephalitis with person-to-person transmission in Bangladesh: implications for prevention and control strategies. Am J Trop Med Hyg 80:96–102
CDC (1999) Update: outbreak of Nipah virus–Malaysia and Singapore, 1999. MMWR Morb Mortal Wkly Rep 48:335–337
Chadha MS, Comer JA, Lowe L, Rota PA, Rollin PE, Bellini WJ, Ksiazek TG, Mishra A (2006) Nipah virus-associated encephalitis outbreak, Siliguri, India. Emerg Infect Dis 12:235–240
Chakraborty A (2011) Nipah outbreak in Lalmonirhat district, 2011. Health and Science Bulletin 9:13–19
Chua KB (2003) Nipah virus outbreak in Malaysia. J Clin Virol 26:265–275
Chua KB, Bellini WJ, Rota PA, Harcourt BH, Tamin A, Lam SK, Ksiazek TG, Rollin PE, Zaki SR, Shieh W, Goldsmith CS, Gubler DJ, Roehrig JT, Eaton B, Gould AR, Olson J, Field H, Daniels P, Ling AE, Peters CJ, Anderson LJ, Mahy BW (2000) Nipah virus: a recently emergent deadly paramyxovirus. Science 288:1432–1435
Chua KB, Goh KJ, Wong KT, Kamarulzaman A, Tan PS, Ksiazek TG, Zaki SR, Paul G, Lam SK, Tan CT (1999) Fatal encephalitis due to Nipah virus among pig-farmers in Malaysia. Lancet 354:1257–1259
Chua KB, Lam SK, Goh KJ, Hooi PS, Ksiazek TG, Kamarulzaman A, Olson J, Tan CT (2001) The presence of Nipah virus in respiratory secretions and urine of patients during an outbreak of Nipah virus encephalitis in Malaysia. J Infect 42:40–43
Field H, Young P, Yob JM, Mills J, Hall L, Mackenzie J (2001) The natural history of Hendra and Nipah viruses. Microbes Infect 3:307–314
Fogarty R, Halpin K, Hyatt AD, Daszak P, Mungall BA (2008) Henipavirus susceptibility to environmental variables. Virus Res 132:140–144
Goh KJ, Tan CT, Chew NK, Tan PS, Kamarulzaman A, Sarji SA, Wong KT, Abdullah BJ, Chua KB, Lam SK (2000) Clinical features of Nipah virus encephalitis among pig farmers in Malaysia. N Engl J Med 342:1229–1235
Gurley ES, Montgomery JM, Hossain MJ, Bell M, Azad AK, Islam MR, Molla MA, Carroll DS, Ksiazek TG, Rota PA, Lowe L, Comer JA, Rollin P, Czub M, Grolla A, Feldmann H, Luby SP, Woodward JL, Breiman RF (2007) Person-to-person transmission of Nipah virus in a Bangladeshi community. Emerg Infect Dis 13:1031–1037
Hadley MB, Blum LS, Mujaddid S, Parveen S, Nuremowla S, Haque ME, Ullah M (2007) Why Bangladeshi nurses avoid 'nursing': social and structural factors on hospital wards in Bangladesh. Soc Sci Med 64:1166–1177
Halim MA, Chowdhury MSH, Muhamed N, Rahman M, Koike M (2008) Sap production from khejur palm (phoenix sylvestris roxb) husbandry: a substantial means of seasonal livelihood in rural Bangladesh. Forests Trees and Livelihoods 18:305–318
Hanna JN, McBride WJ, Brookes DL, Shield J, Taylor CT, Smith IL, Craig SB, Smith GA (2006) Hendra virus infection in a veterinarian. Med J Aust 185:562–564
Harcourt BH, Lowe L, Tamin A, Liu X, Bankamp B, Bowden N, Rollin PE, Comer JA, Ksiazek TG, Hossain MJ, Gurley ES, Breiman RF, Bellini WJ, Rota PA (2005) Genetic characterization of Nipah virus, Bangladesh, 2004. Emerg Infect Dis 11:1594–1597
Hess IM, Massey PD, Walker B, Middleton DJ, Wright TM (2011) Hendra virus: what do we know? N S W Public Health Bull 22:118–122
Homaira N, Rahman M, Hossain MJ, Epstein JH, Sultana R, Khan MS, Podder G, Nahar K, Ahmed B, Gurley ES, Daszak P, Lipkin WI, Rollin PE, Comer JA, Ksiazek TG, Luby SP

(2010) Nipah virus outbreak with person-to-person transmission in a district of Bangladesh, 2007. Epidemiol Infect 138:1630–1636

Hsu VP, Hossain MJ, Parashar UD, Ali MM, Ksiazek TG, Kuzmin I, Niezgoda M, Rupprecht C, Bresee J, Breiman RF (2004) Nipah virus encephalitis reemergence, Bangladesh. Emerg Infect Dis 10:2082–2087

ICDDRB (2003) Outbreaks of encephalitis due to Nipah/Hendra-like Viruses, Western Bangladesh. Health and Science Bulletin 1:1–6

Khan MS, Hossain J, Gurley ES, Nahar N, Sultana R, Luby SP (2010) Use of infrared camera to understand bats' access to date palm sap: implications for preventing Nipah virus transmission. EcoHealth 7:517–525

Luby S, Hossain J, Gurley E, Ahmed B, Banu S, Khan M, Homaira N, Rota P, Rollin P, Comer JA, Kenah E, Ksiazek T, Rahman M (2009a) Recurrent zoonotic transmission of Nipah virus into humans, Bangladesh, 2001–2007. Emerg Infect Dis 15:1229–1235

Luby SP, Gurley ES, Hossain MJ (2009b) Transmission of human infection with Nipah virus. Clin Infect Dis 49:1743–1748

Luby SP, Rahman M, Hossain MJ, Blum LS, Husain MM, Gurley E, Khan R, Ahmed BN, Rahman S, Nahar N, Kenah E, Comer JA, Ksiazek TG (2006) Foodborne transmission of Nipah virus, Bangladesh. Emerg Infect Dis 12:1888–1894

McCormack JG, Allworth AM, Selvey LA, Selleck PW (1999) Transmissibility from horses to humans of a novel paramyxovirus, equine morbillivirus (EMV). J Infect 38:22–23

Middleton DJ, Morrissy CJ, van der Heide BM, Russell GM, Braun MA, Westbury HA, Halpin K, Daniels PW (2007) Experimental Nipah virus infection in Pteropid bats (Pteropus poliocephalus). J Comp Pathol 136:266–272

Middleton DJ, Westbury HA, Morrissy CJ, van der Heide BM, Russell GM, Braun MA, Hyatt AD (2002) Experimental Nipah virus infection in pigs and cats. J Comp Pathol 126:124–136

Mills JN, Alim AN, Bunning ML, Lee OB, Wagoner KD, Amman BR, Stockton PC, Ksiazek TG (2009) Nipah virus infection in dogs, Malaysia, 1999. Emerg Infect Dis 15:950–952

Mohd Nor MN, Gan CH, Ong BL (2000) Nipah virus infection of pigs in peninsular Malaysia. Rev Sci Tech 19:160–165

Montgomery JM, Hossain MJ, Gurley ES, Carroll DS, Croisier A, Bertherat E, Asgari N, Formenty P, Keeler N, Comer J, Bell MR, Akram K, Molla AR, Zaman K, Islam MR, Wagoner K, Mills JN, Rollin PE, Ksiazek TG, Breiman RF (2008) Risk factors for Nipah virus encephalitis in Bangladesh. Emerg Infect Dis 10:1526–1532

Mounts AW, Kaur H, Parashar UD, Ksiazek TG, Cannon D, Arokiasamy JT, Anderson LJ, Lye MS (2001) A cohort study of health care workers to assess nosocomial transmissibility of Nipah virus, Malaysia, 1999. J Infect Dis 183:810–813

Murray K, Selleck P, Hooper P, Hyatt A, Gould A, Gleeson L, Westbury H, Hiley L, Selvey L, Rodwell B et al (1995) A morbillivirus that caused fatal disease in horses and humans. Science 268:94–97

Nahar N, Sultana R, Gurley ES, Hossain MJ, Luby SP (2010) Date palm sap collection: exploring opportunities to prevent Nipah transmission. EcoHealth 7:196–203

NIPORT (2007) Bangladesh demographic and health survey 2007. National institute of population research and training, Mitra and Associates, Dhaka, Bangladesh and Calverton, Maryland [USA}

O'Sullivan JD, Allworth AM, Paterson DL, Snow TM, Boots R, Gleeson LJ, Gould AR, Hyatt AD, Bradfield J (1997) Fatal encephalitis due to novel paramyxovirus transmitted from horses. Lancet 349:93–95

Parashar UD, Sunn LM, Ong F, Mounts AW, Arif MT, Ksiazek TG, Kamaluddin MA, Mustafa AN, Kaur H, Ding LM, Othman G, Radzi HM, Kitsutani PT, Stockton PC, Arokiasamy J, Gary HE Jr, Anderson LJ (2000) Case-control study of risk factors for human infection with a new zoonotic paramyxovirus, Nipah virus, during a 1998–1999 outbreak of severe encephalitis in Malaysia. J Infect Dis 181:1755–1759

Paton NI, Leo YS, Zaki SR, Auchus AP, Lee KE, Ling AE, Chew SK, Ang B, Rollin PE, Umapathi T, Sng I, Lee CC, Lim E, Ksiazek TG (1999) Outbreak of Nipah-virus infection among abattoir workers in Singapore. Lancet 354:1253–1256

Perkins N (2009) Progress audit of biosecurity Queensland response activities at Cawarral in August 2009. AusVet Animal Health Services Pty Ltd, Toowoomba

Playford EG, McCall B, Smith G, Slinko V, Allen G, Smith I, Moore F, Taylor C, Kung YH, Field H (2010) Human Hendra virus encephalitis associated with equine outbreak, Australia, 2008. Emerg Infect Dis 16:219–223

Pulliam JR, Epstein JH, Dushoff J, Rahman SA, Bunning M, Jamaluddin AA, Hyatt AD, Field HE, Dobson AP, Daszak P (2011) Agricultural intensification, priming for persistence and the emergence of Nipah virus: a lethal bat-borne zoonosis. J R Soc Interface

Queensland Horse Council Inc (2012) Hendra Virus. Queensland Horse Council Inc, Queensland, http://www.qldhorsecouncil.com/QHC %20Documents/Notifiable %20Diseases %20Information %20Sheets/Hendra %20Virus.pdf, accessed 2/3/2012

Rahman MA, Hossain MJ, Sultana S, Homaira N, Khan SU, Rahman M, Gurley ES, Rollin PE, Lo MK, Comer JA, Lowe L, Rota PA, Ksiazek TG, Kenah E, Sharker Y, Luby SP (2011) Date Palm Sap Linked to Nipah Virus Outbreak in Bangladesh, 2008. Vector Borne Zoonotic Dis

Rahman MA, Hossain MJ, Sultana S, Homaira N, Khan SU, Rahman M, Gurley ES, Rollin PE, Lo MK, Comer JA, Lowe L, Rota PA, Ksiazek TG, Kenah E, Sharker Y, Luby SP (2012) Date palm sap linked to Nipah virus outbreak in Bangladesh, 2008. Vector Borne Zoonotic Dis 12:65–72

Reynes JM, Counor D, Ong S, Faure C, Seng V, Molia S, Walston J, Georges-Courbot MC, Deubel V, Sarthou JL (2005) Nipah virus in Lyle's flying foxes, Cambodia. Emerg Infect Dis 11:1042–1047

Sazzad HM (2010) Nipah outbreak in Faridpur District, Bangladesh, 2010. Health and Science Bulletin 8:6–11

Selvey L, Sheridan J (1995) Outbreak of severe respiratory disease in humans and horses due to a previously unrecognized Paramyxovirus*. J Travel Med 2:275

Selvey L, Taylor R, Arklay A, Gerrard J (1996) Screening of bat carers for antibodies to equine morbillivirus. Commun Dis Intell 20:477

Selvey LA, Wells RM, McCormack JG, Ansford AJ, Murray K, Rogers RJ, Lavercombe PS, Selleck P, Sheridan JW (1995) Infection of humans and horses by a newly described morbillivirus. Med J Aust 162:642–645

Tan KS, Ahmad Sarji S, Tan CT, Abdullah BJJ, Chong HT, Thayaparan T, Koh CN (2000) Patients with asymptomatic Nipah virus infection may have abnormal cerebral MR imaging. Neurol J Southeast Asia 5:69–73

Tan KSTC, Goh KJ (1999) Epidemiological aspects of Nipah virus infection. Neurol J Southeast Asia 4:77–81

Uppal PK (2000) Emergence of Nipah virus in Malaysia. Ann N Y Acad Sci 916:354–357

Wacharapluesadee S, Boongird K, Wanghongsa S, Ratanasetyuth N, Supavonwong P, Saengsen D, Gongal GN, Hemachudha T (2010) A longitudinal study of the prevalence of Nipah virus in Pteropus lylei bats in Thailand: evidence for seasonal preference in disease transmission. Vector Borne Zoonotic Dis 10:183–190

Wacharapluesadee S, Lumlertdacha B, Boongird K, Wanghongsa S, Chanhome L, Rollin P, Stockton P, Rupprecht CE, Ksiazek TG, Hemachudha T (2005) Bat Nipah virus, Thailand. Emerg Infect Dis 11:1949–1951

Molecular Virology of the Henipaviruses

Michael K. Lo and Paul A. Rota

Abstract Nipah (NiV) and Hendra (HeV) viruses comprise the genus *Henipavirus* and are highly pathogenic paramyxoviruses, which cause fatal encephalitis and respiratory disease in humans. Since their respective initial outbreaks in 1998 and 1994, they have continued to cause sporadic outbreaks resulting in fatal disease. Due to their designation as Biosafety Level 4 pathogens, the level of containment required to work with live henipaviruses is available only to select laboratories around the world. This chapter provides an overview of the molecular virology of NiV and HeV including comparisons to other, well-characterized paramyxoviruses. This chapter also describes the sequence diversity present among the henipaviruses.

Contents

M. K. Lo · P. A. Rota (✉)
MS-C-22, Centers for Disease Control and Prevention,
1600 Clifton Road, Atlanta, GA 30333, USA
e-mail: prota@cdc.gov

Current Topics in Microbiology and Immunology (2012) 359: 41–58

DOI: 10.1007/82_2012_211

Published Online: 3 May 2012

1 Nipah and Hendra Viruses

1.1 Classification

Nipah (NiV) and Hendra (HeV) belong to the family *Paramyxoviridae*, and have nonsegmented negative-stranded (NNS) RNA genomes consisting of helical nucleocapsids encased in an envelope forming relatively spherical, pleomorphic virus particles. There are two subfamilies within the family *Paramyxoviridae*, the *Paramyxovirinae* and the *Pneumovirinae*. The subfamily *Paramyxovirinae* is divided into five genera: *Rubulavirus* (prototype, mumps virus), *Respirovirus* (prototype, human parainfluenza virus 1), *Morbillivirus* (prototype, measles virus), *Avulavirus* (prototype Newcastle disease virus), and with the emergence of NiV and HeV, a new genus *Henipavirus* (Bellini et al. 1998; Mayo 2002).

1.2 Genome Organization

The henipavirus genome consists of six genes encoded from $3'$ to $5'$, the nucleocapsid (N), phosphoprotein (P), matrix (M), fusion (F), attachment glycoprotein (G), and large (L) polymerase, respectively. The prototypical genomes of NiV and HeV are 18,246 nucleotides (nt) and 18,234 nt long, respectively, making them significantly larger than most other paramyxoviruses, with the exception of the rodent-borne J-virus and Beilong virus (Fig. 1) (Jack et al. 2005; Li et al. 2006; Magoffin et al. 2007b). The large genome size is due to a longer open reading frame (ORF) encoding the P and to the large $3'$ untranslated regions (UTR) that flank each gene with the exception of the large polymerase gene (Harcourt et al. 2000, 2001; Mayo 2002; Wang et al. 2000; Yu et al. 1998). The organization of the henipavirus genome indicates that it likely has a similar replication strategy to other paramyxoviruses, which was demonstrated by the requirement of the N, P, and L proteins for minigenome replication (Halpin et al. 2004; Harcourt et al. 2000, 2001; Wang et al. 2000). The genus-specific $3'$ leader and $5'$ trailer regions of paramyxovirus genomes serve as promoters for both transcription and replication (Lamb and Parks 2007). The coding sequence of each gene is flanked by UTRs that contain a gene-start (GS) sequence and a gene-end (GE) sequence. Immediately following the GE sequence is a short non-coding intergenic region that precedes the GS sequence of the following gene. GE signals direct polyadenylation and termination of each mRNA, while GS signals direct initiation and capping (Lamb and Parks 2007). The terminal 12–13 nt of the genome and antigenome are identical, which indicates their critical role as replication promoters. A study of the *cis*-acting signals in the antigenomic promoter indicated a bipartite promoter structure, with two critically conserved regions (1–12 and 79–91 nt). The internal antigenomic promoter element (79–91 nt) while analogous in position to its equivalent in the Sendai virus antigenomic promoter, contained a distinct $5'$-$(\underline{G}NNN\underline{UG})_{14-15}(\underline{G}NNNNN)_{16}$ motif (Tapparel et al. 1998; Walpita and Peters 2007).

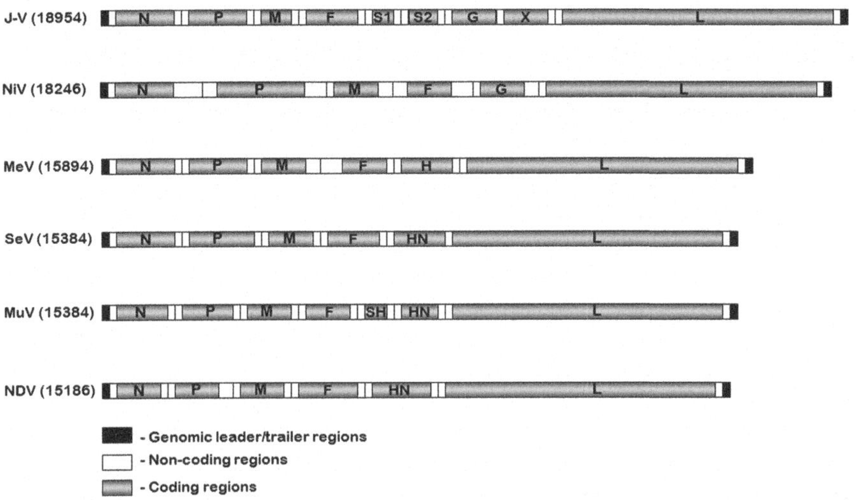

Fig. 1 Comparative genome lengths among viruses in the family *Paramyxovirinae*. NiV has an exceptionally large genome compared to most paramyxoviruses by virtue of its longer non-coding regions and larger P gene (genome lengths in parentheses) (see text). The *solid black segments* indicate leader and trailer sequences. The *clear segments* indicate virus gene non-coding regions, and the *shaded areas* indicate the respective NiV gene coding regions. *MeV* Measles virus, *SeV* Sendai virus, *MuV* Mumps virus, *NDV* Newcastle disease virus

1.3 Replication Cycle

NiV and HeV virions attach to host cell surfaces through the identified receptor molecule ephrin-B2, which is expressed on neurons, smooth muscle, and endothelial cells surrounding small arteries (Bonaparte et al. 2005; Negrete et al. 2005). Ephrin-B3, a related molecule with less affinity for NiV G, has nonetheless also been found to serve as a functional alternative receptor for NiV, but not HeV (Negrete et al. 2006). After receptor binding by G, the F protein then mediates fusion between the viral envelope and the host cell membrane and the viral ribonucleocapsid is then released into the cytoplasm (Lamb and Parks 2007). NiV has also been shown to enter cells via macropinocytosis which is dependent upon a single tyrosine residue in the ephrin-B2 cytoplasmic tail (Pernet et al. 2009). Upon ribonucleocapsid release into the host cell cytoplasm, transcription of viral mRNA begins, with the encapsidated genome serving as the template for mRNA synthesis. The polymerase complex (P and L proteins) initiates transcription at the promoter region in the 3′ leader. Transcription of each gene is regulated by transcriptional start and stop sites at their respective regions in the gene. The polymerase "stutters" at the transcriptional stop signals in order to synthesize poly-A tails of the mRNA. The polymerase pauses at the gene stop signal, crosses the intergenic region, and reinitiates transcription at the next gene start signal. During this process, some of the polymerase complexes will fall off from the template. Given that there is only one promoter at the 3′ leader of the genomic RNA, these polymerase complexes can only reinitiate transcription at the 3′ end. This results in a differential gradient of transcription, in which the genes closer to the 3′ end are transcribed more often than genes closer to the 5′ end of the genome (Bellini et al. 1998). As translation of viral mRNA occurs, viral proteins accumulate in the cell, and an unidentified mechanism triggers the polymerase to stop transcription and initiate genome replication. During genome replication, the polymerase complex enters the 3′ genomic promoter in a similar manner as during transcription, but ignores the mRNA transcriptional signals at gene junctions, causing the polymerase to transcribe a full-length positive-sense antigenome. A promoter in the 3′ non-coding region of the antigenome serves as the location where synthesis of new, negative-sense genomic RNA begins. Newly made genomes can in turn, serve as templates for transcription (Lamb and Parks 2007). To assemble virus particles, the genome replication process must be terminated, newly made genomes must be properly encapsidated, and regions of the cell membrane need to be prepared to accept budding nucleocapsids (Lamb and Parks 2007). Polymerase complexes associated with packaged nucleocapsids are responsible for the next round of infection. The viral glycoproteins are typically synthesized in the endoplasmic reticulum (ER) and mature through the Golgi network to the cell membrane. The henipavirus F glycoproteins are a unique exception to the rule, as F protein processing and maturation occurs in the endosome (Diederich et al. 2005). It is thought that the cytoplasmic tails of paramyxoviral fusion and attachment proteins interact with the matrix protein (Lamb and Parks 2007; Sanderson et al. 1994), which is the protein most likely responsible for driving virus particle assembly and budding (Ciancanelli and Basler 2006; Patch et al. 2007, 2008; Wang et al. 2010).

1.4 Morphology

Electron microscopy studies using both negative stain and thin section preparations depicted extracellular NiV virions as tangled collections of filamentous, helical nucleocapsids encased by the viral envelope (Chua et al. 2000; Goldsmith et al. 2003). While NiV virions share common morphological characteristics with the rest of the family *Paramyxoviridae* (i.e., herringbone-like nucleocapsid structure, nucleocapsid accumulation along plasma membranes during budding), there were several distinct characteristics observed. Unusual cytoplasmic ring-like reticular inclusions (RI) were found in the proximity to membranes of the ER. These inclusions were distinct from the typical nucleocapsid inclusions (NCI) found in paramyxo virus-infected cells. Long cytoplasmic tubules were found in the periphery of NiV infected cells, which were occasionally incorporated into virus particles. NiV virions were on average larger in diameter (500 nm) than typical paramyxo viruses (150–400 nm), with extreme variations in size (180–1,900 nm) (Goldsmith et al. 2003). There are subtle ultrastructural differences that distinguish NiV and HeV. When viewed by negative contrast electron microscopy, the surface projections of HeV virions have a double-fringed appearance, while those of NiV have a single fringe (Hyatt et al. 2001). The localization of the nuclei and viral nucleocapsids within infected cells also differentiates NiV and HeV. Along with the viral nucleocapsids, the nuclei of NiV-infected cells localize to the cell periphery while in HeV infected cells both the nuclei and nucleocapsids are either more centrally located or randomly distributed in the cytoplasm (Hyatt et al. 2001). In situ hybridization studies in cell culture and brain specimens taken from fatal NiV infections indicated that unencapsidated viral mRNA is almost exclusively found in the novel RI mentioned previously, and not in the NCIs (Goldsmith et al. 2003). These results suggest that the abovementioned RI structures play an integral role in of henipavirus transcription and replication.

1.5 Viral Proteins

1.5.1 G Glycoprotein

The henipavirus G proteins are type II membrane proteins that form tetramers upon cell surface expression, and serve primarily in cellular receptor binding (Bossart et al. 2005; Levroney et al. 2005; Wang et al. 2001). The NiV and HeV G proteins are 602 and 604 amino acids long, respectively, and share approximately 83 % amino acid identity (Wang et al. 2001). The N-termini of these proteins constitute the cytoplasmic tail, which is followed by the hydrophobic transmembrane region, a short extracellular stem region, and a large globular head formed by the C-terminal end. Henipavirus G proteins lack both hemagglutinin and neuraminidase activities (Bossart et al. 2001; Tamin et al. 2002; Yu et al. 1998). Two independent groups using different experimental approaches identified the functional henipavirus receptor to be ephrinB2 (Bonaparte et al. 2005; Negrete et al. 2005). EphrinB2 is involved in embryogenic development, vasculogenesis, and axonal guidance; its expression pattern in neurons, smooth muscle, and endothelial cells correlates with the cellular tropism of henipaviruses during infection (Erbar et al. 2008; Palmer and Klein 2003; Poliakov et al. 2004; Thiel et al. 2008). EphrinB3 shares two specific amino acid residues with EphrinB2 through which it can serve as an alternative receptor for NiV infection (Negrete et al. 2006). Identification of the functional residues required for ephrinB2 and ephrinB3 interactions with henipavirus G proteins demonstrated a greater ability of NiV G to bind ephrinB3 than that of HeV G, which may explain the increased neurotropism of NiV compared to HeV in human infections (Negrete et al. 2007). The crystal structures of the ephrinB2 and ephrinB3 interactions with NiV G confirm previously performed functional mapping, and also provide a comprehensive analysis of critical residues composing the hydrophobic binding cleft in NiV G which interacts with high affinity to ephrinB2 and ephrinB3 (Bowden et al. 2008; Xu et al.2008). Analysis of the unbound form of NiV G protein by crystallography and mass spectrometry indicated that NiV G contains highly processed complex-type glycans with negligible amounts of oligomannose-type glycans, and that the *N*-acetylglucosamine (GlcNAc) $\beta 1 \rightarrow 2$Man terminal structures on NiV G were a potential ligand for LSECtin, a C-type lectin expressed on sinusoidal endothelial cells of lymph nodes and liver. Consistent with observations for F and G protein interaction among other paramyxoviruses, a receptor binding activation site in the stalk region of NiV G was shown to trigger fusion by the F protein (Aguilar et al. 2009; Corey and Iorio 2007, 2009; Lamb and Parks 2007; Lee et al. 2008; Melanson and Iorio 2004, 2006). Studies of henipavirus fusion dynamics have demonstrated differential turnover and trafficking rates between the G and F proteins, and have suggested that constant stimulation of the F protein is required by a receptor-engaged G protein (Porotto et al. 2011; Whitman and Dutch 2007; Whitman et al. 2009).

1.5.2 F Glycoprotein

The henipavirus F proteins are 546 amino acid type I transmembrane proteins that form trimers on the cell surface, and are responsible for mediating the fusion of virus and host cell membranes during infection (Bossart et al. 2001; Levroney et al. 2005; Michalski et al. 2000; Tamin et al. 2002; Wang et al. 2001). Similar to other paramyxoviruses, activation of NiV F fusion depends on specific sequences in its cytoplasmic tail (Aguilar et al. 2007; Tong et al. 2002; Waning et al. 2004). The F proteins of paramyxoviruses are synthesized as inactive precursors called F0, and are then typically cleaved by a cellular protease to yield the biologically active F1 and F2 subunits which are linked by disulfide bonds (Lamb and Parks 2007). The membrane-anchored subunit F1 contains a hydrophobic amino terminus that is highly conserved, often referred to as the fusion peptide. Paramyxovirus F proteins with multi-basic consensus sequence (R-X-R/K-R) are activated by ubiquitous host cell proteases such as furin, while F proteins with only one basic residue at the site of cleavage are activated by extracellular trypsin-like proteases upon transport of inactive F0 precursors to the cell membrane. Viruses with multi-basic sequences in their F protein tend to cause systemic infections, while those with single basic cleavage sites such as Sendai virus cause localized infections (Klenk and Garten 1994). While henipavirus infections are systemic in nature, comprehensive mutagenesis studies of the cleavage sites of both HeV and NiV F proteins demonstrated that the lone arginine residue at the cleavage site was not required for F processing, indicating that neither trypsin nor any other furin-like proteases were responsible (Craft and Dutch 2005; Moll et al. 2004a). It was determined that these proteins undergo endocytosis following surface expression, and are processed in acidified endosomes via the endosomal protease cathepsin L (Meulendyke et al. 2005; Pager et al. 2006; Pager and Dutch 2005; Vogt et al. 2005). Specific tyrosine residues in the cytoplasmic domains of the NiV F and G proteins target them to the basolateral membrane in polarized epithelial cells and disruption of these residues abrogated syncytia formation (Weise et al. 2010). N-glycans have been shown to be crucial structural components that facilitate proper folding and transport of viral proteins (Doms et al. 1993; McGinnes et al. 2001; Roberts et al. 1993; Segawa et al. 2000; Tamura et al. 2002; Zimmer et al. 2001). The henipavirus F proteins contain five potential glycosylation sites, with three located in the F_2 subunit, and two in the F_1 subunit. Multiple studies have determined roles for N-linked oligosaccharides in regulating henipavirus F protein fusogenicity, cell surface expression, viral entry, and protection from neutralizing antibodies (Aguilar et al. 2006; Carter et al. 2005; Moll et al. 2004b).

1.5.3 M Protein

The henipavirus matrix (M) proteins have a length of 352 amino acids, and like other paramyxovirus M proteins are very basic, with pIs of 9.99 and 9.44, respectively, and charges of +12 and +14 at neutral pH, respectively (Wang et al. 2001). The paramyxovirus M proteins provide rigidity and structure to the virion through its interactions with the cytoplasmic tail of the F protein, the ribonucleoprotein complex, and the inner surface of the virion envelope (Lamb and Parks 2007; Sanderson et al. 1994; Tashiro et al. 1993, 1996). Late domain motifs found in numerous viral matrix proteins recruit specific host factors to viral assembly sites where they facilitate virus release (Freed 2002). There are several different late domain motifs found among NNS viruses: P(T/S)AP, PPxY, $YP(x)_nL$, and more recently ϕ-P-x-V (Lamb and Parks 2007). Several studies have identified potential late domain motifs (YMYL and YPLGVG) in the NiV M protein, and have shown that mutagenizing either of these motifs drastically reduces virus-like particle budding, and skews the subcellular localization of M toward the nucleus. (Ciancanelli et al. 2009; Patch et al. 2008). A more recent study demonstrated a correlation between ubiquitination of NiV M and budding function, in which a bipartite nuclear localization signal was not only responsible for temporal nuclear localization for the NiV M protein, but also serves as a ubiquitination site which turned out to be required for nuclear export of NiV M (Wang et al. 2010). Once cellular pools of ubiquitin were depleted, NiV M could not be exported from the nucleus, which prevented proper budding. Furthermore, fusing membrane-targeting sequences to an alanine mutant at the putative ubiquitination site at K258 restored budding function.

1.5.4 N Protein

The N proteins of henipaviruses are 532 amino acids long, are responsible for encapsidating the viral genome, and are what cause the visible herringbone-like structures seen in electron micrographs which are hallmarks of paramyxoviruses (Lamb and Parks 2007). As with other paramyxoviruses, henipavirus N proteins have an inherent ability to assemble themselves, and follow the "rule of six" (Halpin et al. 2004; Hausmann et al. 1996; Ong et al. 2009). The N and the P proteins of NiV and HeV were shown to form homologous and heterologous N–P complexes, and have more than one P-binding site on the N protein (Chan et al. 2004). Huang et al. 2011 demonstrated transient phosphorylation at amino acid position 451 of the N protein, which when mutated to an alanine or glutamate significantly reduced NiV minigenome replication.

1.5.5 L Protein

The L protein of paramyxoviruses is the largest and the least abundant viral protein found in an infected cell. The large size of the protein reflects the many different enzymatic functions attributed to it, including initiation, elongation and termination of both mRNA transcription (capping, methylation, and polyadenylation of mRNA) and genome replication (Lamb and Parks 2007). A comparative study done on L proteins of NNS RNA viruses identified six similar domains shared among them (Stec et al. 1991). Henipavirus L proteins are 2,244 amino acids in length, contain all six conserved domains found in NNS viruses, but have a GDNE motif in the putative polymerase catalytic site, as opposed to the GDNQ motif found in most negative sense, non-segmented RNA viruses. (Poch et al. 1989). Site directed mutagenesis of the glutamate residue in the motif GDNE (underlined) of the NiV L protein indicated a robust ability for the NiV L protein to tolerate different substitutions in this motif without completely losing its activity, suggesting that this residue may not be crucial for catalytic polymerase activity (Magoffin et al. 2007a).

1.5.6 P Gene Protein Products

A common characteristic among viruses in the *Paramyxovirinae* subfamily is the ability to generate multiple proteins from the P gene that are distinct from the P protein (Lamb and Parks 2007). Mechanisms by which these accessory gene products are generated include RNA editing, alternative downstream initiation sites, internal ribosomal entry sites, and the use of non-AUG start codons (Curran et al. 1991; Curran and Kolakofsky 1988a, b, 1989; Latorre et al. 1998; Mehdi et al. 1990).

The henipavirus P protein is a 707 (HeV)/709 (NiV) amino acid protein, and is the only essential gene product encoded by the P gene for genome replication (Halpin et al. 2004). A study utilizing mass spectrometry described only serine phosphorylation sites of NiV P, in contrast to both serine and threonine phosphorylation sites in HeV P (Shiell et al. 2003). There are at least two independent N-binding sites in the P protein; one at the N–terminus (amino acids 3–220), and the other at C-terminal end (amino acids 636–709) (Chan et al. 2004). In addition to the P protein, the P gene of henipaviruses encodes at least three non-structural proteins (C, V, and W). The V and W proteins are generated by RNA editing, while the C protein is encoded by a second ORF downstream of the translational initiation site for the P ORF (Harcourt et al. 2000). There is also an additional accessible open reading frame within the P gene of HeV identified as the small basic protein (SB) (Wang et al. 2001). The NiV C, V, and W proteins provided in *trans* via plasmid expression inhibited NiV minigenome transcription and replication in a dose-dependent manner (Sleeman et al. 2008), which is consistent with observations for other paramyxoviruses (Bankamp et al. 2005; Grogan and Moyer 2001; Nishio et al. 2008; Reutter et al. 2001; Witko et al. 2006). Studies using transfected plasmid vectors expressing NiV P, V, and W indicated that the V and P proteins are localized in the cytoplasm, while the W protein is in the nucleus (Rodriguez et al. 2002, 2004; Shaw et al. 2004). A study characterizing the NiV P gene products in infected Vero cells confirmed prior findings, and also showed the C protein being distributed in perinuclear regions of the cell (Lo et al. 2009). The nuclear localization of W was shown to be cell-type specific, as NiV infection of three distinct human primary endothelial cell types demonstrated that W localized predominantly in the cytoplasm, which corresponded with its ability to generate an antiviral response (Lo et al. 2010).

2 Genetics and Reverse Genetics

Yoneda et al. 2006 were the first to develop a T7 polymerase-based full-length reverse genetic system for NiV. An additional gene encoding enhanced green fluorescent protein (EGFP) was inserted between the N and P genes to create a reporter NiV in which replication could be monitored by fluorescence microscopy. The recombinant virus was shown to have similar growth characteristics and pathogenicity in hamsters. The restriction of NiV replication shown in mouse and rat cell lines which expressed ephrinB2 implied that there are other factors besides receptor expression that control NiV tropism (Yoneda et al. 2006). Since then, an independent group developed a similar T7 polymerase-based full-length NiV reverse genetic system to study the P gene (Ciancanelli et al. 2009). There are currently no known full-length HeV reverse genetic systems available.

3 Molecular Diversity

3.1 NiV

Since the initial molecular characterization of NiVs from multiple sources (pigs, humans, and bats) in Malaysia, there have been relatively few full-length NiV ORF sequences from NiV outbreaks in Bangladesh and India to analyze (Chadha et al. 2006; Harcourt et al. 2005; Harit et al. 2006). While the sequences of NiV isolates from five humans and two pigs from the Malaysian outbreak were nearly identical in sequence, the sequence from a third pig isolate was significantly different, suggesting the possibility that there were at least two circulating lineages of NiV present in Malaysia just prior to the outbreak in 1999 (AbuBakar et al. 2004;

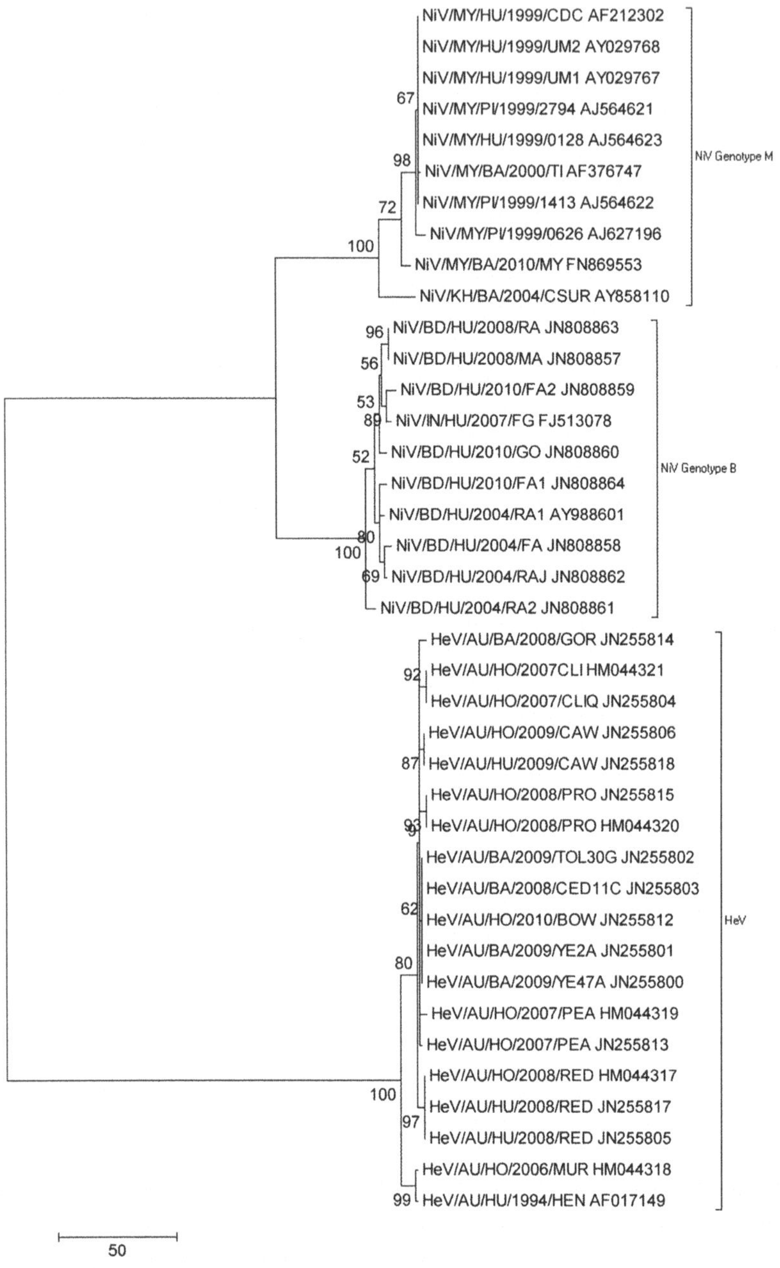

Fig. 2 Phylogenetic analyses of currently available full-length N gene ORF sequences from NiV and HeV using MEGA 5 (Tamura et al. 2011). Maximum Parsimony tree, Close-Neighbor-Interchange algorithm, 1,000 bootstrap replicates. Branch lengths are in units of number of changes over the whole sequence. Available GenBank accession numbers are shown for corresponding sequences. Proposed genotype groupings are indicated by the labeled brackets (NiV Genotype M; NiV Genotype B). *AU* Australia, *BD* Bangladesh, *KH* Cambodia, *IN* India, *MY* Malaysia, *BA* Bat, *HO* Horse, *HU* Human, *PI* Pig

Chan et al. 2001). Interestingly, sequence analysis of the N, P, M, F, and G genes of NiV isolated from Malaysian *Pteropus hypomelanus* on Tioman Island shortly after the outbreak had only 5–6 nt differences when compared with the human isolates (Chua et al. 2002). Higher levels of sequence divergence were observed from the full-length N and G gene ORFs from NiVs isolated from the urine of Cambodian *Pteropus lylei*, although the overall nucleotide and amino acid sequence similarity to the Malaysian sequences was still greater than 98 % (Reynes et al. 2005). Similarly, sequence analysis of a NiV isolate obtained from captive Malaysian *Pteropus vampyrus*, while significantly divergent from other Malaysian sequences at the nucleotide (98 changes) and amino acid (44 changes) levels, still had high levels of sequence similarity across the entire genome (~99 %).Since the NiV sequences from *Pteropus lylei* and *Pteropus vampyrus* grouped more closely compared to other NiV sequences from Malaysia, and because bats of both species have been found to roost together in Thailand and Cambodia, it was suggested that NiV is transmitted between these two species (Rahman et al. 2010).

Sequences of NiV isolates from human outbreaks in India and Bangladesh in 2007 and 2008, along with sequence data obtained from patient specimens during the 2010 outbreak in Bangladesh have supported prior observations of relative heterogeneity among nucleotide sequences obtained from Bangladesh/India as compared with sequences from samples obtained during the initial Malaysian outbreak (AbuBakar et al. 2004; Arankalle et al. 2011; Harcourt et al. 2005; Lo et al. 2012). Phlyogenetic analysis of the sequences obtained from Bangladesh and India from 2007 to 2010 indicated that they group significantly closer to the sequences obtained from Bangladesh in 2004. This observation led to the proposal of a genotyping scheme to classify the two distinct lineages of NiV. Currently available sequences obtained from Malaysia and Cambodia were designated genotype "M", while sequences obtained from Bangladesh and India were designated genotype "B" (Fig. 2). Analysis of the designated 729 nucleotide genotyping window in the N-terminal region of the N gene ORF (N ORF 123–852 nt, NiV genome position 236–964) identified 25 distinct nt which universally differentiated these two genotypes. The topology of the phylogenetic tree and the positions of the branches generated from this smaller nucleotide window were very similar to those of the tree generated with the full-length N, P, M, F, G, and L ORF sequences. Levels of nucleotide variation among full-length ORFs between genotypes M and B ranged from 6 to 9 %. Within each genotype, nucleotide variation ranged from ~0.2 to 3.6 % among full-length ORFs (Lo et al. 2012). Nucleotide variation between genotypes M and B across the complete NiV genome is approximately 8 %.

3.2 HeV

Following a relatively quiet decade since its emergence, there have been 28 documented spillover events of HeV infection, 17 of which occurred over a 3 month span in 2011 (Smith et al. 2011). In contrast to the significant nucleotide sequence differences observed between the two genotypes of NiV, sequence analysis of five HeV isolates from several more recent outbreaks in horses indicated a high level of sequence and amino acid conservation as compared to the original HeV isolate from 1994 ($\sim 1\ \%$ or less) (Marsh et al. 2010) (Fig. 2). There was no discernible correlation found between HeV sequence and time or geographic location of the outbreaks. Sequences of four virus isolates and PCR variants from pooled urine samples of *Pteropus alecto*, *Pteropus poliocephalus*, and *Pteropus conspillicatus* species bats indicated that there was widespread circulation of multiple HeV variants, and that these variants did not necessarily correspond to locations in which spillover events occurred. Phylogenetic analysis of bat HeV sequences indicated that levels of sequence variation were comparable to levels observed among horse and human sequences when compared to the original HeV sequence (Smith et al. 2011).

4 Conclusion

Since their emergence in the 1990s, we have seen significant advances in elucidating the molecular biology of henipaviruses. With the advent of reverse genetics, several animal models for henipavirus infection, and the development of *Pteropus* cell lines, we can look forward to greater insights into the molecular pathogenesis and transmission of these highly pathogenic viruses (Crameri et al. 2009; Weingartl et al. 2009). Technological advances in DNA sequencing as well as improved sample collection techniques during field surveillance and outbreak studies should facilitate the sequencing of additional viral isolates, which will in turn give us a broader perspective of henipavirus molecular diversity (Chua et al. 2002; Marsh et al. 2010; Rahman et al. 2010; Reynes et al. 2005; Smith et al. 2011).

Disclaimer The findings and conclusions of this report are those of the authors and do not necessarily represent the views of the Centers for Disease Control and Prevention.

References

AbuBakar S, Chang LY, Ali AR, Sharifah SH, Yusoff K, Zamrod Z (2004) Isolation and molecular identification of Nipah virus from pigs. Emerg Infect Dis 10(12):2228–2230

Aguilar HC, Matreyek KA, Filone CM, Hashimi ST, Levroney EL, Negrete OA, Bertolotti-Ciarlet A, Choi DY, McHardy I, Fulcher JA, Su SV, Wolf MC, Kohatsu L, Baum LG, Lee B (2006) N-glycans on Nipah virus fusion protein protect against neutralization but reduce membrane fusion and viral entry. J Virol 80(10):4878–4889

Aguilar HC, Matreyek KA, Choi DY, Filone CM, Young S, Lee B (2007) Polybasic KKR motif in the cytoplasmic tail of Nipah virus fusion protein modulates membrane fusion by inside-out signaling. J Virol 81(9):4520–4532

Aguilar HC, Ataman ZA, Aspericueta V, Fang AQ, Stroud M, Negrete OA, Kammerer RA, Lee B (2009) A novel receptor-induced activation site in the Nipah virus attachment glycoprotein (G) involved in triggering the fusion glycoprotein (F). J Biol Chem 284(3):1628–1635

Arankalle VA, Bandyopadhyay BT, Ramdasi AY, Jadi R, Patil DR, Rahman M, Majumdar M, Banerjee PS, Hati AK, Goswami RP, Neogi DK, Mishra AC (2011) Genomic characterization of Nipah virus, West Bengal, India. Emerg Infect Dis 17(5):907–909
Bankamp B, Wilson J, Bellini WJ, Rota PA (2005) Identification of naturally occurring amino acid variations that affect the ability of the measles virus C protein to regulate genome replication and transcription. Virology 336(1):120–129
Bellini WJ, Rota PA, Anderson LJ (1998) Paramyxoviruses. In: Mahy BWJ, Collier L (eds) Topley & Wilson's microbiology and microbial infections, vol 1. 9th edn. VirologyEdward Arnold, London, pp 435–462
Bonaparte MI, Dimitrov AS, Bossart KN, Crameri G, Mungall BA, Bishop KA, Choudhry V, Dimitrov DS, Wang LF, Eaton BT, Broder CC (2005) Ephrin-B2 ligand is a functional receptor for Hendra virus and Nipah virus. Proc Natl Acad Sci U S A 102(30):10652–10657
Bossart KN, Wang LF, Eaton BT, Broder CC (2001) Functional expression and membrane fusion tropism of the envelope glycoproteins of Hendra virus. Virology 290(1): 121–135
Bossart KN, Crameri G, Dimitrov AS, Mungall BA, Feng YR, Patch JR, Choudhary A, Wang LF, Eaton BT, Broder CC (2005) Receptor binding, fusion inhibition, and induction of cross-reactive neutralizing antibodies by a soluble G glycoprotein of Hendra virus. J Virol 79(11): 6690–6702
Bowden TA, Crispin M, Harvey DJ, Aricescu AR, Grimes JM, Jones EY, Stuart DI (2008) Crystal structure and carbohydrate analysis of Nipah virus attachment glycoprotein: a template for antiviral and vaccine design. J Virol 82(23):11628–11636
Carter JR, Pager CT, Fowler SD, Dutch RE (2005) Role of N-linked glycosylation of the Hendra virus fusion protein. J Virol 79(12):7922–7925
Chadha MS, Comer JA, Lowe L, Rota PA, Rollin PE, Bellini WJ, Ksiazek TG, Mishra A (2006) Nipah virus-associated encephalitis outbreak, Siliguri, India. Emerg Infect Dis 12(2):235–240
Chan YP, Chua KB, Koh CL, Lim ME, Lam SK (2001) Complete nucleotide sequences of Nipah virus isolates from Malaysia. J Gen Virol 82(Pt 9):2151–2155
Chan YP, Koh CL, Lam SK, Wang LF (2004) Mapping of domains responsible for nucleocapsid protein–phosphoprotein interaction of Henipaviruses. J Gen Virol 85(Pt 6):1675–1684
Chua KB, Bellini WJ, Rota PA, Harcourt BH, Tamin A, Lam SK, Ksiazek TG, Rollin PE, Zaki SR, Shieh W, Goldsmith CS, Gubler DJ, Roehrig JT, Eaton B, Gould AR, Olson J, Field H, Daniels P, Ling AE, Peters CJ, Anderson LJ, Mahy BW (2000) Nipah virus: a recently emergent deadly paramyxovirus. Science 288(5470):1432–1435
Chua KB, Koh CL, Hooi PS, Wee KF, Khong JH, Chua BH, Chan YP, Lim ME, Lam SK (2002) Isolation of Nipah virus from Malaysian Island flying-foxes. Microbes Infect 4(2):145–151
Ciancanelli MJ, Basler CF (2006) Mutation of YMYL in the Nipah virus matrix protein abrogates budding and alters subcellular localization. J Virol 80(24):12070–12078
Ciancanelli MJ, Volchkova VA, Shaw ML, Volchkov VE, Basler CF (2009) Nipah virus sequesters inactive STAT1 in the nucleus via a P gene-encoded mechanism. J Virol 83(16): 7828–7841
Corey EA, Iorio RM (2007) Mutations in the stalk of the measles virus hemagglutinin protein decrease fusion but do not interfere with virus-specific interaction with the homologous fusion protein. J Virol 81(18):9900–9910
Corey EA, Iorio RM (2009) Measles virus attachment proteins with impaired ability to bind CD46 interact more efficiently with the homologous fusion protein. Virology 383(1):1–5
Craft WW Jr, Dutch RE (2005) Sequence motif upstream of the Hendra virus fusion protein cleavage site is not sufficient to promote efficient proteolytic processing. Virology 341(1): 130–140
Crameri G, Todd S, Grimley S, McEachern JA, Marsh GA, Smith C, Tachedjian M, De Jong C, Virtue ER, Yu M, Bulach D, Liu JP, Michalski WP, Middleton D, Field HE, Wang LF (2009) Establishment, immortalisation and characterisation of pteropid bat cell lines. PLoS ONE 4(12):e8266
Curran J, Kolakofsky D (1988a) Ribosomal initiation from an ACG codon in the Sendai virus P/C mRNA. EMBO J 7(1):245–251

Curran J, Kolakofsky D (1988b) Scanning independent ribosomal initiation of the Sendai virus X protein. EMBO J 7(9):2869–2874
Curran J, Kolakofsky D (1989) Scanning independent ribosomal initiation of the Sendai virus Y proteins in vitro and in vivo. EMBO J 8(2):521–526
Curran J, Boeck R, Kolakofsky D (1991) The Sendai virus P gene expresses both an essential protein and an inhibitor of RNA synthesis by shuffling modules via mRNA editing. EMBO J 10(10):3079–3085
Diederich S, Moll M, Klenk HD, Maisner A (2005) The Nipah virus fusion protein is cleaved within the endosomal compartment. J Biol Chem 280(33):29899–29903
Doms RW, Lamb RA, Rose JK, Helenius A (1993) Folding and assembly of viral membrane proteins. Virology 193(2):545–562
Erbar S, Diederich S, Maisner A (2008) Selective receptor expression restricts Nipah virus infection of endothelial cells. Virol J 5:142
Freed EO (2002) Viral late domains. J Virol 76(10):4679–4687
Goldsmith CS, Whistler T, Rollin PE, Ksiazek TG, Rota PA, Bellini WJ, Daszak P, Wong KT, Shieh WJ, Zaki SR (2003) Elucidation of Nipah virus morphogenesis and replication using ultrastructural and molecular approaches. Virus Res 92(1):89–98
Grogan CC, Moyer SA (2001) Sendai virus wild-type and mutant C proteins show a direct correlation between L polymerase binding and inhibition of viral RNA synthesis. Virology 288(1):96–108
Halpin K, Bankamp B, Harcourt BH, Bellini WJ, Rota PA (2004) Nipah virus conforms to the rule of six in a minigenome replication assay. J Gen Virol 85(Pt 3):701–707
Harcourt BH, Tamin A, Ksiazek TG, Rollin PE, Anderson LJ, Bellini WJ, Rota PA (2000) Molecular characterization of Nipah virus, a newly emergent paramyxovirus. Virology 271(2):334–349
Harcourt BH, Tamin A, Halpin K, Ksiazek TG, Rollin PE, Bellini WJ, Rota PA (2001) Molecular characterization of the polymerase gene and genomic termini of Nipah virus. Virology 287(1):192–201
Harcourt BH, Lowe L, Tamin A, Liu X, Bankamp B, Bowden N, Rollin PE, Comer JA, Ksiazek TG, Hossain MJ, Gurley ES, Breiman RF, Bellini WJ, Rota PA (2005) Genetic characterization of Nipah virus, Bangladesh, 2004. Emerg Infect Dis 11(10):1594–1597
Harit AK, Ichhpujani RL, Gupta S, Gill KS, Lal S, Ganguly NK, Agarwal SP (2006) Nipah/Hendra virus outbreak in Siliguri, West Bengal, India in 2001. Indian J Med Res 123(4):553–560
Hausmann S, Jacques JP, Kolakofsky D (1996) Paramyxovirus RNA editing and the requirement for hexamer genome length. RNA 2(10):1033–1045
Huang M, Sato H, Hagiwara K, Watanabe A, Sugai A, Ikeda F, Kozuka-Hata H, Oyama M, Yoneda M, Kai C (2011) Determination of a phosphorylation site in Nipah virus nucleoprotein and its involvement in virus transcription. J Gen Virol 92(Pt 9):2133–2141
Hyatt AD, Zaki SR, Goldsmith CS, Wise TG, Hengstberger SG (2001) Ultrastructure of Hendra virus and Nipah virus within cultured cells and host animals. Microbes Infect 3(4):297–306
Jack PJ, Boyle DB, Eaton BT, Wang LF (2005) The complete genome sequence of J virus reveals a unique genome structure in the family *Paramyxoviridae*. J Virol 79(16):10690–10700
Klenk HD, Garten W (1994) Host cell proteases controlling virus pathogenicity. Trends Microbiol 2(2):39–43
Lamb RA, Parks GD (2007) Paramyxoviridae: the viruses and their replication. In: Knipe DM, Howley PM (eds) Fields virology, 5th edn. vols 1 and 2. Lippincott Williams & Wilkins, Philadelphia, pp 1449–1496
Latorre P, Kolakofsky D, Curran J (1998) Sendai virus Y proteins are initiated by a ribosomal shunt. Mol Cell Biol 18(9):5021–5031
Lee JK, Prussia A, Paal T, White LK, Snyder JP, Plemper RK (2008) Functional interaction between paramyxovirus fusion and attachment proteins. J Biol Chem 283(24):16561–16572
Levroney EL, Aguilar HC, Fulcher JA, Kohatsu L, Pace KE, Pang M, Gurney KB, Baum LG, Lee B (2005) Novel innate immune functions for galectin-1: galectin-1 inhibits cell fusion by

Nipah virus envelope glycoproteins and augments dendritic cell secretion of proinflammatory cytokines. J Immunol 175(1):413–420

Li Z, Yu M, Zhang H, Magoffin DE, Jack PJ, Hyatt A, Wang HY, Wang LF (2006) Beilong virus, a novel paramyxovirus with the largest genome of non-segmented negative-stranded RNA viruses. Virology 346(1):219–228

Lo MK, Harcourt BH, Mungall BA, Tamin A, Peeples ME, Bellini WJ, Rota PA (2009) Determination of the Henipavirus phosphoprotein gene mRNA editing frequencies and detection of the C, V and W proteins of Nipah virus in virus-infected cells. J Gen Virol 90(Pt 2):398–404

Lo MK, Miller DM, Aljofan M, Mungall BA, Rollin PE, Bellini WJ, Rota PA (2010) Characterization of the antiviral and inflammatory responses against Nipah virus in endothelial cells and neurons. Virology 404(1):78–88

Lo MK, Lowe L, Hummel KB, Sazzad HM, Gurley ES, Hossain MJ, Luby SP, Miller DM, Comer JA, Rollin PE, Bellini WJ, Rota PA (2012) Characterization of Nipah virus from outbreaks in Bangladesh, 2008–2010. Emerg Infect Dis 18(2):248–255

Magoffin DE, Halpin K, Rota PA, Wang LF (2007a) Effects of single amino acid substitutions at the E residue in the conserved GDNE motif of the Nipah virus polymerase (L) protein. Arch Virol 152(4):827–832

Magoffin DE, Mackenzie JS, Wang LF (2007b) Genetic analysis of J-virus and Beilong virus using minireplicons. Virology 364(1):103–111

Marsh GA, Todd S, Foord A, Hansson E, Davies K, Wright L, Morrissy C, Halpin K, Middleton D, Field HE, Daniels P, Wang LF (2010) Genome sequence conservation of Hendra virus isolates during spillover to horses, Australia. Emerg Infect Dis 16(11):1767–1769

Mayo MA (2002) A summary of taxonomic changes recently approved by ICTV. Arch Virol 147(8):1655–1663

McGinnes L, Sergel T, Reitter J, Morrison T (2001) Carbohydrate modifications of the NDV fusion protein heptad repeat domains influence maturation and fusion activity. Virology 283(2):332–342

Mehdi H, Ono E, Gupta KC (1990) Initiation of translation at CUG, GUG, and ACG codons in mammalian cells. Gene 91(2):173–178

Melanson VR, Iorio RM (2004) Amino acid substitutions in the F-specific domain in the stalk of the newcastle disease virus HN protein modulate fusion and interfere with its interaction with the F protein. J Virol 78(23):13053–13061

Melanson VR, Iorio RM (2006) Addition of N-glycans in the stalk of the newcastle disease virus HN protein blocks its interaction with the F protein and prevents fusion. J Virol 80(2):623–633

Meulendyke KA, Wurth MA, McCann RO, Dutch RE (2005) Endocytosis plays a critical role in proteolytic processing of the Hendra virus fusion protein. J Virol 79(20):12643–12649

Michalski WP, Crameri G, Wang L, Shiell BJ, Eaton B (2000) The cleavage activation and sites of glycosylation in the fusion protein of Hendra virus. Virus Res 69(2):83–93

Moll M, Diederich S, Klenk HD, Czub M, Maisner A (2004a) Ubiquitous activation of the Nipah virus fusion protein does not require a basic amino acid at the cleavage site. J Virol 78(18):9705–9712

Moll M, Kaufmann A, Maisner A (2004b) Influence of N-glycans on processing and biological activity of the Nipah virus fusion protein. J Virol 78(13):7274–7278

Negrete OA, Levroney EL, Aguilar HC, Bertolotti-Ciarlet A, Nazarian R, Tajyar S, Lee B (2005) EphrinB2 is the entry receptor for Nipah virus, an emergent deadly paramyxovirus. Nature 436(7049):401–405

Negrete OA, Wolf MC, Aguilar HC, Enterlein S, Wang W, Muhlberger E, Su SV, Bertolotti-Ciarlet A, Flick R, Lee B (2006) Two key residues in ephrinB3 are critical for its use as an alternative receptor for Nipah virus. PLoS Pathog 2(2):e7

Negrete OA, Chu D, Aguilar HC, Lee B (2007) Single amino acid changes in the Nipah and Hendra virus attachment glycoproteins distinguish ephrinB2 from ephrinB3 usage. J Virol 81(19):10804–10814

Nishio M, Ohtsuka J, Tsurudome M, Nosaka T, Kolakofsky D (2008) Human parainfluenza virus type 2 V protein inhibits genome replication by binding to the L protein: possible role in promoting viral fitness. J Virol 82(13):6130–6138
Ong ST, Yusoff K, Kho CL, Abdullah JO, Tan WS (2009) Mutagenesis of the nucleocapsid protein of Nipah virus involved in capsid assembly. J Gen Virol 90(Pt 2):392–397
Pager CT, Dutch RE (2005) Cathepsin L is involved in proteolytic processing of the Hendra virus fusion protein. J Virol 79(20):12714–12720
Pager CT, Craft WW Jr, Patch J, Dutch RE (2006) A mature and fusogenic form of the Nipah virus fusion protein requires proteolytic processing by cathepsin L. Virology 346(2):251–257
Palmer A, Klein R (2003) Multiple roles of ephrins in morphogenesis, neuronal networking, and brain function. Genes Dev 17(12):1429–1450
Patch JR, Crameri G, Wang LF, Eaton BT, Broder CC (2007) Quantitative analysis of Nipah virus proteins released as virus-like particles reveals central role for the matrix protein. Virol J 4:1
Patch JR, Han Z, McCarthy SE, Yan L, Wang LF, Harty RN, Broder CC (2008) The YPLGVG sequence of the Nipah virus matrix protein is required for budding. Virol J 5(1):137
Pernet O, Pohl C, Ainouze M, Kweder H, Buckland R (2009) Nipah virus entry can occur by macropinocytosis. Virology 395(2):298–311
Poch O, Sauvaget I, Delarue M, Tordo N (1989) Identification of four conserved motifs among the RNA-dependent polymerase encoding elements. EMBO J 8(12):3867–3874
Poliakov A, Cotrina M, Wilkinson DG (2004) Diverse roles of eph receptors and ephrins in the regulation of cell migration and tissue assembly. Dev Cell 7(4):465–480
Porotto M, Devito I, Palmer SG, Jurgens EM, Yee JL, Yokoyama CC, Pessi A, Moscona A (2011) Spring-loaded model revisited: paramyxovirus fusion requires engagement of a receptor binding protein beyond initial triggering of the fusion protein. J Virol 85(24):12867–12880
Rahman SA, Hassan SS, Olival KJ, Mohamed M, Chang LY, Hassan L, Saad NM, Shohaimi SA, Mamat ZC, Naim MS, Epstein JH, Suri AS, Field HE, Daszak P (2010) Characterization of Nipah virus from naturally infected Pteropus vampyrus bats Malaysia. Emerg Infect Dis 16(12):1990–1993
Reutter GL, Cortese-Grogan C, Wilson J, Moyer SA (2001) Mutations in the measles virus C protein that up regulate viral RNA synthesis. Virology 285(1):100–109
Reynes JM, Counor D, Ong S, Faure C, Seng V, Molia S, Walston J, Georges-Courbot MC, Deubel V, Sarthou JL (2005) Nipah virus in Lyle's flying foxes Cambodia. Emerg Infect Dis 11(7):1042–1047
Roberts PC, Garten W, Klenk HD (1993) Role of conserved glycosylation sites in maturation and transport of influenza A virus hemagglutinin. J Virol 67(6):3048–3060
Rodriguez JJ, Parisien JP, Horvath CM (2002) Nipah virus V protein evades alpha and gamma interferons by preventing STAT1 and STAT2 activation and nuclear accumulation. J Virol 76(22):11476–11483
Rodriguez JJ, Cruz CD, Horvath CM (2004) Identification of the nuclear export signal and STAT-binding domains of the Nipah virus V protein reveals mechanisms underlying interferon evasion. J Virol 78(10):5358–5367
Sanderson CM, Wu HH, Nayak DP (1994) Sendai virus M protein binds independently to either the F or the HN glycoprotein in vivo. J Virol 68(1):69–76
Segawa H, Yamashita T, Kawakita M, Taira H (2000) Functional analysis of the individual oligosaccharide chains of Sendai virus fusion protein. J Biochem 128(1):65–72
Shaw ML, Garcia-Sastre A, Palese P, Basler CF (2004) Nipah virus V and W proteins have a common STAT1-binding domain yet inhibit STAT1 activation from the cytoplasmic and nuclear compartments, respectively. J Virol 78(11):5633–5641
Shiell BJ, Gardner DR, Crameri G, Eaton BT, Michalski WP (2003) Sites of phosphorylation of P and V proteins from Hendra and Nipah viruses: newly emerged members of Paramyxoviridae. Virus Res 92(1):55–65

Sleeman K, Bankamp B, Hummel KB, Lo MK, Bellini WJ, Rota PA (2008) The C, V and W proteins of Nipah virus inhibit minigenome replication. J Gen Virol 89(Pt 5):1300–1308

Smith I, Broos A, de Jong C, Zeddeman A, Smith C, Smith G, Moore F, Barr J, Crameri G, Marsh G, Tachedjian M, Yu M, Kung YH, Wang LF, Field H (2011) Identifying Hendra virus diversity in pteropid bats. PLoS ONE 6(9):e25275

Stec DS, Hill MG 3rd, Collins PL (1991) Sequence analysis of the polymerase L gene of human respiratory syncytial virus and predicted phylogeny of nonsegmented negative-strand viruses. Virology 183(1):273–287

Tamin A, Harcourt BH, Ksiazek TG, Rollin PE, Bellini WJ, Rota PA (2002) Functional properties of the fusion and attachment glycoproteins of Nipah virus. Virology 296(1): 190–200

Tamura T, Yamashita T, Segawa H, Taira H (2002) N-linked oligosaccharide chains of Sendai virus fusion protein determine the interaction with endoplasmic reticulum molecular chaperones. FEBS Lett 513(2–3):153–158

Tamura K, Peterson D, Peterson N, Stecher G, Nei M, Kumar S (2011) MEGA5: Molecular Evolutionary Genetics Analysis using maximum likelihood, evolutionary distance, and maximum parsimony methods. Mol Biol Evol 28(10):2371–2379

Tapparel C, Maurice D, Roux L (1998) The activity of Sendai virus genomic and antigenomic promoters requires a second element past the leader template regions: a motif (GNNNNN)3 is essential for replication. J Virol 72(4):3117–3128

Tashiro M, Seto JT, Klenk HD, Rott R (1993) Possible involvement of microtubule disruption in bipolar budding of a Sendai virus mutant, F1-R, in epithelial MDCK cells. J Virol 67(10):5902–5910

Tashiro M, McQueen NL, Seto JT, Klenk HD, Rott R (1996) Involvement of the mutated M protein in altered budding polarity of a pantropic mutant, F1-R, of Sendai virus. J Virol 70(9): 5990–5997

Thiel L, Diederich S, Erbar S, Pfaff D, Augustin HG, Maisner A (2008) Ephrin-B2 expression critically influences Nipah virus infection independent of its cytoplasmic tail. Virol J 5:163

Tong S, Li M, Vincent A, Compans RW, Fritsch E, Beier R, Klenk C, Ohuchi M, Klenk HD (2002) Regulation of fusion activity by the cytoplasmic domain of a paramyxovirus F protein. Virology 301(2):322–333

Vogt C, Eickmann M, Diederich S, Moll M, Maisner A (2005) Endocytosis of the Nipah virus glycoproteins. J Virol 79(6):3865–3872

Walpita P, Peters CJ (2007) Cis-acting elements in the antigenomic promoter of Nipah virus. J Gen Virol 88(Pt 9):2542–2551

Wang LF, Yu M, Hansson E, Pritchard LI, Shiell B, Michalski WP, Eaton BT (2000) The exceptionally large genome of Hendra virus: support for creation of a new genus within the family Paramyxoviridae. J Virol 74(21):9972–9979

Wang L, Harcourt BH, Yu M, Tamin A, Rota PA, Bellini WJ, Eaton BT (2001) Molecular biology of Hendra and Nipah viruses. Microbes Infect 3(4):279–287

Wang YE, Park A, Lake M, Pentecost M, Torres B, Yun TE, Wolf MC, Holbrook MR, Freiberg AN, Lee B (2010) Ubiquitin-regulated nuclear-cytoplasmic trafficking of the Nipah virus matrix protein is important for viral budding. PLoS Pathog 6(11):e1001186

Waning DL, Russell CJ, Jardetzky TS, Lamb RA (2004) Activation of a paramyxovirus fusion protein is modulated by inside-out signaling from the cytoplasmic tail. Proc Natl Acad Sci U S A 101(25):9217–9222

Weingartl HM, Berhane Y, Czub M (2009) Animal models of Henipavirus infection: a review. Vet J 181(3):211–220

Weise C, Erbar S, Lamp B, Vogt C, Diederich S, Maisner A (2010) Tyrosine residues in the cytoplasmic domains affect sorting and fusion activity of the Nipah virus glycoproteins in polarized epithelial cells. J Virol 84(15):7634–7641

Whitman SD, Dutch RE (2007) Surface density of the Hendra G protein modulates Hendra F protein-promoted membrane fusion: role for Hendra G protein trafficking and degradation. Virology 363(2):419–429

Whitman SD, Smith EC, Dutch RE (2009) Differential rates of protein folding and cellular trafficking for the Hendra virus F and G proteins: implications for F-G complex formation. J Virol 83(17):8998–9001

Witko SE, Kotash C, Sidhu MS, Udem SA, Parks CL (2006) Inhibition of measles virus minireplicon-encoded reporter gene expression by V protein. Virology 348(1):107–119

Xu K, Rajashankar KR, Chan YP, Himanen JP, Broder CC, Nikolov DB (2008) Host cell recognition by the Henipaviruses: crystal structures of the Nipah G attachment glycoprotein and its complex with ephrin-B3. Proc Natl Acad Sci U S A 105(29):9953–9958

Yoneda M, Guillaume V, Ikeda F, Sakuma Y, Sato H, Wild TF, Kai C (2006) Establishment of a Nipah virus rescue system. Proc Natl Acad Sci U S A 103(44):16508–16513

Yu M, Hansson E, Langedijk JP, Eaton BT, Wang LF (1998) The attachment protein of Hendra virus has high structural similarity but limited primary sequence homology compared with viruses in the genus *Paramyxovirus*. Virology 251(2):227–233

Zimmer G, Trotz I, Herrler G (2001) N-glycans of F protein differentially affect fusion activity of human respiratory syncytial virus. J Virol 75(10):4744–4751

Henipavirus Receptor Usage and Tropism

Olivier Pernet, Yao E Wang and Benhur Lee

Abstract Nipah (NiV) and Hendra (HeV) viruses are the deadliest human pathogens within the *Paramyxoviridae* family, which include human and animal pathogens of global biomedical importance. NiV and HeV infections cause respiratory and encephalitic illness with high mortality rates in humans. Henipaviruses (HNV) are the only Paramyxoviruses classified as biosafety level 4 (BSL4) pathogens due to their extreme pathogenicity, potential for bioterrorism, and lack of licensed vaccines and therapeutics. HNV use ephrin-B2 and ephrin-B3, highly conserved proteins, as viral entry receptors. This likely accounts for their unusually broad species tropism, and also provides opportunities to study how receptor usage, cellular tropism, and end-organ pathology relates to the pathobiology of HNV infections. The clinical and pathologic manifestations of NiV and HeV virus infections are reviewed in the chapters by Wong et al. and Geisbert et al. in this issue. Here, we will review the biology of the HNV receptors, and how receptor usage relates to HNV cell tropism in vitro and in vivo.

Contents

O. Pernet · Y. EWang · B. Lee (✉)
Department of Microbiology, Immunology and Molecular Genetics,
David Geffen School of Medicine at UCLA, BSRB 251a,
615 Charles E. Young Drive East, Los Angeles, CA 90095-1489, USA
e-mail: bleebhl@ucla.edu

Current Topics in Microbiology and Immunology (2012) 359: 59–78
DOI: 10.1007/82_2012_222

Published Online: 15 June 2012

1 The Receptors

1.1 The Molecular Biology of Ephrin-B2

Ephrin-B2 (also known as EPLG5, HTKL, Htk-L, and LERK5) and ephrin-B3 are the cellular receptors for HNV (Bonaparte et al. 2005; Negrete et al. 2005; Negrete et al. 2006). They belong to the ephrin family of receptor tyrosine kinases (RTKs) comprising six GPI-anchored ephrin-As and three transmembrane ephrin-Bs (Eph Nomenclature Committee 1997). Ephrin-B2 and -B3 are type I transmembrane proteins of ~330 amino acids encoded on human chromosome 13 and 17, respectively (Bennett et al. 1995; Bergemann et al. 1995) Ephrin family members are highly conserved: human and murine ephrin-B2 differs by only 3 % at the amino acid level and many mammalian ephrin-B2 homlogs have been shown to bind HNV-G with similar affinities (Bossart et al. 2008).

By convention, ephrins (A and B) are designated as ligands for EphA and/or EphB receptors, which are also RTKs. Thus, cognate ephrin-Eph interactions result in bidirectional signaling cascades: ephrins induce "forward" signaling through the Eph receptors on the opposing cell, while Eph receptors initiate "reverse" signaling through the ephrin ligands (Cowan and Henkemeyer 2001; Zhao et al. 2006). Ephrin-Eph interactions are promiscuous, although the promiscuity, with a few exceptions, is limited to members within the same class (A or B). Structural evidence indicates that ephrin-B2 binds to its cognate receptors EphB4 and EphB2 via critical residues in a flexible "G-H loop" (amino acids 120–125) that fits into a shallow cleft on the opposing ephB receptors (Chrencik et al. 2006; Füller et al. 2003; Kobayashi et al. 2007). The functional signaling cascade seemingly arises from oligomeric interactions between clusters of ephrinB-EphB molecules at the point of cell–cell contact (reviewed in Wilkinson 2003).

Due to its function in mediating cell adhesion/repulsion, ephrin-B2 plays critical roles in chemotaxis and cell migration (Meyer et al. 2005), as evidenced by the fact that ephrin-B2 homozygous knock-out mice are embryonic lethals with defects such as primitive and uniformally sized vasculature, underdeveloped heart, and poor organization of the intersomitic vessels (Adams et al. 1999; Gerety and Anderson

2002). During neurogenesis, ephrin-B2 guides the migration of neuron precursors, contributing to the extremely precise organization pattern of brain cells (Zimmer et al. 2003); during vasculogenesis, the repulsion between EphB4-expressing vein precursors and ephrin-B2-expressing arterial precursors results in the formation of defined junctions between veins and arteries (Wang et al. 1998). Additionally, the function of ephrin-B2 has been extended to immune activation and bone formation in the adult body. EphB4-ephrin-B2-interaction has been shown to mediate the attachment of monocytes (EphB4 positive) to the endothelium (ephrin-B2 positive) during extravasation. Quite interestingly, while cytoplasmic tail deletion of ephrin-B2 does not affect monocyte attachment *per se*, it abolishes extravasation, suggesting that the "reverse" signaling through ephrin-B2 is required for this process (Pfaff et al. 2008). Similarly, during bone formation, "reverse" signaling downstream of ephrin-B2 in osteoclasts upon contact with EphB4-expressing osteoblasts regulates the proliferation of osteoclasts (Edwards and Mundy 2008; Zhao et al. 2006). A role for ephrin-B2 in lymph node remodeling has also been suggested (Mäkinen et al. 2005).

Similar to ephrin-B2, the alternative receptor for NiV, ephrin-B3, is also involved in axon guidance during neurogenesis. For example, it acts as a midline repellent for axons of the corticospinal tract (Benson et al. 2005; Bergemann et al. 1998; Kullander et al. 2001).

1.2 Surface Expression and Regulation

In some cases, ephrin-B2 protein is hardly detectable at the cell surface although its mRNA is present in these cells (Yoneda et al. 2006). One potential explanation could be that intramembrane proteases remove ephrins from the cell surface. For example, in cells that express rhomboid protease RHBDL2, ephrin-B3 is not detectable on the cell surface due to the efficient cleavage by this protease (Pascall and Brown 2004). In the case of ephrin-B2, its extracellular domain can be cleaved by the Kuzbanian metalloproteinase upon interaction with EphB4 (Holmberg and Frisén 2002). It has also been demonstrated that the reverse signaling is dependent on the presence of Presenillin-1, an aspartyl protease from the γ-secretase complex. Presenillin-1 cleaves ephrin-B2 cytoplasmic tail into a C-terminal fragment (CTF) (Georgakopoulos et al. 2006), which regulates c-Src at two levels: first, CTF translocates to the nucleus where it activates c-Src transcription; second, CTF binds c-Src and therefore releases it from its inhibitor Csk. The c-Src-CTF complex then phosphorylates the cytoplasmic tail of uncleaved ephrin-B2 on tyrosine residues 304, 330, and 331 (Georgakopoulos et al. 2011; Su et al. 2004). The phosphorylated ephrin-B2, in turn, activates a kinase cascade that induces membrane ruffling and membrane exchange, leading to cell–cell repulsion (Fig. 1) (Holmberg and Frisén 2002; Wilkinson 2003). This kinase cascade may also intersect with the signals involved in cytoskeleton rearrangement and membrane extensions involved in macropinocytosis, a process that is thought to be involved in NiV uptake upon viral attachment protein interaction with ephrin-B2.

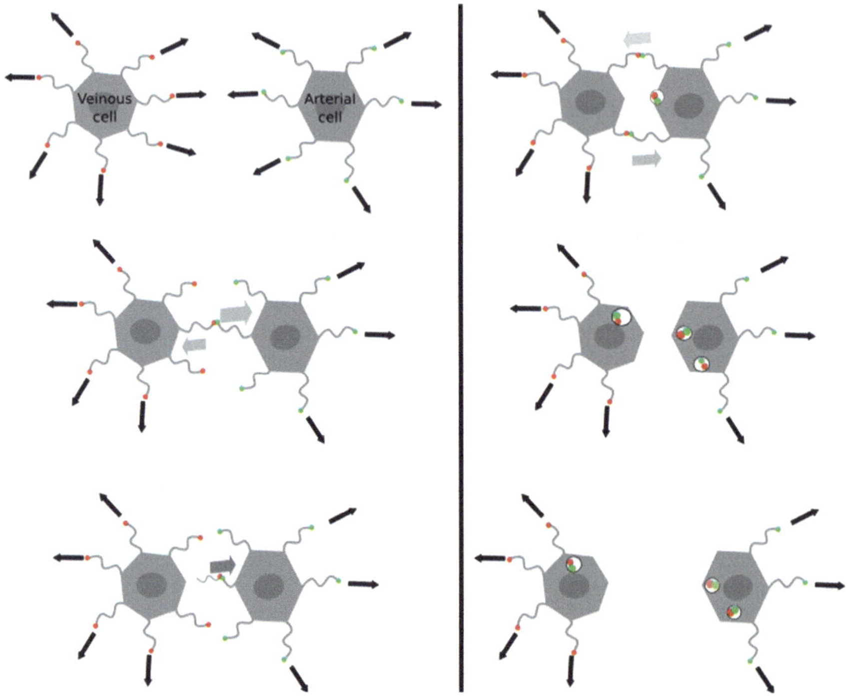

Fig. 1 Schematic illustration of EphB4/phrin-B2-mediated cell–cell repulsion. *Green dots* represent ephrin-B2. Red dots represent EphB4. *Arrows* indicate the direction of the filopodia extensions. Migrating cells move along the filopods they project. When ephrin-B2 on the surface of an arterial cell interacts with EphB4 on the surface of a venous cell, bidirectional signaling induces membrane ruffling and internalization of the EphB4–ephrin-B2 complex. Continued filopodia growth in opposite directions leads to the repulsion of these two cells. However, context-dependent signaling can also lead to cell attraction and forward cell propulsion

1.3 Distribution

1.3.1 Ephrin-B2

Consistent with its role in vasculogenesis, ephrin-B2 is expressed in endothelial and smooth muscle cells in arterial vessels and at angiogenesis sites (Gale et al. 2001). The brain is a main site of ephrin-B2 expression, especially during the fetal stage. Within the brain, the cortex (especially prefrontal) and neuroepithelial cells contain the highest levels of ephrin-B2, while its expression can also be detected in the olfactory bulb and the amygdala, albeit at lower levels. Outside of the brain, the placenta, lungs, and the prostate contain abundant ephrin-B2 protein, and high levels of ephrin-B2 mRNA have also been detected in cardiomyocytes and bronchial epithelial cells (Liebl et al. 2003; Su et al. 2004) (BioGPS—probe ref.202668).

1.3.2 Ephrin-B3

With the possible exception of the prostate and heart, ephrin-B3 expression is mostly restricted to the central nervous system (CNS). It can be detected in the spinal cord and throughout the brain, with the highest levels found in the occipital lobe, the prefrontal cortex, and the amygdala. Lower levels can be detected in the pons, the globus pallidus, the subthalamic nucleus, the temporal lobe, the hypothalamus, the corpus callosum, and the hippocampus (Benson et al. 2005; Liebl et al. 2003; Su et al. 2004) (BioGPS—probe ref.205031).

1.4 Receptor and Host Range

Due to its critical role during embryogenesis, ephrin-B2 (and B3) is evolutionarily conserved. It is found from fish and amphibians to mammals with relatively few changes. For example, the amino acid sequence similarity between human ephrin-B2 and ephrin-B2 from mice, pigs (as well as cats, horses and dogs), and fruit bats reaches 97, 96, and 95 %, respectively (Bossart et al. 2008). Even zebrafish ephrin-B2 (*Danio rerio*) is 65 % similar to human ephrin-B2 at the amino acid level, and can even serve as entry receptors for NiV (unpublished observations). Thus, the conservation of ephrin-B2 may explain the unusually broad host range of HNV compared to most other Paramyxoviruses. Henipaviruses can infect several orders of the Class *Mammalia* under natural or experimental settings (see below). Remarkably, NiV innoculated into chicken embryos (Class *Aves*) also resulted in a histopathological picture that resembles that found in humans: severe lesions in the CNS with high viral antigenic load found in the vasculature and neurons (Tanimura et al. 2006). The conservation of cross Class tissue tropism is unprecedented for a Paramyxovirus, and it remains to be determined if HNV tropism extends outside the Superclass *Tetrapoda* to which Mammalia and Aves belong, to the Superclass *Osteichthyes* (bony fish), which includes the above-mentioned Zebrafish.

The primary host for HNV has been identified as fruit bats (family *Megachiroptera*) in the genera of *Pteropus, Eidolon,* and *Rousettus* (Hayman et al. 2008; 2011; Olson et al. 2002; Young et al. 1996). Although *Microchiroptera* species are generaly not infected (Hasebe et al. 2012), some have been detected seropositive for HNV (Li et al. 2008). Given that the bats comprise the second largest order in the class *Mammalia* (after rodents), with more than 1,240 species constituting 20 % of all known mammals, the finding of HNV in both *Megachiroptera* and *Microchiroptera* suggests that HNV may be more widely distributed than currently appreciated (Kunz et al. 2011).

During the HeV outbreaks in Australia, and the initial NiV outbreaks in Malaysia, horses and pigs, respectively, have been shown to act as intermediate amplifying hosts, transmitting the viruses from bats to humans (Chua 2010; Chua et al. 1999; Selvey et al. 1995). During the Meherpur outbreak (Bangladesh 2001), cows have been suspected to be the amplifying host (Hsu et al. 2004). However, much epidemiological evidence suggests that, in Bangladesh, bats can transmit

NiV directly to humans, and human-to-human transmission has also been documented (Homaira et al. 2010; Tan and Chua 2008).

Under laboratory conditions, the host range of HNV can be extended to rodents (including golden hamsters (Wong et al. 2003)), guinea pigs (Williamson et al. 2001; Williamson and Torres-Velez 2010), and ferrets (Bossart et al. 2009), and nonhuman primates (including African green monkeys (Old World monkey) (Geisbert et al. 2010; Rockx et al. 2010) and Saimiri/Squirell monkeys (New World monkey) (Marianneau et al. 2010)). Notably, while hamsters and ferrets are readily infectable, and recapitulate many symptoms reflective of human infections (for details, see the review on animal models for HNV in this issue), mice are resistant to NiV infection (Wong et al. 2003) despite the fact that murine ephrin-B2 shares 97 % sequence similarity with human ephrin-B2. Other animals that are susceptible to HNV infection, at least under laboratory conditions, include cats, dogs (Hooper et al. 2001; Mills et al. 2009; Mungall et al. 2006) (class *Mammalia)*, and chicken embryos (Tanimura et al. 2006) (class *Aves*) as discussed above. The curious exception of murine resistance to HNV infection likely occurs at a post-entry step, as the HNV attachment glycoprotein binds to murine ephrinB2 just as well as human ephrinB2 (Bossart et al. 2008).

1.5 Ephrins and Henipavirus Cellular Tropism In Vitro

The susceptibility of a given cell line to HNV infection in vitro largely depends on its ephrin-B2/B3 expression. For example, the hamster cell line CHO lacks any endogenous ephrinB expression, and expression of exogenous ephrin-B2 and -B3 (but not -B1) renders CHO cells highly permissive to NiV infection (Negrete et al. 2006). Most of the common cell lines used in the laboratory (including HEK293T, vero, HeLa-CCL2, but not the HeLa-USU subclone) can support HNV infection (see Table 1). On the other hand, although known to be endothelial-tropic, NiV is not able to infect all the endothelial cell lines. Endothelial cells from the capillaries and the brain such as PBMECs (porcine microvascular endothelial cells) and HBMECs (human brain endothelial cells), which express high levels of ephrin-B2, are susceptible to NiV infection, whereas PAECs (porcine aorta endothelial cells) and MyEnd (murine myocard), which do not express detectable levels of the receptors, are resistant to infection (Erbar et al. 2008).

So far, most cell lines with detectable ephrinB2 expression have been shown to be permissive to NiV infection with the possible exception of P815 (mouse mast cells) and 208f (rat embryonic fibroblasts). It is unclear why these cell lines are unable to support NiV replication, but in the case of 208f, interestingly, the amount of ephrinB2 at the cells surface as detected by flow cytometry is low although the mRNA is present at high levels. It is possible that the receptor may be down-regulated by intramembrane proteases similar to what has been shown for ephrin-B3/B2 and Rhomboid proteases as discussed above. A list of cell lines with their ephrin-B2/B3 expression profiles and infectability phenotypes is shown in Table 1

Table 1 Cell line tropism of Henipaviruses and receptor expression

Cell line	Species/Type	Ephrin-B2	Ephrin-B3	Infectability	References
CHO	Chinese Hamster Ovary	–	–	–	Negrete et al. (2005)
CHO B2	Stably expressing ephrin-B2	+	–	+	Negrete et al. (2005)
CHO B3	Stably expressing ephrin-B3	–	+	+	Negrete et al. (2005)
HeLa-CCL2	Human Uterus	+		+	Bonaparte et al. (2005)
HeLa-USU	Human Uterus	–	–	–	Bonaparte et al. (2005)
Vero	Vervet kidney epithelial	+	+	+	Bossart et al. (2005); Erbar et al. (2008)
MyEnd	Myocard Endothelial	–	–	–	Erbar et al. (2008)
HBMEC	Human brain endothelial	+	–	+	Erbar et al. (2008)
PBMEC	Porcine brain microvascular endothelial	+	+	+	Erbar et al. (2008)
PAEC	Pig Aorta	–	–	–	Erbar et al. (2008)
PK13	Pig Fibroblast	–	–	–	Negrete et al. (2005)
BHK	Hamster Kidney	+		+	Yoneda et al. (2006)
293T	Human Kidney (endothelial)	+		+	Yoneda et al. (2006)
4/4RM4	Rat epithelial	+	ND	+	Yoneda et al. (2006)
L2	Rat lung epithelia	+	ND	+	Yoneda et al. (2006)
208f	Rat embryonic fibroblast	+	ND	–	Yoneda et al. (2006)
P815	Mouse mast cell	+	ND	–	Yoneda et al. (2006)
PCI 13	Human head and neck carcinoma	ND[a]	ND[a]	+	Bossart et al. (2005)
U373-MG	Human Glioblastoma	ND[a]	ND[a]	+	Bossart et al. (2005)

[a] These cells have been tested positive for "a receptor". However, this receptor has not been identified in the above mentioned reference

2 Glycoprotein-Receptor Interaction

2.1 The Attachment Glycoprotein G

The HNV attachment glycoprotein is a type II transmembrane protein (602 and 604 amino acids for NiV-G and HeV-G, respectively). It shares structural similarities with the attachment glycoproteins of other Paramyxoviruses (Bowden et al. 2008a; Xu et al. 2008). These include an N-terminal cytoplasmic tail, a single transmembrane domain, a stalk region, and a globular head that folds as a six-bladed β-propeller barrel with a central canyon on the membrane-distal face (Bowden et al. 2008a; Xu et al. 2008). Despite these structural similarities, HNV-G does not possess hemagglutination (H) or neurominidase (N) activities, features that are common to most other Paramyxoviruses (reviewed in Lee and Ataman 2011). In addition, the stalk region of HNV-G is about 40 amino acids longer that the HN stalk from Paramyxoviruses that use glycan-based receptors (sialic acids) (Respiroviruses, Rubulaviruses, and Avulaviruses), but is similar in length to the H stalk in Morbilliviruses (the only other genus that uses protein-based receptors) (Maar et al. 2012). On the other hand, structural phylogeny analysis of the globular head domain indicates that HNV-G is more closely related to the HN attachment glycoproteins from glycan-using Paramyxoviruses than to the H attachment glycoprotein from Morbilliviruses (using measles H as the representative example) (Bowden et al. 2008a; 2010b). HN is thought to be the ancestral protein—thus the hybrid features of HNV-G suggest that HNV-G independently evolved to use protein-based receptors after MeV-H's own switch from using glycan-based to protein-based receptors. The more recent evolutionary history of HNV-G may contribute to the relatively broad species tropism of HNV, unlike Measles virus, which has had more time to adapt to using only the cognate receptors (CD150/SLAM) from humans and certain nonhuman primates (Hashiguchi et al. 2011a).

There is biochemical evidence that HNV-G oligomerizes as a dimer of a dimers (Bishop et al. 2008; Maar et al. 2012), consistent with the tetrameric assembly found for other paramyxoviral attachment proteins (Lamb et al. 2006; Yuan et al. 2008). Positional analysis of glycosylation sites and other structural modeling criteria suggest that HNV-G has a dimerization interface that is conserved amongst *Paramyxoviridae* (Bowden et al. 2008b; 2010a) although the relative angle of dimer association and the area of the dimer interface differs among HNVes, Morbilliviruses, and other glycan-using Paramyxoviruses (reviewed in Lee and Ataman 2011). Recent work on the structure of the NDV globular domain crystallized with part of its stalk region indicates that the stalk region adopts a four helix bundle that holds the tetrameric assembly together (Yuan et al. 2008; Yuan et al. 2011b). Strikingly, each pair of globular head domains was titled almost at right angles with respect to the other pair, exposing the four receptor-binding sites in four different directions. It is unclear whether HNV-G will also adopt this configuration—if it does, these putative structural features might place highly restrictive constraints on how the ephrin-B2/B3 "G-H" loop can access the receptor binding cleft in the globular head of HNV-G.

2.2 Henipavirus-G-receptor Interaction

Upon binding to its cellular receptor ephrin-B2/B3, conformational changes in NiV-G contribute to the "triggering" of the fusion protein F, leading to the fusion between viral and cellular membranes (for details, please see review by Aguilar and Iorio in this issue). The solvent exposed "G-H" loop of ephrin-B2/B3 is important for interacting with the endogenous EphB receptors, as well as the HNV-G proteins (Bowden et al. 2008a; Himanen et al. 2007; Xu et al. 2008). However, compared to EphB4 (or EphB2), both NiV-G and HeV-G have ~tenfold higher affinity for ephrin-B2 (K_d in the picomolar range) (Negrete et al. 2007). This is the highest

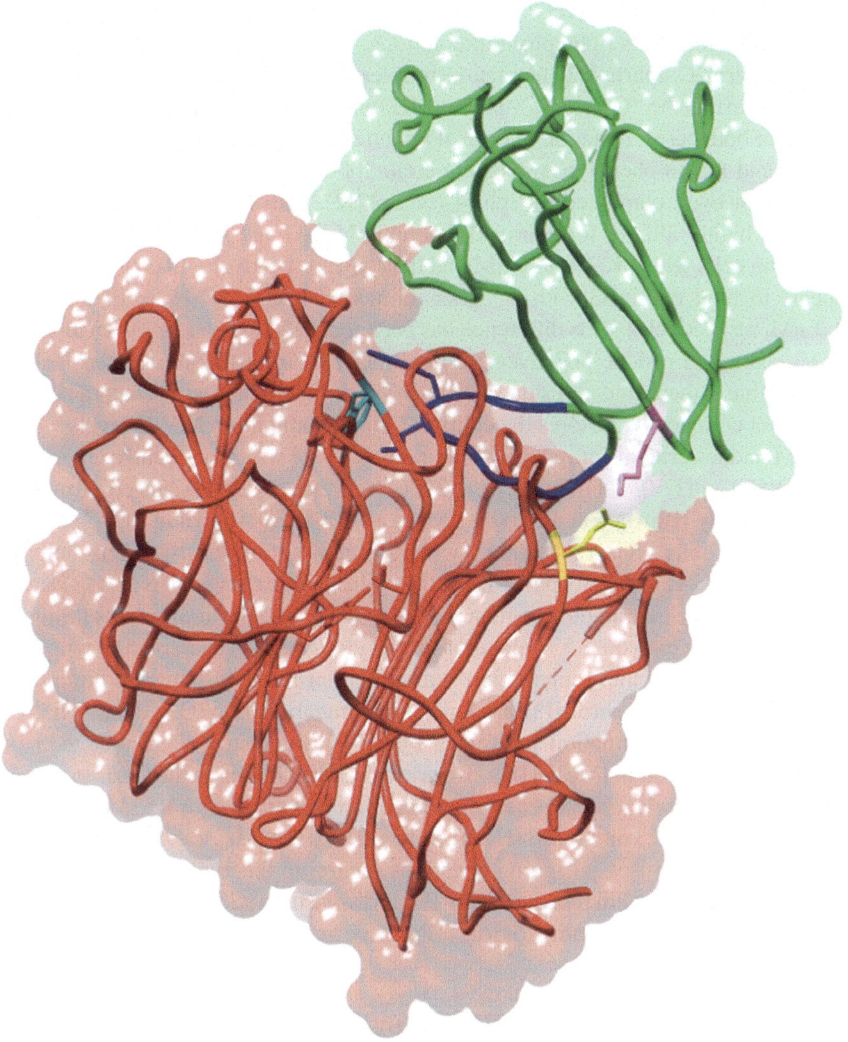

Fig. 2 Interaction between NiV-G and ephrin-B2. Basic amino acid residue K60 (*pink*) of ephrin-B2 (*green*) forms a salt bridge with acidic amino acid E533 (*yellow*) of NiV-G (*red*), guiding the GH-Loop of ephrin-B2 (*blue*) into the hydrophobic canyon located at the top of NiV-G globular head. Amino acid W504 (*cyan*) of NiV-G, located in the hydrophobic canyon, is important for differential binding to ephrin-B2 versus ephrinB3. Based on PDB ID:2VSM

affinity viral envelope-receptor interaction known to date, and likely reflects selective pressures under which HNV have evolved to use ephrin-B2/B3 as a receptor. For example, Eph/ephrin interactions, though highly promiscuous, are regulated in part by the strength of the signaling that results from clusters of Eph/ephrin interactions at the cell–cell interface. The high affinity of HNV-G for ephrin-B2/B3 may have evolved to compete effectively with the high avidity interactions that result from clusters of endogenous Eph/ephrin interactions.

Henipavirus-G-ephrin-B2 interaction mimics the endogenous EphB4-ephrin-B2 interaction in that the "G-H loop" of ephrin-B2 (amino acids 120–125) is important for binding in both cases; mutation of critical amino acids in this region results in altered binding and fusion phenotypes (Negrete et al. 2006; Pernet et al. 2009; Yuan et al. 2011a). However, compared to EphB4 (or EphB2), both NiV-G and HeV-G have ~tenfold higher affinity for ephrin-B2: soluble NiV-G and HeV-G bind ephrinB2 expressing cells with a K_d of 0.27 nM and 0.57 nM, respectively. In contrast, NiV-G appears to bind ephrinB3 with a higher affinity than HeV-G ($K_d = 0.78$ nM vs. 24.3 nM, respectively). This difference has been localized to a valine (NiV) to serine (HeV) change in residue 507 (Negrete et al. 2007). Curiously, an updated HeV-G sequence has a theronine at residue 507, and this serine to theronine change largely, though not completely, restores the ephrinB3 binding affinity to NiV-G levels.

The residues critical for receptor binding on NiV-G was functionally mapped to E533, W504 and E505 (Guillaume et al. 2006; Negrete et al. 2007), and confirmed by crystallography (Bowden et al. 2008a; Xu et al. 2008). As shown in Fig. 2, NiV-G E533 makes contact (via a salt bridge) with residue K60 on ephrin-B2. This helps to orientate the "G-H loop" of ephrin-B2 in the hydrophobic canyon of NiV-G in a way that facilitates the interaction between the "G-H loop" and the amino acids lining up the surface of the canyon. E533 makes a similar charge–charge interaction with R57 on ephrin-B3. Since these interactions are important for stabilizing the binding of both ephrin-B2 and ephrin-B3, a single point mutation E533Q reduces NiV-G-mediated syncytia formation and entry by 90 %. Interestingly, residue 533 on Measles virus H protein (MeV-H) is also critical for its binding to the CD150/SLAM receptor, although the charge–charge interaction is reversed with R533 on MeV-H interacting with E123 on SLAM (Hashiguchi et al. 2011b; Massé et al. 2004; Tatsuo et al. 2000). This raises intriguing questions regarding the evolution of two independent genera of protein receptor using Paramyxoviruses (G and H) from the likely ancestral glycan-using Paramyxovirus (HN) (Lee and Ataman 2011).

2.3 The Internalization of Viral Particles

The membrane fusion mediated by paramyxoviral glycoproteins is known to be pH independent and can presumably occur on the plasma membrane of the target cell. However, NiV has also been shown to exploit the cellular macropinocytosis pathway, potentially to its own advantage (Pernet et al. 2009).

Similar to the cognate receptor EphB4, NiV-G triggers the activation of ephrin-B2 by phosphorylation on specific tyrosine residues in the cytoplasmic tail. However, while residues Y304, Y330, and Y331 of ephrin-B2 are all involved in EphB4-triggered downstream events, only Y304 phosphorylation seems critical for "reverse signaling" during NiV-G binding (Pernet et al. 2009; Su et al. 2004). Subsequently, phospho-tyrosines 304 recruit the adaptor protein Grb4 (also known as Nck2) (Cowan and Henkemeyer 2001) which, in turn, recruits Ras and activates multiple downstream pathways involved in intracellular vesicle trafficking (Arf6), actin remodeling, and membrane ruffling (Rab5/RNtre and PI3K/PKC/Rab34), and regulating the switch between macropinocytosis and filopodia formation (Rac1/Cdc42) (Bucci et al. 1995; Marston et al. 2003; Radhakrishna et al. 1999; Sun et al. 2003). A panoply of assays using inhibitors of macropinocytosis and actin depolymerization, suggest the NiV-ephinB2 complex is internalized (Pernet et al. 2009) in a manner not unlike the bidirectional endocytosis seen for EphB4-ephrin-B2 complexes when they interact on opposing cells (reviewed in Pitulescu and Adams 2010).

Many viruses appear to use macropinocytosis to facilitate entry. Examples include vaccinia (*Poxviridae*) (Mercer and Helenius 2008), Dengue virus (*Flaviviridae*) (Suksanpaisan et al. 2009; Zamudio-Meza et al. 2009), and Ebola virus (*Filioviridae*) (Saeed et al. 2010). Measles virus, another Paramyxovirus that uses protein-based receptors for entry, can cross-link CD46 (a cognate receptor for the vaccine strain) and induce pseudopodia engulfment in a process similar to macropinocytosis (Crimeen-Irwin et al. 2003). Rac1/Cdc42-mediated internalization triggered by ephrin-B2 signaling is a well-established pathway in endothelial cells (Vandenbroucke et al. 2008). Co-opting this pathway in its natural target cell may have advantages for the virus: (1) rapid internalization potentially helps the virus to evade host immune recognition and (2) downregulation of ephrin-B2 from the cell surface (Pernet et al. 2009) may also prevent super-infection and facilitate the release of progeny virions.

3 Receptor Usage and Henipavirus Pathogenesis

3.1 Pathology and Symptoms

Henipavirus infection leads to vasculitis, necrosis, thrombosis, as well as brain parenchyma lesion associated with the formation of giant multi-nucleated cells (for details, see the review on "Clinical and pathological manifestations of human HNV infection" in this issue). These cytopathic effects are largely due to the syncytia-inducing ability of NiV and HeV glycoproteins, and the sites of lesion match ephrin-B2/B3 expression profiles.

The clinical presentations following HNV infection include neurological symptoms (e.g. headaches, drowsiness, disorientation, myoclonus, motor, and sensory loss), respiratory disorders (observed in HeV-infected horses, NiV-infected pigs, and 25–40 % of NiV-infected humans), unstable blood pressure, and, in one case, vision loss (Chua et al. 1999; Lim et al. 1999; Lim et al. 2003; Wong et al. 2002). Totally 40–92 % of NiV-infected humans succumb to acute encephalitis with an average time of 10 days from fever onset to death, while 3–7 % of infected patients exhibit a late onset or relapsed encephalitis months to years after the initial infection (Chua et al. 1999; Luby et al. 2006).

3.2 Ephrin-B2/B3 and NiV Pathogenesis

The pathology and symptoms associated with NiV infection can, in some cases, be explained by the expression pattern of ephrin-B2/B3 in vivo and the resultant cellular tropism of the virus (see Table 2). The more generalized symptomology may be a result of the inflammatory response directed against the replicating virus once it has established a primary infection.

A hallmark of HNV infection is the cellular pathology that results from the infection of microvascular endothelial cells (Hooper et al. 1997a, b; Wong et al. 2002). Syncytia formation induced by infection results in endothelial cell dysfunction and apoptosis; in the end-organ microvasculature, this can lead to vascular inflammation and thrombosis (Lee 2007). At autopsy, widespread vasculitis is seen in the lung (62 %), heart (31 %), kidney (24 %), and the CNS (80 %) (Wong et al. 2002), concordant with the high expression levels of ephrin-B2 in these tissues (Hafner et al. 2006). Focal perivascular necrosis was observed in most of the highly vascular organs, with the spleen being the most striking example. Furthermore, viral antigen staining was particularly evident in the periarteriolar sheaths in the white pulp (Wong et al. 2002), which is the only region in the spleen where ephrin-B2 expression can be detected (Gale et al. 2001).

In the CNS, where severe damage to vessels was seen in 80 % of the patients, vasculitis presumably results in the disruption of the blood–brain barrier, allowing the virus access to the CNS where it can infect ephrin-B2 and B3-positive cells in the brain parenchyma (Wong et al. 2002). However, it is also possible that infected brain microvascular endothelial cells produce virus that bud out from the basolateral surface and therefore gain access to the CNS without a breach in the BBB. There is evidence that the cytoplasmic domain of the Nipah F and G proteins contain basolateral targeting signals, at least when examined in polarized epithelial cells (Weise et al. 2010). Of course, both these pathological models are not mutually exclusive, and in all likelihood, occur together, and contribute to the devastating sequalae of the disease.

Consistent with the diffuse expression pattern of ephrin-B2 in the brain, MRI scans of acutely infected patients reveal multiple hyperintense lesions in the cortex, pons, putamen, and the cerebral and cerebellar peduncles. These lesions may underlie some of the neurological symptoms seen in Nipah infected patients (see the chapter by Wong and Tan in this issue). Interestingly, ephrin-B3, but not ephrin-B2, is also expressed in the brainstem (Negrete et al. 2007). Neurological symptoms reflective of brainstem involvement such as coma, abnormal doll's-eye reflex, abnormal pupils, and segmental myoclonus are bad prognostic factors, and brainstem neuronal dysfunction is thought to be the major cause of death in NiV encephalitis (Goh et al. 2000). In the African green monkey model of HeV infection, localized lesions of intense viral antigen staining can also be seen in the brainstem (Bossart et al. 2011). Thus, although ephrin-B2 likely serves as the receptor that allows for establishment of a primary HNV infection, ephrin-B3 mediated cellular pathology may also contribute significantly to the ultimate cause of death.

Table 2 Examples of infected sites and the related pathobiology

Site	Pathology	Symptoms	Receptor
Lungs	Microvascular destruction	Cough and other respiratory disorder	ephrin-B2
Retinal artery	Artery occlusion	Vision loss	ephrin-B2
CNS	BBB rupture	Motor and sensory loss, encephalitis	ephrin-B2
	Syncytia in brain parenchyma		ephrin-B2/B3
Liver	–	None	–

Henipavirus transmission from mother to fetus has been observed in the fruit bats, the natural reservoir for HNV, consistent with the expression of ephrin-B2 in the placenta (Williamson et al. 2000). In addition, vertical transmission and fetal replication of NiV has been documented in an experimentally infected cat (Mungall et al. 2007). There is some speculation that the seasonal uptick in HNV spillover events is associated with breeding cycle of the Asian fruit bats, which give birth in the winter (Jan-Mar in Bangladesh, May-Oct in Australia) (McFarlane et al. 2011; Plowright et al. 2008; Wacharapluesadee et al. 2010).

3.3 Alternative Tropism and Trans-Infection

Blood cells (including lymphocytes and monocytes) in general are not permissive to NiV infection (with the possible exception of dendritic cells, in which low levels of viral replication have been detected) (Mathieu et al. 2011). However, they can efficiently transport the virus from one infection site to another or even to a new host, as suggested by a recent study (Mathieu et al. 2011). This lymphocyte-mediated *trans*-infection is likely due to the binding of the virus to the cell surface via its glycoproteins without establishing a productive infection. Indeed, the virus can bind to the cell surface in a receptor-independent manner, since binding can occur in CHO cells which do not express detectable levels of ephrin-B2 or B3, and trans-infection can be abolished by C-type lectin inhibitors (Mathieu et al. 2011). These results are consistent with an earlier report by Bowden et al. showing that NiV-G can act as a putative ligand for the endothelial cell lectin, LSECtin (Bowden et al. 2008b).

4 Perspectives

Ephrin-B2 and B3, the cellular receptors for HNV entry, are mainly expressed in neuronal and endothelial cells. Receptor expression largely determines the cellular tropism of HNV, which is underscored by the high level of concordance between ephrin-B2/B3 distribution pattern in vivo and HNV-induced cellular pathology.

Although much has been done to elucidate the receptor interactions and tropism of HNV, the entry route of the virus into the human body still remains elusive. In Bangladesh, it has been documented that people can acquire NiV infection by drinking contaminated date palm sap (Luby et al. 2006; Rahman et al. 2012). Since the cells in the epithelium of the digestive tract do not express the HNV receptors, and the virus is sensitive to low pH exposure (Fogarty et al. 2008), how the virus manages to reach the susceptible cell types to establish a primary infection in this case is not clear. It is possible the virus may gain access to tissue macrophages and/or dendritic cells in the oropharyngeal submucosa through microscopic lesions commonly found in the mouth and throat. Macrophages in inflamed tissues such as the gingiva in periodontal disease, and the tonsils in general, express high levels of ephrin-B2 (Yuan et al. 2000, 2004) even if cultured monocyte-derived macrophages do not (Bossart et al. 2001). Binding to macrophages (or dendritic cells) may also allow the virus can be carried to permissive cells via "trans-infection" as discussed above.

An alternative explanation is that NiV is able to use another receptor for entry in the absence of ephrin-B2 and B3, similar to MeV. It has been known for a while that epithelial cells, which lack CD150/SLAM and CD46, are permissive for MeV infection. This elusive epithelial cell receptor was recently identified as nectin-4 (Noyce et al. 2011; Mühlebach et al. 2011). Since nectin-4 is expressed in epithelial cells (especially tracheal cells), it may facilitate the initial infection through the respiratory tract in addition to the well-established route of host entry via transmigrating alveolar macrophages, although it has been proposed as the exit receptor used by MeV to cross the tracheal epithelium and emerge in the airways (Mühlebach et al. 2011). Given the structural similarity between NiV-G and MeV-H, it is possible that similar mechanisms might exist for NiV-G as well.

Although HNV infections have a high mortality rate, spillover events into the human population remain rare. Preventive measures such as the use of universal precautions to protect against contaminated fluids from infected horses (Australia) or public education campaigns against drinking raw palm date juice appears to be the most cost-effective interim measures for preventing spread in the human population (Luby et al. 2009; Nahar et al. 2010). Many therapeutic strategies have been investigated (reviewed in Aguilar and Lee 2011; Vigant and Lee 2011), and promising vaccine candidates are being advanced through the pipeline (see chapter by C. Broder in this issue). The high affinity for HNV-G for its ephrin-B receptors makes it difficult to directly antagonize this molecular interaction via traditional small molecule therapeutics. However, passive immunotherapy with a monoclonal antibody that blocks HNV-G-receptor interactions appears promising, and even exhibits post-exposure efficacy when administered up to 72 h post-infection in the African green monkey model (Bossart et al. 2011). Nonetheless, the considerable knowledge gained from the study of the HNV-G proteins and their receptors can potentially be harnessed for use in therapies that target the ephrin-B2-EphB4 axis that is implicated in the tumorigenesis or tumor angiogenesis in certain cancers (reviewed in Pasquale 2008).

References

Adams RH, Wilkinson GA, Weiss C, Diella F, Gale NW et al (1999) Roles of ephrinb ligands and ephb receptors in cardiovascular development: demarcation of arterial/venous domains, vascular morphogenesis, and sprouting angiogenesis. Genes Dev 13:295–306

Aguilar HC, Lee B (2011) Emerging Paramyxoviruses: molecular mechanisms and antiviral strategies. Expert Rev Mol Med 13:e6

Bennett BD, Zeigler FC, Gu Q, Fendly B, Goddard AD et al (1995) Molecular cloning of a ligand for the eph-related receptor protein-tyrosine kinase htk. Proc Natl Acad Sci USA 92:1866–1870

Benson MD, Romero MI, Lush ME, Lu QR, Henkemeyer M et al (2005) Ephrin-b3 is a myelin-based inhibitor of neurite outgrowth. Proc Natl Acad Sci USA 102:10694–10699

Bergemann AD, Cheng HJ, Brambilla R, Klein R, Flanagan JG (1995) Elf-2, a new member of the eph ligand family, is segmentally expressed in mouse embryos in the region of the hindbrain and newly forming somites. Mol Cell Biol 15:4921–4929

Bergemann AD, Zhang L, Chiang MK, Brambilla R, Klein R et al (1998) Ephrin-B3, a ligand for the receptor EphB3, expressed at the midline of the developing neural tube. Oncogene 16:471–480

Bishop KA, Hickey AC, Khetawat D, Patch JR, Bossart KN et al (2008) Residues in the stalk domain of the Hendra virus g glycoprotein modulate conformational changes associated with receptor binding. J Virol 82:11398–11409

Bonaparte MI, Dimitrov AS, Bossart KN, Crameri G, Mungall BA et al (2005) Ephrin-b2 ligand is a functional receptor for Hendra virus and Nipah virus. Proc Natl Acad Sci USA 102:10652–10657

Bossart KN, Crameri G, Dimitrov AS, Mungall BA, Feng Y et al (2005) Receptor binding, fusion inhibition, and induction of cross-reactive neutralizing antibodies by a soluble g glycoprotein of Hendra virus. J Virol 79:6690–6702

Bossart KN, Geisbert TW, Feldmann H, Zhu Z, Feldmann F et al (2011) A neutralizing human monoclonal antibody protects African green monkeys from Hendra virus challenge. Sci Transl Med 3:105ra103

Bossart KN, Tachedjian M, McEachern JA, Crameri G, Zhu Z et al (2008) Functional studies of host-specific ephrin-b ligands as Henipavirus receptors. Virology 372:357–371

Bossart KN, Wang LF, Eaton BT, Broder CC (2001) Functional expression and membrane fusion tropism of the envelope glycoproteins of Hendra virus. Virology 290:121–135

Bossart KN, Zhu Z, Middleton D, Klippel J, Crameri G et al (2009) A neutralizing human monoclonal antibody protects against lethal disease in a new ferret model of acute Nipah virus infection. PLoS Pathog 5:e1000642

Bowden TA, Aricescu AR, Gilbert RJC, Grimes JM, Jones EY et al (2008a) Structural basis of Nipah and Hendra virus attachment to their cell-surface receptor ephrin-B2. Nat Struct Mol Biol 15:567–572

Bowden TA, Crispin M, Harvey DJ, Aricescu AR, Grimes JM et al (2008b) Crystal structure and carbohydrate analysis of Nipah virus attachment glycoprotein: a template for antiviral and vaccine design. J Virol 82:11628–11636

Bowden TA, Crispin M, Harvey DJ, Jones EY, Stuart DI (2010a) Dimeric architecture of the Hendra virus attachment glycoprotein: evidence for a conserved mode of assembly. J Virol 84:6208–6217

Bowden TA, Crispin M, Jones EY, Stuart DI (2010b) Shared paramyxoviral glycoprotein architecture is adapted for diverse attachment strategies. Biochem Soc Trans 38:1349–1355

Bucci C, Lütcke A, Steele-Mortimer O, Olkkonen VM, Dupree P et al (1995) Co-operative regulation of endocytosis by three rab5 isoforms. FEBS Lett 366:65–71

Chrencik JE, Brooun A, Kraus ML, Recht MI, Kolatkar AR et al (2006) Structural and biophysical characterization of the EphB4*ephrinB2 protein-protein interaction and receptor specificity. J Biol Chem 281:28185–28192

Chua KB (2010) Epidemiology, surveillance and control of Nipah virus infections in Malaysia. Malays J Pathol 32:69–73

Chua KB, Goh KJ, Wong KT, Kamarulzaman A, Tan PS et al (1999) Fatal encephalitis due to Nipah virus among pig-farmers in Malaysia. Lancet 354:1257–1259

Cowan CA, Henkemeyer M (2001) The sh2/sh3 adaptor grb4 transduces B-ephrin reverse signals. Nature 413:174–179

Crimeen-Irwin B, Ellis S, Christiansen D, Ludford-Menting MJ, Milland J et al (2003) Ligand binding determines whether cd46 is internalized by clathrin-coated pits or macropinocytosis. J Biol Chem 278:46927–46937

Edwards CM, Mundy GR (2008) Eph receptors and ephrin signaling pathways: a role in bone homeostasis. Int J Med Sci 5:263–272

Eph Nomenclature Committee (1997) Unified nomenclature for eph family receptors and their ligands, the ephrins. Eph nomenclature committee. Cell 90:403–404

Erbar S, Diederich S, Maisner A (2008) Selective receptor expression restricts Nipah virus infection of endothelial cells. Virol. J 5:142

Fogarty R, Halpin K, Hyatt AD, Daszak P, Mungall BA (2008) Henipavirus susceptibility to environmental variables. Virus Res 132:140–144

Füller T, Korff T, Kilian A, Dandekar G, Augustin HG (2003) Forward ephb4 signaling in endothelial cells controls cellular repulsion and segregation from ephrinB2 positive cells. J Cell Sci 116:2461–2470

Gale NW, Baluk P, Pan L, Kwan M, Holash J et al (2001) Ephrin-B2 selectively marks arterial vessels and neovascularization sites in the adult, with expression in both endothelial and smooth-muscle cells. Dev Biol 230:151–160

Geisbert TW, Daddario-DiCaprio KM, Hickey AC, Smith MA, Chan Y et al (2010) Development of an acute and highly pathogenic nonhuman primate model of Nipah virus infection. PLoS ONE 5:e10690

Georgakopoulos A, Litterst C, Ghersi E, Baki L, Xu C et al (2006) Metalloproteinase/presenilin1 processing of ephrinb regulates ephb-induced src phosphorylation and signaling. EMBO J 25:1242–1252

Georgakopoulos A, Xu J, Xu C, Mauger G, Barthet G et al (2011) Presenilin1/gamma-secretase promotes the ephb2-induced phosphorylation of ephrinB2 by regulating phosphoprotein associated with glycosphingolipid-enriched microdomains/csk binding protein. FASEB J 25:3594–3604

Gerety SS, Anderson DJ (2002) Cardiovascular ephrinB2 function is essential for embryonic angiogenesis. Development 129:1397–1410

Goh KJ, Tan CT, Chew NK, Tan PS, Kamarulzaman A et al (2000) Clinical features of Nipah virus encephalitis among pig farmers in Malaysia. N Engl J Med 342:1229–1235

Guillaume V, Aslan H, Ainouze M, Guerbois M, Wild TF et al (2006) Evidence of a potential receptor-binding site on the Nipah virus g protein (NiV-G): identification of globular head residues with a role in fusion promotion and their localization on an NiV-G structural model. J Virol 80:7546–7554

Hafner C, Becker B, Landthaler M, Vogt T (2006) Expression profile of Eph receptors and ephrin ligands in human skin and downregulation of epha1 in nonmelanoma skin cancer. Mod Pathol 19:1369–1377

Hasebe F, Thuy NTT, Inoue S, Yu F, Kaku Y et al (2012) Serologic evidence of Nipah virus infection in bats, Vietnam. Emerg Infect Dis 18:536–537

Hashiguchi T, Maenaka K, Yanagi Y (2011a) Measles virus hemagglutinin: structural insights into cell entry and measles vaccine. Front Microbiol 2:247

Hashiguchi T, Ose T, Kubota M, Maita N, Kamishikiryo J et al (2011b) Structure of the Measles virus hemagglutinin bound to its cellular receptor slam. Nat Struct Mol Biol 18:135–141

Hayman DTS, Suu-Ire R, Breed AC, McEachern JA, Wang L et al (2008) Evidence of Henipavirus infection in West African fruit bats. PLoS ONE 3:e2739

Hayman DTS, Wang L, Barr J, Baker KS, Suu-Ire R et al (2011) Antibodies to Henipavirus or Henipa-like viruses in domestic pigs in Ghana, West Africa. PLoS ONE 6:e25256

Himanen J, Saha N, Nikolov DB (2007) Cell-cell signaling via Eph receptors and ephrins. Curr Opin Cell Biol 19:534–542

Holmberg J, Frisén J (2002) Ephrins are not only unattractive. Trends Neurosci 25:239–243
Homaira N, Rahman M, Hossain MJ, Epstein JH, Sultana R et al (2010) Nipah virus outbreak with person-to-person transmission in a district of Bangladesh, 2007. Epidemiol Infect 138:1630–1636
Hooper P, Zaki S, Daniels P, Middleton D (2001) Comparative pathology of the diseases caused by Hendra and Nipah viruses. Microbes Infect 3:315–322
Hooper PT, Ketterer PJ, Hyatt AD, Russell GM (1997a) Lesions of experimental equine Morbillivirus pneumonia in horses. Vet Pathol 34:312–322
Hooper PT, Westbury HA, Russell GM (1997b) The lesions of experimental equine Morbillivirus disease in cats and guinea pigs. Vet Pathol 34:323–329
Hsu VP, Hossain MJ, Parashar UD, Ali MM, Ksiazek TG et al (2004) Nipah virus encephalitis reemergence, Bangladesh. Emerg Infect Dis 10:2082–2087
Kobayashi H, Kitamura T, Sekiguchi M, Homma MK, Kabuyama Y et al (2007) Involvement of EphB1 receptor/ephrinB2 ligand in neuropathic pain. Spine 32:1592–1598
Kullander K, Croll SD, Zimmer M, Pan L, McClain J et al (2001) Ephrin-B3 is the midline barrier that prevents corticospinal tract axons from recrossing, allowing for unilateral motor control. Genes Dev 15:877–888
Kunz TH, Braun de Torrez E, Bauer D, Lobova T, Fleming TH (2011) Ecosystem services provided by bats. Ann NY Acad Sci 1223:1–38
Lamb RA, Paterson RG, Jardetzky TS (2006) Paramyxovirus membrane fusion: lessons from the F and HN atomic structures. Virology 344:30–37
Lee B (2007) Envelope-receptor interactions in Nipah virus pathobiology. Ann NY Acad Sci 1102:51–65
Lee B, Ataman ZA (2011) Modes of Paramyxovirus fusion: a Henipavirus perspective. Trends Microbiol 19:389–399
Li Y, Wang J, Hickey AC, Zhang Y, Li Y et al (2008) Antibodies to nipah or nipah-like viruses in bats, China. Emerg Infect Dis 14:1974–1976
Liebl DJ, Morris CJ, Henkemeyer M, Parada LF (2003) mRNA expression of ephrins and Eph receptor tyrosine kinases in the neonatal and adult mouse central nervous system. J Neurosci Res 71:7–22
Lim CC, Sitoh YY, Lee KE, Kurup A, Hui F (1999) Meningoencephalitis caused by a novel Paramyxovirus: an advanced MRI case report in an emerging disease. Singap Med J 40:356–358
Lim CCT, Lee WL, Leo YS, Lee KE, Chan KP et al (2003) Late clinical and magnetic resonance imaging follow up of Nipah virus infection. J Neurol Neurosurg Psychiatr 74:131–133
Luby SP, Gurley ES, Hossain MJ (2009) Transmission of human infection with Nipah virus. Clin Infect Dis 49:1743–1748
Luby SP, Rahman M, Hossain MJ, Blum LS, Husain MM et al (2006) Foodborne transmission of Nipah virus, Bangladesh. Emerg Infect Dis 12:1888–1894
Mäkinen T, Adams RH, Bailey J, Lu Q, Ziemiecki A et al (2005) PDZ interaction site in ephrinB2 is required for the remodeling of lymphatic vasculature. Genes Dev 19:397–410
Marianneau P, Guillaume V, Wong T, Badmanathan M, Looi RY et al (2010) Experimental infection of squirrel monkeys with Nipah virus. Emerg Infect Dis 16:507–510
Maar D, Harmon B, Chu D et al (2012) Cysteines in the stalk of the Nipah virus G glycoprotein are located in a distinct subdomain critical for fusion activation. J Virol (in press)
Marston DJ, Dickinson S, Nobes CD (2003) Rac-dependent trans-endocytosis of ephrinbs regulates Eph–ephrin contact repulsion. Nat Cell Biol 5:879–888
Massé N, Ainouze M, Néel B, Wild TF, Buckland R et al (2004) Measles virus (mv) hemagglutinin: evidence that attachment sites for mv receptors slam and CD46 overlap on the globular head. J Virol 78:9051–9063
Mathieu C, Pohl C, Szecsi J, Trajkovic-Bodennec S, Devergnas S et al (2011) Nipah virus uses leukocytes for efficient dissemination within a host. J Virol 85:7863–7871
McFarlane R, Becker N, Field H (2011) Investigation of the climatic and environmental context of Hendra virus spillover events 1994–2010. PLoS ONE 6:e28374

Mercer J, Helenius A (2008) Vaccinia virus uses macropinocytosis and apoptotic mimicry to enter host cells. Science 320:531–535

Meyer S, Hafner C, Guba M, Flegel S, Geissler EK et al (2005) Ephrin-B2 overexpression enhances integrin-mediated ECM-attachment and migration of B16 melanoma cells. Int J Oncol 27:1197–1206

Mills JN, Alim ANM, Bunning ML, Lee OB, Wagoner KD et al (2009) Nipah virus infection in dogs, Malaysia, 1999. Emerg Infect Dis 15:950–952

Mühlebach MD, Mateo M, Sinn PL, Prüfer S, Uhlig KM et al (2011) Adherens junction protein nectin-4 is the epithelial receptor for Measles virus. Nature 480:530–533

Mungall BA, Middleton D, Crameri G, Bingham J, Halpin K et al (2006) Feline model of acute Nipah virus infection and protection with a soluble glycoprotein-based subunit vaccine. J Virol 80:12293–12302

Mungall BA, Middleton D, Crameri G, Halpin K, Bingham J et al (2007) Vertical transmission and fetal replication of Nipah virus in an experimentally infected cat. J Infect Dis 196:812–816

Nahar N, Sultana R, Gurley ES, Hossain MJ, Luby SP (2010) Date palm sap collection: exploring opportunities to prevent nipah transmission. Ecohealth 7:196–203

Negrete OA, Chu D, Aguilar HC, Lee B (2007) Single amino acid changes in the nipah and Hendra virus attachment glycoproteins distinguish ephrinB2 from ephrinB3 usage. J Virol 81:10804–10814

Negrete OA, Levroney EL, Aguilar HC, Bertolotti-Ciarlet A, Nazarian R et al (2005) EphrinB2 is the entry receptor for Nipah virus, an emergent deadly Paramyxovirus. Nature 436:401–405

Negrete OA, Wolf MC, Aguilar HC, Enterlein S, Wang W et al (2006) Two key residues in ephrinB3 are critical for its use as an alternative receptor for Nipah virus. PLoS Pathog 2:e7

Noyce RS, Bondre DG, Ha MN, Lin LT, Sisson G, Tsao MS, Richardson CD (2011) Tumor cell marker PVRL4 (nectin 4) is an epithelial cell receptor for Measles virus. PLoS Pathog 7:e1002240

Olson JG, Rupprecht C, Rollin PE, An US, Niezgoda M et al (2002) Antibodies to nipah-like virus in bats (*pteropus lylei*), Cambodia. Emerg Infect Dis 8:987–988

Pascall JC, Brown KD (2004) Intramembrane cleavage of ephrinB3 by the human rhomboid family protease, RHBDl2. Biochem Biophys Res Commun 317:244–252

Pasquale EB (2008) Eph–ephrin bidirectional signaling in physiology and disease. Cell 133:38–52

Pernet O, Pohl C, Ainouze M, Kweder H, Buckland R (2009) Nipah virus entry can occur by macropinocytosis. Virology 395:298–311

Pfaff D, Héroult M, Riedel M, Reiss Y, Kirmse R et al (2008) Involvement of endothelial ephrin-B2 in adhesion and transmigration of EphB-receptor-expressing monocytes. J Cell Sci 121:3842–3850

Pitulescu ME, Adams RH (2010) Eph/ephrin molecules—a hub for signaling and endocytosis. Genes Dev 24:2480–2492

Plowright RK, Field HE, Smith C, Divljan A, Palmer C et al (2008) Reproduction and nutritional stress are risk factors for Hendra virus infection in little red flying foxes (*Pteropus scapulatus*). Proc Biol Sci 275:861–869

Radhakrishna H, Al-Awar O, Khachikian Z, Donaldson JG (1999) ARF6 requirement for Rac ruffling suggests a role for membrane trafficking in cortical actin rearrangements. J Cell Sci 112(Pt 6):855–866

Rahman MA, Hossain MJ, Sultana S, Homaira N, Khan SU et al (2012) Date palm sap linked to Nipah virus outbreak in Bangladesh, 2008. Vector Borne Zoonotic Dis 12:65–72

Rockx B, Bossart KN, Feldmann F, Geisbert JB, Hickey AC et al (2010) A novel model of lethal Hendra virus infection in African green monkeys and the effectiveness of ribavirin treatment. J Virol 84:9831–9839

Saeed MF, Kolokoltsov AA, Albrecht T, Davey RA (2010) Cellular entry of Ebola virus involves uptake by a macropinocytosis-like mechanism and subsequent trafficking through early and late endosomes. PLoS Pathog 6:e1001110

Selvey LA, Wells RM, McCormack JG, Ansford AJ, Murray K et al (1995) Infection of humans and horses by a newly described Morbillivirus. Med J Aust 162:642–645

Su Z, Xu P, Ni F (2004) Single phosphorylation of TYR304 in the cytoplasmic tail of ephrin B2 confers high-affinity and bifunctional binding to both the SH2 domain of GRB4 and the PDZ domain of the PDZ-RGS3 protein. Eur J Biochem 271:1725–1736

Suksanpaisan L, Susantad T, Smith DR (2009) Characterization of Dengue virus entry into HEPG2 cells. J Biomed Sci 16:17

Sun P, Yamamoto H, Suetsugu S, Miki H, Takenawa T et al (2003) Small GTPase Rah/Rab34 is associated with membrane ruffles and macropinosomes and promotes macropinosome formation. J Biol Chem 278:4063–4071

Tan C, Chua K (2008) Nipah virus encephalitis. Curr Infect Dis Rep 10:315–320

Tanimura N, Imada T, Kashiwazaki Y, Sharifah SH (2006) Distribution of viral antigens and development of lesions in chicken embryos inoculated with Nipah virus. J Comp Pathol 135:74–82

Tatsuo H, Ono N, Tanaka K, Yanagi Y (2000) Slam (CDW150) is a cellular receptor for Measles virus. Nature 406:893–897

Vandenbroucke E, Mehta D, Minshall R, Malik AB (2008) Regulation of endothelial junctional permeability. Ann NY Acad Sci 1123:134–145

Vigant F, Lee B (2011) Hendra and nipah infection: pathology, models and potential therapies. Infect Disord Drug Targets 11:315–336

Wacharapluesadee S, Boongird K, Wanghongsa S, Ratanasetyuth N, Supavonwong P et al (2010) A longitudinal study of the prevalence of Nipah virus in *Pteropus lylei* bats in Thailand: evidence for seasonal preference in disease transmission. Vector Borne Zoonotic Dis 10:183–190

Wang HU, Chen ZF, Anderson DJ (1998) Molecular distinction and angiogenic interaction between embryonic arteries and veins revealed by ephrin-B2 and its receptor Eph-B4. Cell 93:741–753

Weise C, Erbar S, Lamp B, Vogt C, Diederich S et al (2010) Tyrosine residues in the cytoplasmic domains affect sorting and fusion activity of the Nipah virus glycoproteins in polarized epithelial cells. J Virol 84:7634–7641

Wilkinson DG (2003) How attraction turns to repulsion. Nat Cell Biol 5:851–853

Williamson MM, Hooper PT, Selleck PW, Westbury HA, Slocombe RF (2000) Experimental Hendra virus infectionin pregnant guinea-pigs and fruit bats (*Pteropus poliocephalus*). J Comp Pathol 122:201–207

Williamson MM, Hooper PT, Selleck PW, Westbury HA, Slocombe RF (2001) A guinea-pig model of Hendra virus encephalitis. J Comp Pathol 124:273–279

Williamson MM, Torres-Velez FJ (2010) Henipavirus: a review of laboratory animal pathology. Vet Pathol 47:871–880

Wong KT, Grosjean I, Brisson C, Blanquier B, Fevre-Montange M et al (2003) A golden hamster model for human acute Nipah virus infection. Am J Pathol 163:2127–2137

Wong KT, Shieh W, Kumar S, Norain K, Abdullah W et al (2002) Nipah virus infection: pathology and pathogenesis of an emerging Paramyxoviral zoonosis. Am J Pathol 161:2153–2167

Xu K, Rajashankar KR, Chan Y, Himanen JP, Broder CC et al (2008) Host cell recognition by the Henipaviruses: crystal structures of the nipah g attachment glycoprotein and its complex with ephrin-B3. Proc Natl Acad Sci USA 105:9953–9958

Yoneda M, Guillaume V, Ikeda F, Sakuma Y, Sato H et al (2006) Establishment of a Nipah virus rescue system. Proc Natl Acad Sci USA 103:16508–16513

Young PL, Halpin K, Selleck PW, Field H, Gravel JL et al (1996) Serologic evidence for the presence in pteropus bats of a Paramyxovirus related to equine Morbillivirus. Emerg Infect Dis 2:239–240

Yuan J, Marsh G, Khetawat D, Broder CC, Wang L et al (2011a) Mutations in the G–H loop region of ephrin-B2 can enhance Nipah virus binding and infection. J Gen Virol 92:2142–2152

Yuan K, Hong T, Chen JJW, Tsai WH, Lin MT (2004) Syndecan-1 up-regulated by ephrinB2/EphB4 plays dual roles in inflammatory angiogenesis. Blood 104:1025–1033

Yuan K, Jin YT, Lin MT (2000) Expression of tie-2, angiopoietin-1, angiopoietin-2, ephrinb2 and ephb4 in pyogenic granuloma of human gingiva implicates their roles in inflammatory angiogenesis. J Periodont Res 35:165–171

Yuan P, Leser GP, Demeler B, Lamb RA, Jardetzky TS (2008) Domain architecture and oligomerization properties of the Paramyxovirus PIV 5 hemagglutinin-neuraminidase (HN) protein. Virology 378:282–291

Yuan P, Swanson KA, Leser GP, Paterson RG, Lamb RA et al (2011b) Structure of the newcastle disease virus hemagglutinin–neuraminidase (HN) ectodomain reveals a four-helix bundle stalk. Proc Natl Acad Sci USA 108:14920–14925

Zamudio-Meza H, Castillo AM, González-Bonilla C et al (2009) Rac1 and CDC42 GTPases cross-talk regulates formation of filopodia required for Dengue virus type-2 entry into HMEC-1 cells. J Gen Virol 90(pt 12):2902–2911.Epub 2009 Aug 26

Zhao C, Irie N, Takada Y, Shimoda K, Miyamoto T et al (2006) Bidirectional ephrinb2–Ephb4 signaling controls bone homeostasis. Cell Metab 4:111–121

Zimmer M, Palmer A, Köhler J, Klein R (2003) EphB-ephrinB bi-directional endocytosis terminates adhesion allowing contact mediated repulsion. Nat Cell Biol 5:869–878

Henipavirus Membrane Fusion and Viral Entry

Hector C. Aguilar and Ronald M. Iorio

Abstract Nipah (NiV) and Hendra (HeV) viruses cause cell–cell fusion (syncytia) in brain, lung, heart, and kidney tissues, leading to encephalitis, pneumonia, and often death. Membrane fusion is essential to both viral entry and virus-induced cell–cell fusion, a hallmark of henipavirus infections. Elucidiation of the mechanism(s) of membrane fusion is critical to understanding henipavirus pathobiology and has the potential to identify novel strategies for the development of antiviral therapeutic agents. Henipavirus membrane fusion requires the coordinated actions of the viral attachment (G) and fusion (F) glycoproteins. Current henipavirus fusion models posit that attachment of NiV or HeV G to its cell surface receptors releases F from its metastable pre-fusion conformation to mediate membrane fusion. The identification of ephrinB2 and ephrinB3 as henipavirus receptors has paved the way for recent advances in our understanding of henipavirus membrane fusion. These advances highlight mechanistic similarities and differences between membrane fusion for the henipavirus and other genera within the *Paramyxoviridae* family. Here, we review these mechanisms and the current gaps in our knowledge in the field.

H. C. Aguilar (✉)
Department of Veterinary Microbiology and Pathology,
Paul G. Allen School for Global Animal Health,
College of Veterinary Medicine, Washington State University,
P.O. Box 617010, Bustad 423, 100 Grimes Way,
Pullman, WA 99164-7010, USA
e-mail: haguilar@vetmed.wsu.edu

R. M. Iorio
Department of Microbiology and Physiological Systems,
University of Massachusetts Medical School,
Worcester, MA 01655, USA

Current Topics in Microbiology and Immunology (2012) 359: 79–94
DOI: 10.1007/82_2012_200

Published Online: 17 March 2012

Contents

1 Introduction

Henipavirus membrane fusion is essential to both viral entry (viral–cell fusion) and virally induced syncytium formation (cell–cell fusion) in infected tissues. Elucidation of the mechanism by which these viruses promote these processes is key to understanding virus pathobiology. In addition, the intricate sequence of events that culminate in fusion constitute attractive antiviral targets (Aguilar and Lee 2011; McKinnell and Saag 2009; Murineddu et al. 2010).

The paramyxovirus fusion (F) protein has canonical structural features of Class I fusion proteins such as the retroviral Moloney murine leukemia virus (MoMuLV) p15E, the lentiviral human immunodeficiency virus (HIV-1) gp41, the Ebola virus GP2 and the influenza virus HA. All share similar trimeric coiled-coil core structures (Baker et al. 1999; Carr and Kim 1993; Fass et al. 1996; Lou et al. 2006; Yin et al. 2005) and several conserved functional domains (Lamb and Jardetzky 2007; Lamb et al. 2006; White et al. 2008). Class I fusion proteins are synthesized as trimeric precursors, which are cleaved for activation into a metastable pre-fusion conformation (Fig. 1a) and poised for the promotion of membrane fusion. Cleavage generates a new hydrophobic N-terminal fusion peptide (FP) that is buried intra-molecularly until F-triggering and pre-hairpin intermediate (PHI) formation, when the FP is harpooned into the target cell membrane (Fig. 1b) (White et al. 2008). The PHI contains two helical regions (HR1 and HR2) with high propensity to bind each other to form a six-helix bundle (6HB), enabling membrane fusion (Fig. 1c) (Lamb et al. 2006; White et al. 2008; Xu et al. 2004).

The current understanding is that, for the majority of paramyxoviruses, binding of the attachment protein to cellular receptors is the trigger for F to undergo this fusion cascade (Fig. 1a). Thus, paramyxovirus membrane fusion generally requires a cooperative effort of the viral attachment and fusion (F) glycoproteins, likely mediated by a virus-specific interaction between the two proteins. This complex interplay between the two surface glycoproteins not only introduces an

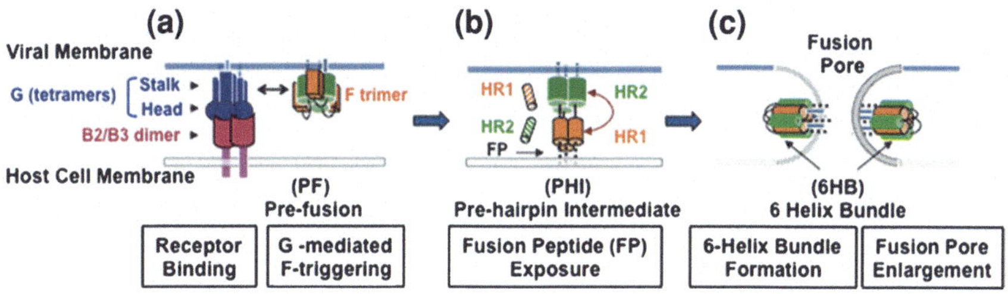

Fig. 1 The henipavirus fusion cascade. A schematic of the henipavirus fusion cascade divided into its three major stages: **a** EphrinB2 or ephrinB3 binding to henipavirus G triggers F through allosteric mechanisms that are still to be elucidated; **b** After F is triggered, it forms a pre-hairpin intermediate (PHI), in which a fusion peptide is harpooned into the host cell membrane. The PHI can be captured by peptides that mimic the henipavirus F HR1 (*orange striped cylinders*) or HR2 regions (*green striped cylinders*) and bind the F HR2 or HR1 regions, respectively; **c** The HR1 and HR2 region in the PHI coalesce to form the six-helix bundle (6HB) conformation, bringing the viral and cell membranes together and facilitating viral-host membrane fusion and viral entry. The viral membrane can be substituted by a cell membrane expressing the F and G glycoproteins, in cell–cell fusion, resulting in syncytium formation

added level of complexity to the fusion process, but also renders it amenable to novel antiviral intervention strategies.

All paramyxovirus attachment proteins are type II homotetrameric membrane glycoproteins, having a receptor-binding "head" domain connected to its transmembrane anchor via a "stalk" domain, as well as a short cytoplasmic tail (Fig. 1a) (Bowden et al. 2008a; Xu et al. 2008). While the hemagglutinin-neuraminidase (HN) proteins of viruses such as Newcastle disease virus (NDV), parainfluenza virus 5 (PIV5) and hPIV3 recognize sialic acid-containing receptors (Lamb and Parks 2007), the attachment proteins of other members of the family recognize specific protein receptors. The measles virus (MV) hemagglutinin (H) binds to CD46 (Dorig et al. 1993), SLAM (Tatsuo et al. 2000) or tumor cell marker nectin-4, the latter on epithelial cells (Muhlebach et al. 2011; Noyce et al. 2011). The Henipavirus G attachment protein recognizes ephrinB2 (B2) or ephrinB3 (B3), the membrane-bound ligands for the EphB class of receptor tyrosine kinases (Bonaparte et al. 2005; Negrete et al. 2005, 2006). B2 appears to be the higher affinity receptor for both viruses with NiV G having higher affinity than HeV G for both receptors (Negrete et al. 2006, 2007).

Similar to most other paramyxoviruses, current NiV and HeV membrane fusion models propose that binding of G to the cell surface receptor B2/B3 triggers a fusion cascade in F. However, the intricacies of the cascade of events by which the binding of receptor by the head of the G protein is transduced to the F protein to trigger the conformational change in F that is integral to the fusion process are still not well understood.

2 The Role(s) of Henipavirus G in Membrane Fusion

2.1 Paramyxovirus Attachment Proteins

With only a few exceptions, including respiratory syncytial virus (RSV) and human metapneumovirus (HMPV), the attachment proteins of paramyxoviruses are essential for cell–cell fusion and viral entry. Although membrane fusion is enhanced in the presence of the attachment glycoprotein for RSV (Kahn et al. 1999), low pH, and not the presence of the attachment protein, is believed to be the fusion trigger for some HMPV strains (Herfst et al. 2008; Schowalter et al. 2006a). Furthermore, for HMPV, no increase in viral entry was observed in the presence of the attachment protein (Kahn et al. 1999; Schowalter et al. 2006a and reviewed in Dutch 2010; Lamb et al. 2006).

For all paramyxoviruses for which fusion requires both surface glycoproteins, the specific role(s) of the attachment protein in the process is of great interest (Lamb and Jardetzky 2007; Lamb et al. 2006; White et al. 2008). It is commonly accepted that the binding of the attachment protein to cell surface receptors is the trigger for the conformational cascade in the F protein that culminates in membrane fusion (Fig. 1a). Even though the determinants of F-triggering on the attachment protein have not been completely defined, evidence indicates that determinants in the stalk of NDV HN and measles H mediate the interactions with the homologous F proteins ((Bishop et al. 2008; Corey and Iorio 2007; Lee et al. 2008b; Melanson and Iorio 2004, 2006). Moreover, the recently solved structures of the stalks of NDV and PIV-5 HN demonstrate them to be four-helix bundles (Bose et al. 2011; Yuan et al. 2011) and the NDV HN structure reveals that the putative F-interactive residues (Melanson and Iorio 2004) are, indeed, exposed on the surface of the helix. Similarly, recent evidence indicates an interaction between an analogous domain in the stalk region of the measles virus attachment protein and the globular head domain of the cognate F protein (Paal et al. 2009). However, despite this progress for other viruses, the domains on the henipavirus G and F proteins that mediate complex formation between them remain to be identified.

2.2 Henipavirus G and Membrane Fusion

For the henipaviruses, it is widely accepted that the attachment of G to B2 or B3 initiates F-triggering (Fig. 1). However, exactly how G links receptor-binding to F-triggering is not well understood. Biochemical and biophysical evidence strongly suggests that a receptor-induced conformational change in NiV-G is integral to the fusion cascade. Moreover, this conformational change was reported to involve critical residues at the base of the NiV-G head domain (Aguilar et al. 2009) and was marked by receptor-attachment enhancement of the binding of a monoclonal antibody (Mab45) and by secondary structural changes detected by circular dichroism. In addition, fusion-null HeV-G stalk domain mutants were differentially recognized by a panel of monoclonal antibodies when bound to B2 relative to the wild-type HeV-G (Bishop et al. 2008). Taken together, these NiV and HeV data are consistent with a requirement for receptor-induced conformational changes in G for F-triggering.

An alternate (and not mutually exclusive) possibility derives from analogy with recent findings for the MV H protein revealed by the structures of co-crystals of soluble H head domains with receptor (Hashiguchi et al. 2011). Two different organizations of the dimers in the H tetramer were identified with one thought to be pre-fusion and the other post-fusion H. Although the dimers themselves are unchanged in the two forms, in the post-fusion form, the tetrameric stalk is dissociated into two widely separated dimers, which may result in the release of F to assume its fusion-active conformation. Thus, the possibility exists that the fusion triggering relevant changes in G could also involve the association between the dimers rather than only changes in the conformation of the monomers themselves.

However, one can envision an even more passive scenario. For example, one possibility that has not completely been eliminated is that, on receptor-binding, G may serve simply to position F in close proximity to the target cell membrane, which itself may trigger F without the need of major structural rearrangements or conformational changes in G. A piece of evidence that is consistent with such a "passive" mechanism of F-triggering by G is that no significant differences were found between the nonreceptor-bound and receptor-bound soluble NiV G head domain crystal structures (Bowden et al. 2008a; Xu et al. 2008). However, the absence of the NiV-G stalk domain in these experiments may account for these results. Perhaps, the presence of the stalk allows for proper disassembly or rearrangement of higher ordered oligomers on receptor engagement, as the stalks of other paramyxovirus attachment proteins have been shown to be involved in oligomerization (Gravel and Morrison 2003; Melanson and Iorio 2004, 2006; Stone-Hulslander and Morrison 1999; Yuan et al. 2011). For example, de-oligomerization may lead to exposure of neoepitopes that trigger F. The stalk domain may also influence G's conformational changes. Indeed, the lack of the stalk domain of soluble NiV G abrogated its receptor-induced conformational changes observed by circular dichroism (Aguilar et al. 2009).

Nonetheless, whatever the mechanism used by the henipaviruses to regulate the activation of their F proteins, it is highly likely that this process is finely tuned and thus vulnerable to disruption. A better understanding of this process may lead to the development of therapeutics that target domains that limit the development of resistance. For example, anti-henipavirus G antibodies that recognize receptor-induced exposed epitopes may be suitable for passive immunization strategies. Reagents that target henipavirus receptor binding have shown success in disrupting henipavirus infections. For example, a monoclonal antibody that competes with B2 for binding to henipavirus G protected ferrets from NiV infection (Bossart et al. 2009) and green monkeys from HeV infection (Bossart et al. 2011).

Additionally, the structure of the B2- or B3-bound henipavirus G complex shows a large protein–protein interface and reveals a lock-and-key binding pocket that may be targeted by small molecules (Bowden et al. 2008b; Xu et al. 2008). For example, Trp125 and Phe120 in the G-H loop of ephrinB2 interact differently with EphB4 than with HNV-G, suggesting a "druggable" pocket to disrupt B2/B3-G interactions specifically (Lee et al. 2008a), provided that a drug can be found that competes for the strong B2-G interactions (Negrete et al. 2006). The identification of new roles of G in the membrane fusion cascade may also result in the discovery of new antiviral targets. For example, a recent report suggests that G-HN/receptor engagement may be important beyond the F-triggering step for NiV and hPIV3 (Porotto et al. 2011).

3 F/G Interactions

The role of G/H/HN-F interactions in membrane fusion are highlighted by several studies of paramyxoviruses and henipaviruses. For the paramyxoviruses, heterotypic attachment and fusion glycoprotein combinations are fusion-null, or nearly so, compared to homotypic combinations (Lamb et al. 2006; White et al. 2008). Even for the closely related NiV and HeV, significantly lower levels of cell–cell fusion were observed when heterotypic F and G were expressed in receptor-positive cell lines relative to when homotypic F and G were expressed (Bossart et al. 2001, 2002). The simplest explanation for homotypic henipavirus glycoprotein combinations being significantly more efficient than heterotypic ones in promoting cell–cell fusion is that the F/G interactions are highly specific and are likely to play an important role in F-triggering.

Additional evidence for the importance of F/G interactions in henipavirus fusion is the inverse correlation between the fusion capacity of F (fusogenicity) and the avidity of F/G interactions for F fusion mutants (Aguilar et al. 2006, 2007). F mutants with hyperfusogenic or hypofusogenic phenotypes displayed lower or higher avidities of F/G interactions, respectively. These data also suggest that F/G interactions must have optimal avidities to result in optimal levels of fusion, and that F/G dissociation might be a critical and rate limiting step in henipavirus membrane fusion.

There are two models for the role of F/G interactions in membrane fusion for the paramyxoviruses. The association (or provocateur) and dissociation (or clamp) models propose that association or dissociation, respectively, of the fusion and attachment glycoproteins is a rate limiting step in membrane fusion (reviewed in (Iorio and Mahon 2008; Iorio et al. 2009)). Several lines of evidence support the idea that the henipaviruses promote fusion via the dissociation mechanism: (1) the amount of fusion promoted by NiV G and mutated F proteins is inversely proportional to G-F complex avidity (Aguilar et al. 2006, 2007); (2) decreased receptor binding increases the amount of the HeV complex (Bishop et al. 2007); and, (3) receptor binding lowers the avidity of NiV F-G interactions in viral-like particles (Aguilar et al. 2009). These findings are consistent with henipavirus fusion being promoted by receptor binding-triggered dissociation of a pre-formed complex. Interestingly, whether a particular paramyxovirus uses the association or dissociation mechanism to promote fusion seems to segregate with receptor type (Iorio and Mahon 2008; Iorio et al. 2009). Paramyxoviruses (i.e. MV and NiV) that recognize protein receptors seem to utilize the dissociation model, whereas those i.e. NDV) that recognize sialic acid receptors utilize the association model, though more work is needed to determine whether there is, indeed, a causal relationship between receptor type and fusion mechanism.

Although the F/G interactive domains are likely highly specific, at least some aspects of F-triggering must be conserved among the paramyxoviruses. Recent evidence showed the NDV F protein can be triggered for fusion by a chimeric attachment protein that binds NiV receptors (Mirza et al. 2011). Subsequent construction of a converse chimera showed NiV F can be triggered by a chimeric attachment protein that binds the NDV sialic acid-containing receptor, though for reasons that are not clear, fusion was dependent on treatment with a neuraminidase inhibitor (Porotto et al. 2011). These conserved features of F-triggering are likely important targets for interfering with henipavirus and paramyxovirus membrane fusion.

In addition, fusion-deficient F and G mutants have been identified that retain the ability to interact with their respective homotypic partners (F mutants with wild-type G and vice versa) (Aguilar et al. 2006, 2007, 2009). These mutants must be blocked at a step in the fusion cascade that is distinct from glycoprotein complex formation, suggesting that there are additional steps that, once elucidated, will not only serve to dissect the steps in the triggering cascade, but are also likely to reveal new strategies for blocking virus entry. Indeed, similar fusion deficient, F-interactive mutations have been identified in the measles H protein (Corey and Iorio 2007) and NDV HN protein (Mahon et al. 2011), suggesting that this phenomenon likely exists across the paramyxovirus family.

4 The Role of Henipavirus F in Membrane Fusion

4.1 Paramyxovirus Fusion Proteins

The determinants of membrane fusion triggering on paramyxovirus F glycoproteins have not been clearly defined. The HR1, HR2, and fusion peptide must be released from the pre-fusion conformation into the pre-hairpin intermediate, but what regulates this release is still unknown (Lamb and Jardetzky 2007; Yin et al. 2006). An additional gap in knowledge in paramyxovirus fusion is what happens after 6HB formation (Fig. 1c). An understanding of the subsequent fusion pore formation and the determinants of its modulation in F are still to be addressed (Lamb and Jardetzky 2007; White et al. 2008).

The henipavirus G and F signaling cascade appears to be even more complex than that of the other paramyxoviruses based on the unique trafficking pathway of the henipavirus F glycoprotein. Most F glycoprotein precursors (F_0) are cleaved ($F_1 + F_2$) by furin-like cellular proteases during transport through the *trans* Golgi network to the cell surface (Begona Ruiz-Arguello et al. 2002; Garten et al. 1994; Gonzalez-Reyes et al. 2001; Ortmann et al. 1994; Watanabe et al. 1995) or by tissue-specific extracellular proteases such as tryptase Clara or mini-plasmin (Murakami et al. 2001; van den Hoogen et al. 2001). However, the henipavirus F proteins are cleaved by cathepsin-L in endosomal compartments (Diederich et al. 2005, 2008; Pager et al. 2006; Pager and Dutch 2005). First, the precursor F_0 is expressed at the plasma membrane prior to endocytosis into the endosomal compartment, where it is cleaved by cathepsin L, generating mature F comprised of disulfide-linked F_1 and F_2 subunits. The mature F protein is then trafficked back to the cell surface (Diederich et al. 2005; Meulendyke et al. 2005), likely independently of G (Whitman et al. 2009). This complex trafficking pathway must be reconciled with the likelihood that F and G cell surface interactions regulate receptor-induced F-triggering and membrane fusion (Fig. 1; Aguilar et al. 2006, 2007, 2009; Iorio and Mahon 2008; Iorio et al. 2009; Lee et al. 2008a).

4.2 Henipavirus F and Membrane Fusion

A few recent studies have shed some light on the roles the henipavirus F glycoproteins play in modulating membrane fusion. The study of the roles of N-glycans in the henipavirus F glycoproteins has identified important differences between the role of F in membrane fusion for the henipavirus and other paramyxovirus F glycoproteins. Additionally, the analysis of several protein domains in the ectodomain and cytoplasmic tail of the F glycoprotein has provided illuminative and interesting results.

For NiV and HeV, one N-glycan (F4) was found to be critical for cell surface expression (Aguilar et al. 2006; Carter et al. 2005; Moll et al. 2004). In contrast, removal of other N-glycans in the henipavirus F glycoproteins by mutation of the $N^1X^2(S/T)^3$ N-glycosylation sites resulted in wild type or increased levels of cell–cell fusion and viral entry (Aguilar et al. 2006; Carter et al. 2005; Moll et al. 2004). These results were quite different from those obtained with other paramyxoviral F proteins, for which similar N-glycan removal generally resulted in aberrant cell surface expression or loss of fusogenicity (Hu et al. 1995; McGinnes et al. 2001; Segawa et al. 2000; von Messling and Cattaneo 2003). These results not only highlight the differences in the role of N-glycosylation for the henipaviruses as compared to other paramyxoviruses, but also provide evidence for a potentially higher capacity of the henipavirus F glycoproteins to cause cell–cell fusion. It should be noted that N–glycan removal also increased the sensitivity of NiV F to antibody neutralization, suggesting that the henipavirus F N-glycans may be retained at least in part to shield against antibody neutralization (Aguilar et al. 2006).

NiV-F N-glycans were also found to be able to mediate binding to galectin-1, an innate immune lectin that binds specific galactose-containing carbohydrates on the surface of pathogens or mammalian cells (reviewed by Camby et al. 2006). Galectin-1 inhibited NiV-mediated cell–cell fusion (Levroney et al. 2005). Interestingly, the individual N-glycan in NiV-F (F3) whose removal resulted in the highest levels of hyperfusogenicity (Aguilar et al. 2006) also was the N-glycan that most optimally mediated galectin-1 binding to NiV F. Galectin-1 binding to this N-glycan inhibited maturation, mobility, and triggering of F (Garner et al. 2010). Because 14 single nucleotide polymorphisms have been identified at the genetic locus of galectin-1 (Iida et al. 2005), the prospect of galectin-1 genetic variability contributing to henipavirus pathology is intriguing.

Among paramyxovirus F proteins, the F3 N-glycan in NiV F is uniquely located in the third major helical region (HR3) in the F_2 subunit of the protein (Aguilar et al. 2006). Mutation of two residues within this region, Y79A and P89A, resulted in folding and expression defects, and lower levels of fusion (Gardner and Dutch 2007). The HR3 is closely packed to the fusion peptide in the pre-fusion PIV5 F structure (Yin et al. 2006). However, the role of the HR3 region in modulating release of the fusion peptide from the pre-fusion conformation into the PHI has not been formally demonstrated. It is also interesting that the F5N-glycan in NiV and HeV F is located in the center of the HR2 helix and that, among paramyxoviruses, only NDV and PIV5 share that in common with the henipaviruses. Given that HR1 and HR2 are known to come together to form the 6HB, it does not seem possible that the F5 N-glycan would reside on the face of HR2 that is involved in this interaction, but rather on the opposite side of the helix. Nonetheless, the ability of these viruses to form the 6HB in the presence of a bulky N-glycan seems somewhat impaired, at least for NiV, as removal of the F5 N-glycan resulted in hyperfusogenicity (Aguilar et al. 2006).

Other residues in the henipavirus F ectodomain that have been studied are two histidines positioned near the cathepsin-L cleavage site (H102 and H372). Mutation of these residues, particularly of H372, suggested that the side chain packing beneath the fusion peptide is an important modulator of henipavirus membrane fusion. Studies for other paramyxovirus F proteins implicate fusion peptide and surrounding residues as modulators of membrane fusion (Lee et al. 2008b; Rawling et al. 2008; Russell et al. 2004). Additionally, mutations in the region of HeV F between HR1 and HR2 resulted in defects in homotrimer formation, proteolytic processing, and transport to the cell surface (Gardner et al. 2007).

Fusion-modulatory determinants have also been found in the cytoplasmic tail of henipavirus F glycoproteins. A polybasic motif (KKR) in the cytoplasmic tail of NiV F was found to modulate cell–cell fusion and viral entry (Aguilar et al. 2007). Interestingly, hyperfusogenic and hypofusogenic phenotypes within this motif correlated with conformational changes in the NiV F ectodomain, detected by conformation-dependent monoclonal antibodies. These data imply that, similar to other class I fusion proteins (Aguilar et al. 2003; Spies et al. 1994; Waning et al. 2004; Wyss et al. 2005), there is important inside-out signaling across the transmembrane domain of F that can modulate fusion for the henipavirus

F glycoproteins. This conserved feature of class I fusion proteins represents a yet unexplored target for antiviral therapeutic intervention.

Blocking viral entry by trapping one of the fusion protein intermediates during the membrane fusion cascade has been a therapeutic approach pursued and utilized for class I fusion protein enveloped viruses. For example, enfuvirtide, sifuvirtide, and other analogs are peptides that mimic the C-terminal heptad repeat region (HR2) of class I fusion proteins, and are approved for HIV-1 treatment (He et al. 2008; Makinson and Reynes 2009; Poveda et al. 2005). The paramyxovirus HR2 (a.k.a. HRC) peptide has also been used to trap the henipavirus F pre-hairpin intermediate conformation (Aguilar et al. 2006, 2007, 2009, 2010; Bossart et al. 2005; Lambert et al. 1996; Porotto et al. 2006, 2007; Russell et al. 2001) (Fig. 1b). Although the N-terminal HR1 mimetic peptide also inhibits fusion, it generally does so less efficiently (Bossart et al. 2005), even when artificially trimerized to mimic the HR1 core (Aguilar et al. 2009). However, the addition of cholesterol to a paramyxovirus F HR2 peptide has been shown to increase its efficiency against henipavirus infections, making this potential therapeutic approach promising (Porotto et al. 2010).

The rate of fusion kinetics for HIV-1 seems to correlate inversely with relative sensitivities to HR2 peptide inhibition (Miyauchi et al. 2009). For the paramyxoviruses, a peptide containing the HR2 sequence of the hPIV3 F protein inhibited NiV and HeV viral infection more efficiently than that of the homologous virus, despite a stronger HR1/HR2 interaction for the latter (Porotto et al. 2006, 2007). This unexpected finding was attributed to relatively slower fusion kinetics for NiV and HeV compared to hPIV3 (Porotto et al. 2009). It would be interesting to determine if slower fusion kinetics induced by B3 (Aguilar et al. 2010) correlate with higher sensitivities to HR2 peptide inhibition, and if differential usage of B2/B3 receptors by NiV and HeV result in differential sensitivities to HR2 peptide inhibition in vitro and in vivo.

Another tactic to inhibit membrane fusion has been the use of monoclonal antibodies to block henipavirus F conformational changes important for the fusion cascade. Two anti-NiV-F antibodies were reported to neutralize NiV and HeV in vitro (0.27–20 ng/2.5 $\times$ 10^4 cells) and in hamsters (180–520 μg/animal) (Guillaume et al. 2006). Although the epitopes recognized by these antibodies have not been identified, their cross-reactivity with HeV is attractive, as they may target a conserved region in henipavirus F that may limit the origination of viable escape variants. Indeed, antibodies that target epitopes that change conformation during the membrane fusion cascade are desirable, since mutations that annul both antibody binding and the need of conformational changes would be relatively more rare.

Lastly, new F-triggering assays have been developed that are capable of measuring the half-lives of the NiV F fusion intermediates. "Time of addition" and "time of stopping" assays are based on HR2 peptide trapping of the pre-hairpin intermediate and can measure pre-fusion to 6HB and pre-fusion to pre-hairpin intermediates, respectively (Aguilar et al. 2010). These assays are promising for deciphering the roles of distinct henipavirus F domains in

F-triggering and in subsequent steps in membrane fusion, and in revealing new potential antiviral targets. In addition, the discovery of the roles of cellular factors in Henipavirus membrane fusion has only begun (Schowalter et al. 2006b) and may offer additional antiviral targets.

5 Epilogue

The more complete our understanding of the mechanisms of membrane fusion, the greater the possibility that we can develop strategies to block NiV and HeV entry and pathobiology. For example, based on our current knowledge of henipavirus membrane fusion, one can envision that henipavirus entry could be blocked by reagents that block B2/B3 binding to G, conformational changes in G, F/G interactions, F-triggering by G, fusion pore formation, or other individual steps in the G and F fusion conformational cascade yet to be identified. Additionally, one must not assume that cell–cell fusion and viral-cell fusion are equivalently modulated events. Although both processes share membrane fusion as a common theme, important differences in the determinants that modulate them may need to be considered. These determinants may be present in the G and F glycoproteins, in other viral proteins, and/or in the host cell.

References

Aguilar HC, Anderson WF, Cannon PM (2003) Cytoplasmic tail of Moloney murine leukemia virus envelope protein influences the conformation of the extracellular domain: implications for mechanism of action of the R Peptide. J Virol 77:1281–1291

Aguilar HC, Matreyek KA, Filone CM, Hashimi ST, Levroney EL, Negrete OA, Bertolotti-Ciarlet A, Choi DY, McHardy I, Fulcher JA, Su SV, Wolf MC, Kohatsu L, Baum LG, Lee B (2006) N-glycans on Nipah virus fusion protein protect against neutralization but reduce membrane fusion and viral entry. J Virol 80:4878–4889

Aguilar HC, Matreyek KA, Choi DY, Filone CM, Young S, Lee B (2007) Polybasic KKR motif in the cytoplasmic tail of Nipah virus fusion protein modulates membrane fusion by inside-out signaling. J Virol 81:4520–4532

Aguilar HC, Ataman ZA, Aspericueta V, Fang AQ, Stroud M, Negrete OA, Kammerer RA, Lee B (2009) A novel receptor-induced activation site in the nipah virus attachment glycoprotein (G) involved in triggering the fusion glycoprotein (F). J Biol Chem 284:1628–1635

Aguilar HC, Aspericueta V, Robinson LR Aanensen KE, Lee B (2010) A quantitative and kinetic fusion protein-triggering assay can discern distinct steps in the nipah virus membrane fusion cascade. J Virol 84:8033–8041. doi:10.1128/JVI.00469-10, JVI.00469-10 [pii]

Aguilar HC, Lee B (2011) Emerging paramyxoviruses: molecular mechanisms and antiviral strategies. Expert Rev Mol Med 13:e6. doi:10.1017/S1462399410001754,S1462399410001754 [pii]

Baker KA, Dutch RE, Lamb RA, Jardetzky TS (1999) Structural basis for paramyxovirus-mediated membrane fusion. Mol Cell 3:309–319

Begona Ruiz-Arguello M, Gonzalez-Reyes L, Calder LJ, Palomo C, Martin D, Saiz MJ, Garcia-Barreno B, Skehel JJ, Melero JA (2002) Effect of proteolytic processing at two distinct sites on shape and aggregation of an anchorless fusion protein of human respiratory syncytial virus and fate of the intervening segment. Virology 298:317–326. S0042682202914972, [pii]

Bishop KA, Stantchev TS, Hickey AC, Khetawat D, Bossart KN, Krasnoperov V, Gill P, Feng YR, Wang L, Eaton BT, Wang LF, Broder CC (2007) Identification of Hendra virus G glycoprotein residues that are critical for receptor binding. J Virol 81:5893–5901

Bishop KA, Hickey AC, Khetawat D, Patch JR, Bossart KN, Zhu Z, Wang LF, Dimitrov DS, Broder CC (2008) Residues in the stalk domain of the Hendra virus G glycoprotein modulate conformational changes associated with receptor binding. J Virol 82:11398–11409

Bonaparte MI, Dimitrov AS, Bossart KN, Crameri G, Mungall BA, Bishop KA, Choudhry V, Dimitrov DS, Wang LF, Eaton BT, Broder CC (2005) From the cover: ephrin-B2 ligand is a functional receptor for Hendra virus and Nipah virus. Proc Natl Acad Sci U S A 102: 10652–10657

Bose S, Welch BD, Kors CA, Yuan P, Jardetzky TS, Lamb RA (2011) Structure and mutagenesis of the parainfluenza virus 5 hemagglutinin-neuraminidase stalk domain reveals a four-helix bundle and the role of the stalk in fusion promotion. J Virol 85:12855–12866. doi:10.1128/JVI.06350-11

Bossart KN, Wang LF, Eaton BT, Broder CC (2001) Functional expression and membrane fusion tropism of the envelope glycoproteins of Hendra virus. Virology 290:121–135

Bossart KN, Wang LF, Flora MN, Chua KB, Lam SK, Eaton BT, Broder CC (2002) Membrane fusion tropism and heterotypic functional activities of the Nipah virus and Hendra virus envelope glycoproteins. J Virol 76:11186–11198

Bossart KN, Mungall BA, Crameri G, Wang LF, Eaton BT, Broder CC (2005) Inhibition of Henipavirus fusion and infection by heptad-derived peptides of the Nipah virus fusion glycoprotein. Virol J 2:57

Bossart KN, Zhu Z, Middleton D, Klippel J, Crameri G, Bingham J, McEachern JA, Green D, Hancock TJ, Chan YP, Hickey AC, Dimitrov DS, Wang LF, Broder CC (2009) A neutralizing human monoclonal antibody protects against lethal disease in a new ferret model of acute nipah virus infection. PLoS Pathog 5:e1000642. doi:10.1371/journal.ppat.1000642

Bossart KN, Geisbert TW, Feldmann H, Zhu Z, Feldmann F, Geisbert JB, Yan L, Feng YR, Brining D, Scott D, Wang Y, Dimitrov AS, Callison J, Chan YP, Hickey AC, Dimitrov DS, Border CC, Rockx B (2011) A neutralizing human monoclonal antibody protects african green monkeys from hendra virus challenge. Sci Transl Med 3:105ra103. doi: 10.1126/scitranslmed.3002901

Bowden TA, Aricescu AR, Gilbert RJ, Grimes JM, Jones EY, Stuart DI (2008a) Structural basis of Nipah and Hendra virus attachment to their cell-surface receptor ephrin-B2. Nat Struct Mol Biol 15:567–572

Bowden TA, Crispin M, Harvey DJ, Aricescu AR, Grimes JM, Jones EY, Stuart DI (2008b) Crystal structure and carbohydrate analysis of Nipah virus attachment glycoprotein: a template for antiviral and vaccine design. J Virol 82:11628–11636

Camby I, Le Mercier M, Lefranc F, Kiss R (2006) Galectin-1: a small protein with major functions. Glycobiology 16:137R-157R. doi:10.1093/glycob/cwl025, cwl025 [pii]

Carr CM, Kim PS (1993) A spring-loaded mechanism for the conformational change of influenza hemagglutinin. Cell 73:823–832

Carter JR, Pager CT, Fowler SD, Dutch RE (2005) Role of N-linked glycosylation of the Hendra virus fusion protein. J Virol 79:7922–7925

Corey EA, Iorio RM (2007) Mutations in the stalk of the measles virus hemagglutinin protein decrease fusion but do not interfere with virus-specific interaction with the homologous fusion protein. J Virol 81:9900–9910

Diederich S, Moll M, Klenk HD, Maisner A (2005) The Nipah virus fusion protein is cleaved within the endosomal compartment. J Biol Chem 280:29899–29903

Diederich S, Thiel L, Maisner A (2008) Role of endocytosis and cathepsin-mediated activation in Nipah virus entry. Virology 375:391–400

Dorig RE, Marcil A, Chopra A, Richardson CD (1993) The human CD46 molecule is a receptor for measles virus (Edmonston strain). Cell 75:295–305

Dutch RE (2010) Entry and fusion of emerging paramyxoviruses. PLoS Pathog 6:e1000881. doi:10.1371/journal.ppat.1000881

Fass D, Harrison SC, Kim PS (1996) Retrovirus envelope domain at 1.7 Angstrom resolution. Nat Struct Biol 3:465–469

Gardner AE, Dutch RE (2007) A conserved region in the F(2) subunit of paramyxovirus fusion proteins is involved in fusion regulation. J Virol 81:8303–8314. doi: 10.1128/JVI.00366-07, JVI.00366-07 [pii]

Gardner AE, Martin KL, Dutch RE (2007) A conserved region between the heptad repeats of paramyxovirus fusion proteins is critical for proper F protein folding. Biochemistry 46:5094–5105. doi:10.1021/bi6025648

Garner OB, Aguilar HC, Fulcher JA, Levroney EL, Harrison R, Wright L, Robinson LR, Aspericueta V, Panico M, Haslam SM, Morris HR, Dell A, Lee B, Baum LG (2010) Endothelial galectin-1 binds to specific glycans on nipah virus fusion protein and inhibits maturation, mobility, and function to block syncytia formation. PLoS Pathog 6:e1000993. doi:10.1371/journal.ppat.1000993

Garten W, Hallenberger S, Ortmann D, Schafer W, Vey M, Angliker H, Shaw E, Klenk HD (1994) Processing of viral glycoproteins by the subtilisin-like endoprotease furin and its inhibition by specific peptidylchloroalkylketones. Biochimie 76: 217–25. 0300-9084(94) 90149-X [pii]

Gonzalez-Reyes L, Ruiz-Arguello MB, Garcia-Barreno B, Calder L, Lopez JA, Albar JP, Skehel JJ, Wiley DC, Melero JA (2001) Cleavage of the human respiratory syncytial virus fusion protein at two distinct sites is required for activation of membrane fusion. Proc Natl Acad Sci U S A 98:9859–9864. doi:10.1073/pnas.151098198, 151098198 [pii]

Gravel KA, Morrison TG (2003) Interacting domains of the HN and F proteins of newcastle disease virus. J Virol 77:11040–11049

Guillaume V, Contamin H, Loth P, Grosjean I, Courbot MC, Deubel V, Buckland R, Wild TF (2006) Antibody prophylaxis and therapy against Nipah virus infection in hamsters. J Virol 80:1972–1978. doi:10.1128/JVI.80.4.1972-1978.2006, 80/4/1972 [pii]

Hashiguchi T, Ose T, Kubota M, Maita N, Kamishikiryo J, Maenaka K, Yanagi Y (2011) Structure of the measles virus hemagglutinin bound to its cellular receptor SLAM. Nat struct mol biol 18:135–141. doi:10.1038/nsmb.1969

He Y, Xiao Y, Song H, Liang Q, Ju D, Chen X, Lu H, Jing W, Jiang S, Zhang L (2008) Design and evaluation of sifuvirtide, a novel HIV-1 fusion inhibitor. J Biol Chem 283:11126–11134. doi: 10.1074/jbc.M800200200, M800200200 [pii]

Herfst S, Mas V, Ver LS, Wierda RJ, Osterhaus AD, Fouchier RA, Melero JA (2008) Low-pH-induced membrane fusion mediated by human metapneumovirus F protein is a rare, strain-dependent phenomenon. J Virol 82:8891–8895. doi:10.1128/JVI.00472-08

Hu A, Cathomen T, Cattaneo R, Norrby E (1995) Influence of N-linked oligosaccharide chains on the processing, cell surface expression and function of the measles virus fusion protein. J Gen Virol 76(Pt 3):705–710

Iida A, Ozaki K, Tanaka T, Nakamura Y (2005) Fine-scale SNP map of an 11-kb genomic region at 22q13.1 containing the galectin-1 gene. J Hum Genet 50:42–45. doi:10.1007/s10038-004-0218-4

Iorio RM, Mahon PJ (2008) Paramyxoviruses: different receptors—different mechanisms of fusion. Trends Microbiol 16:135–137

Iorio RM, Melanson VR, Mahon PJ (2009) Glycoprotein interactions in paramyxovirus fusion. Future Virol 4:335–351. doi:10.2217/fvl.09.17

Kahn JS, Schnell MJ, Buonocore L, Rose JK (1999) Recombinant vesicular stomatitis virus expressing respiratory syncytial virus (RSV) glycoproteins: RSV fusion protein can mediate infection and cell fusion. Virology 254:81–91. doi:10.1006/viro.1998.9535

Lamb RA, Paterson RG, Jardetzky TS (2006) Paramyxovirus membrane fusion: lessons from the F and HN atomic structures. Virology 344:30–37

Lamb RA, Jardetzky TS (2007) Structural basis of viral invasion: lessons from paramyxovirus F. Curr Opin Struct Biol 17:427–436

Lamb RA, Parks GD (2007) Paramyxoviridae: the viruses and their replication. In: Knipe DM, Howley PM (eds) Fields virology, 5th edn, vol 1. Lippincott Williams & Wilkins, Philadelphia, pp 1449–1496

Lambert DM, Barney S, Lambert AL, Guthrie K, Medinas R, Davis DE, Bucy T, Erickson J, Merutka G, Petteway SR Jr (1996) Peptides from conserved regions of paramyxovirus fusion (F) proteins are potent inhibitors of viral fusion. Proc Natl Acad Sci U S A 93:2186–2191

Lee B, Ataman ZA, Jin L (2008a) Evil versus 'eph-ective' use of ephrin-B2. Nat Struct Mol Biol 15:540–542

Lee JK, Prussia A, Paal T, White LK, Snyder JP, Plemper RK (2008b) Functional interaction between paramyxovirus fusion and attachment proteins. J Biol Chem 283:16561–16572

Levroney EL, Aguilar HC, Fulcher JA, Kohatsu L, Pace KE, Pang M, Gurney KB, Baum LG, Lee B (2005) Novel innate immune functions for galectin-1: galectin-1 inhibits cell fusion by Nipah virus envelope glycoproteins and augments dendritic cell secretion of proinflammatory cytokines. J Immunol 175:413–420

Lou Z, Xu Y, Xiang K, Su N, Qin L, Li X, Gao GF, Bartlam M, Rao Z (2006) Crystal structures of Nipah and Hendra virus fusion core proteins. FEBS J 273:4538–4547

Mahon PJ, Mirza AM, Iorio RM (2011) Role of the two sialic acid binding sites on the newcastle disease virus HN protein in triggering the interaction with the F protein required for the promotion of fusion. J Virol 85:12079–12082. doi:10.1128/JVI.05679-11

Makinson A, Reynes J (2009) The fusion inhibitor enfuvirtide in recent antiretroviral strategies. Curr Opin HIV AIDS 4:150–158. doi:10.1097/COH.0b013e32832498d8, 01222929-200903000-00013 [pii]

McGinnes L, Sergel T, Reitter J, Morrison T (2001) Carbohydrate modifications of the NDV fusion protein heptad repeat domains influence maturation and fusion activity. Virology 283:332–342

McKinnell JA, Saag MS (2009) Novel drug classes: entry inhibitors [enfuvirtide, chemokine (C–C motif) receptor 5 antagonists]. Curr Opin HIV AIDS 4: 513–517. doi:10.1097/COH.0b013e328331d3d0, 01222929-200911000-00010 [pii]

Melanson VR, Iorio RM (2004) Amino acid substitutions in the F-specific domain in the stalk of the newcastle disease virus HN protein modulate fusion and interfere with its interaction with the F protein. J Virol 78:13053–13061

Melanson VR, Iorio RM (2006) Addition of N-glycans in the stalk of the Newcastle disease virus HN protein blocks its interaction with the F protein and prevents fusion. J Virol 80:623–633

Meulendyke KA, Wurth MA, McCann RO, Dutch RE (2005) Endocytosis plays a critical role in proteolytic processing of the Hendra virus fusion protein. J Virol 79:12643–12649

Mirza AM, Aguilar HC, Zhu Q, Mahon PJ, Rota PA, Lee B, Iorio RM (2011) Triggering of the newcastle disease virus fusion protein by a chimeric attachment protein that binds to Nipah virus receptors. J Biol Chem 286:17851–17860. doi:10.1074/jbc.M111.233965

Miyauchi K, Kozlov MM, Melikyan GB (2009) Early steps of HIV-1 fusion define the sensitivity to inhibitory peptides that block 6-helix bundle formation. PLoS Pathog 5:e1000585. doi:10.1371/journal.ppat.1000585

Moll M, Kaufmann A, Maisner A (2004) Influence of N-glycans on processing and biological activity of the nipah virus fusion protein. J Virol 78:7274–7278

Muhlebach MD, Mateo M, Sinn PL, Prufer S, Uhlig KM, Leonard VH, Navaratnarajah CK, Frenzke M, Wong XX, Sawatsky B, Ramachandran S, McCray PB, Cichutek K, von Messling V, Lopez M, Cattaneo R (2011) Adherens junction protein nectin-4 is the epithelial receptor for measles virus. Nature 480(7378):530. doi:10.1038/nature10639

Murakami M, Towatari T, Ohuchi M, Shiota M, Akao M, Okumura Y, Parry MA, Kido H (2001) Mini-plasmin found in the epithelial cells of bronchioles triggers infection by broad-spectrum influenza A viruses and Sendai virus. Eur J Biochem 268:2847–55. ejb2166 [pii]

Murineddu G, Murruzzu C, Pinna GA (2010) An overview on different classes of viral entry and respiratory syncitial virus (RSV) fusion inhibitors. Curr Med Chem 17:1067–1091. BSP/CMC/E-Pub/067 [pii]

Negrete OA, Levroney EL, Aguilar HC, Bertolotti-Ciarlet A, Nazarian R, Tajyar S, Lee B (2005) EphrinB2 is the entry receptor for Nipah virus, an emergent deadly paramyxovirus. Nature 436:401–405

Negrete OA, Wolf MC, Aguilar HC, Enterlein S, Wang W, Muhlberger E, Su SV, Bertolotti-Ciarlet A, Flick R, Lee B (2006) Two key residues in ephrinB3 are critical for its use as an alternative receptor for Nipah virus. PLoS Pathog 2:e7

Negrete OA, Chu D, Aguilar HC, Lee B (2007) Single amino acid changes in the Nipah and Hendra virus attachment glycoproteins distinguish ephrinB2 from ephrinB3 usage. J Virol 81:10804–10814

Noyce RS, Bondre DG, Ha MN, Lin LT, Sisson G, Tsao MS, Richardson CD (2011) Tumor cell marker PVRL4 (nectin 4) is an epithelial cell receptor for measles virus. PLoS pathogens 7:e1002240. doi:10.1371/journal.ppat.1002240

Ortmann D, Ohuchi M, Angliker H, Shaw E, Garten W, Klenk HD (1994) Proteolytic cleavage of wild type and mutants of the F protein of human parainfluenza virus type 3 by two subtilisin-like endoproteases, furin and Kex2. J Virol 68:2772–2776

Paal T, Brindley MA, St Clair C, Prussia A, Guas D, Krumm SA, Synder JP, Plemper RK (2009) Probing the spatial organization of measles virus fusion complexes. J Virol 83: 10480-93. doi:10.1128/JVI.01195-09, JVI.01195-09 [pii]

Pager CT, Dutch RE (2005) Cathepsin L is involved in proteolytic processing of the Hendra virus fusion protein. J Virol 79:12714–12720

Pager CT, Craft WW Jr, Patch J, Dutch RE (2006) A mature and fusogenic form of the Nipah virus fusion protein requires proteolytic processing by cathepsin L. Virology 346:251–257

Porotto M, Doctor L, Carta P, Fornabaio M, Greengard O, Kellogg GE, Moscona A (2006) Inhibition of hendra virus fusion. J Virol 80:9837–9849

Porotto M, Carta P, Deng Y, Kellogg GE, Whitt M, Lu M, Mungall BA, Moscona A (2007) Molecular determinants of antiviral potency of paramyxovirus entry inhibitors. J Virol 81:10567–10574. doi:10.1128/JVI.01181-07, JVI.01181-07 [pii]

Porotto M, Yokoyama CC, Orefice G, Kim HS, Aljofan M, Mungall BA, Moscona A (2009) Kinetic dependence of paramyxovirus entry inhibition. J Virol 83:6947–6951. doi:10.1128/JVI.00416-09, JVI.00416-09 [pii]

Porotto M, Rockx B, Yokoyama CC, Talekar A, Devito I, Palermo LM, Liu J, Cortese R, Lu M, Feldmann H, Pessi A, Moscona A (2010) Inhibition of Nipah Virus Infection In Vivo: Targeting an Early Stage of Paramyxovirus Fusion Activation during Viral Entry. PLoS Pathog 6:e1001168. doi:10.1371/journal.ppat.1001168

Porotto M, Devito I, Palmer SG, Jurgens EM, Yee JL, Yokoyama CC, Pessi A, Moscona A (2011) Spring-loaded model revisited: Paramyxovirus fusion requires engagement of a receptor binding protein beyond initial triggering of the fusion protein. J Virol. doi:10.1128/JVI.05873-11

Poveda E, Briz V, Soriano V (2005) Enfuvirtide, the first fusion inhibitor to treat HIV infection. AIDS Rev 7:139–147

Rawling J, Garcia-Barreno B, Melero JA (2008) Insertion of the two cleavage sites of the respiratory syncytial virus fusion protein in Sendai virus fusion protein leads to enhanced cell–cell fusion and a decreased dependency on the HN attachment protein for activity. J Virol 82:5986–5998. doi:10.1128/JVI.00078-08

Russell CJ, Jardetzky TS, Lamb RA (2001) Membrane fusion machines of paramyxoviruses: capture of intermediates of fusion. EMBO J 20:4024–4034

Russell CJ, Jardetzky TS, Lamb RA (2004) Conserved glycine residues in the fusion peptide of the paramyxovirus fusion protein regulate activation of the native state. J Virol 78:13727–13742. doi:10.1128/JVI.78.24.13727-13742.2004

Schowalter RM, Smith SE, Dutch RE (2006a) Characterization of human metapneumovirus F protein-promoted membrane fusion: critical roles for proteolytic processing and low pH. J Virol 80:10931–10941

Schowalter RM, Wurth MA, Aguilar HC, Lee B, Moncman CL, McCann RO, Dutch RE (2006b) Rho GTPase activity modulates paramyxovirus fusion protein-mediated cell–cell fusion. Virology 350:323–334

Segawa H, Yamashita T, Kawakita M, Taira H (2000) Functional analysis of the individual oligosaccharide chains of sendai virus fusion protein. J Biochem 128:65–72
Spies CP, Ritter GD Jr, Mulligan MJ, Compans RW (1994) Truncation of the cytoplasmic domain of the simian immunodeficiency virus envelope glycoprotein alters the conformation of the external domain. J Virol 68:585–591
Stone-Hulslander J, Morrison TG (1999) Mutational analysis of heptad repeats in the membrane-proximal region of Newcastle disease virus HN protein. J Virol 73:3630–3637
Tatsuo H, Ono N, Tanaka K, Yanagi Y (2000) SLAM (CDw150) is a cellular receptor for measles virus. Nature 406:893–897. doi:10.1038/35022579
van den Hoogen BG, de Jong JC, Groen J, Kuiken T, de Groot R, Fouchier RA, Osterhaus AD (2001) A newly discovered human pneumovirus isolated from young children with respiratory tract disease. Nat Med 7:719–724. doi: 10.1038/89098, 89098 [pii]
von Messling V, Cattaneo R (2003) N-linked glycans with similar location in the fusion protein head modulate paramyxovirus fusion. J Virol 77:10202–10212
Waning DL, Russell CJ, Jardetzky TS, Lamb RA (2004) Activation of a paramyxovirus fusion protein is modulated by inside-out signaling from the cytoplasmic tail. Proc Natl Acad Sci U S A 101:9217–9222
Watanabe M, Hirano A, Stenglein S, Nelson J, Thomas G, Wong TC (1995) Engineered serine protease inhibitor prevents furin-catalyzed activation of the fusion glycoprotein and production of infectious measles virus. J Virol 69:3206–3210
White JM, Delos SE, Brecher M, Schornberg K (2008) Structures and mechanisms of viral membrane fusion proteins: multiple variations on a common theme. Crit Rev Biochem Mol Biol 43:189–219
Whitman SD, Smith EC, Dutch RE (2009) Differential rates of protein folding and cellular trafficking for the Hendra virus F and G proteins: implications for F-G complex formation. J Virol 83:8998–9001. doi:10.1128/JVI.00414-09, JVI.00414-09 [pii]
Wyss S, Dimitrov AS, Baribaud F, Edwards TG, Blumenthal R, Hoxie JA (2005) Regulation of human immunodeficiency virus type 1 envelope glycoprotein fusion by a membrane-interactive domain in the gp41 cytoplasmic tail. J Virol 79:12231–12241
Xu K, Rajashankar KR, Chan YP, Himanen JP, Broder CC, Nikolov DB (2008) Host cell recognition by the henipaviruses: crystal structures of the Nipah G attachment glycoprotein and its complex with ephrin-B3. Proc Natl Acad Sci U S A 105:9953–9958
Xu Y, Lou Z, Liu Y, Cole DK, Su N, Qin L, Li X, Bai Z, Rao Z, Gao GF (2004) Crystallization and preliminary crystallographic analysis of the fusion core from two new zoonotic paramyxoviruses, Nipah virus and Hendra virus. Acta Crystallogr D Biol Crystallogr 60:1161–1164. doi:10.1107/S0907444904009515
Yin HS, Paterson RG, Wen X, Lamb RA, Jardetzky TS (2005) Structure of the uncleaved ectodomain of the paramyxovirus (hPIV3) fusion protein. Proc Natl Acad Sci U S A 102: 9288–9293
Yin HS, Wen X, Paterson RG, Lamb RA, Jardetzky TS (2006) Structure of the parainfluenza virus 5 F protein in its metastable, prefusion conformation. Nature 439:38–44
Yuan P, Swanson KA, Leser GP, Paterson RG, Lamb RA, Jardetzky TS (2011) Structure of the Newcastle disease virus hemagglutinin-neuraminidase (HN) ectodomain reveals a four-helix bundle stalk. Proc Natl Acad Sci U S A 108:14920–14925. doi:10.1073/pnas.1111691108

Clinical and Pathological Manifestations of Human Henipavirus Infection

K. T. Wong and C. T. Tan

Abstract The clinicopathological features of human Nipah virus and Hendra virus infections appear to be similar. The clinical manifestations may be mild, but if severe, includes acute encephalitic and pulmonary syndromes with a high mortality. The pathological features in human acute henipavirus infections comprise vasculopathy (vasculitis, endothelial multinucleated syncytia, thrombosis), microinfarcts and parenchymal cell infection in the central nervous system, lung, kidney and other major organs. Viral inclusions, antigens, nucleocapsids and RNA are readily demonstrated in blood vessel wall and numerous types of parenchymal cells. Relapsing henipavirus encephalitis is a rare complication reported in less than 10% of survivors of the acute infection and appears to be distinct from the acute encephalitic syndrome. Pathological evidence suggests viral recrudescence confined to the central nervous system as the cause.

Contents

K. T. Wong (✉)
Deptartment of Pathology, Faculty of Medicine,
University of Malaya, 50603 Kuala Lumpur, Malaysia
e-mail: wongkt@ummc.edu.my

C. T. Tan
Department of Medicine, Faculty of Medicine,
University of Malaya, 50603 Kuala Lumpur, Malaysia

Current Topics in Microbiology and Immunology (2012) 359: 95–104
DOI: 10.1007/82_2012_205

Published Online: 18 March 2012

1 Introduction

Henipavirus genus, a recently established group of paramyxoviruses (Wang et al. 2000) comprising the Hendra virus (HeV) and Nipah virus (NiV), has emerged to cause severe infection in humans and animals. There are several previous reviews on NiV or henipavirus infections (Wong 2000; Wong et al. 2002a, b; Tan and Wong 2003; Wong 2010; Wong and Ong 2011) but the present one focuses on updated clinical features and pathology in infected patients.

In Australia starting in 1994, HeV was reported to have infected numerous horses and at least seven persons with four fatalities so far (Murray et al. 1995; Selvey et al. 1995; Rogers et al. 1996; O'Sullivan et al. 1997; Field et al. 2000; Hanna et al. 2006; Playford et al. 2010; Williamson and Torres-Velez 2010). The first known NiV outbreak in Malaysia occurred mainly among pig farmers and other people who had close contact with pigs from 1989 to 1999 (Chua et al. 1999). Although a prevalence of 265 Malaysian cases of acute NiV encephalitis with 105 fatalities have been reported (Parashar et al. 2000), if milder non-encephalitic cases are included (Tan et al. 2002), the total number infected was estimated to exceed 350 cases (Wong et al. 2002a, b). Eleven cases were reported from Singapore (Paton et al. 1999) at about the same time as the Malaysian outbreak. Beginning 2001, several on-going and recurrent NiV outbreaks were reported from Bangladesh and the neighbouring Bengal region in India involving more than 120 people (Harit et al. 2006; Hossain et al. 2008). Hence, worldwide there may be more than 500 human henipavirus infections to date.

2 Clinical Aspects and Sequelae of Henipavirus Infection

Although there is some variation, the general clinical manifestation of henipavirus infection appears to be similar. After a short incubation, an acute illness manifesting as acute encephalitis or pulmonary disease with a high mortality occurs. In some patients relapsing encephalitis or late-onset encephalitis could complicate initial recovery from the acute illness months or years later. Some infected people may remain asymptomatic.

The incubation period for henipavirus infection ranges from a few days to 2 weeks (Selvey et al. 1995; Chong et al. 2000; Goh et al. 2000; Hanna et al. 2006; Playford et al. 2010). Milder symptoms could include fever, headache, influenza-like illness and drowsiness. Severe HeV infection like NiV, may present either as a neurological or pulmonary syndrome but since there were very few patients, the clinical features are far less well characterised than the latter. Neurological signs could include confusion, motor deficits and seizures. The pulmonary syndrome comprises an influenza-like illness, hypoxaemia and diffuse alveolar shadowing in chest X-rays (Selvey et al. 1995; Playford et al. 2010).

Severe NiV encephalitic syndrome presents mainly with fever, headache, dizziness, vomiting and reduced consciousness (Chong et al. 2000; Goh et al. 2000;

Chadha et al. 2006; Hossain et al. 2008). The clinical signs included areflexia, hypotonia, abnormal pupillary and doll's eye reflex, tachycardia, hypertension, meningism, myoclonus and seizures. A pulmonary syndrome has been described in some patients who present with cough, atypical pneumonia, acute respiratory distress syndrome and abnormal chest X-ray findings (Paton et al. 1999; Chong et al. 2000; Goh et al. 2000). Vomiting and diarrhoea has also been reported in some patients (Hossain et al. 2008).

MR scans in acute henipavirus encephalitis showed disseminated, small discrete hyperintense lesions in both grey and white matter of the brain (Sarji et al. 2000; Lim et al. 2002; Playford et al. 2010). These features were thought to be sufficiently characteristic to be a useful adjunct for diagnosis (Fig. 3a). Interestingly, some cases from Bangladesh and Australia appear to show more confluent high signal lesions in grey and white matter as well (Quddus et al. 2004; Playford et al. 2010).

Cerebrospinal fluid (CSF) examination showed elevated protein levels and/or white cell counts in more than 75% of NiV patients but glucose levels were normal (Lee et al. 1999; Goh et al. 2000). Electroencephalography most commonly demonstrated continuous diffuse, symmetrical slowing with or without focal discharges in acute NiV encephalitis (Chew et al. 1999).

Specific anti-henipavirus antibodies detectable in most patients in serum and CSF are important for diagnosis. Again, more is known about NiV than HeV seroconversion. Anti-NiV IgM appears early and by about 2 weeks, it is found in 100% of patients and may persist for more than 3 months. IgG seroconversion was 100% by about 3 weeks and may persist for several years (Ramasundram et al. 2000; Chong and Tan 2003; Siva et al. 2009). Specific neutralising IgM or IgG antibodies have been reported in HeV infection (Selvey et al. 1995; O'Sullivan et al. 1997; Hanna et al. 2006).

In general, mortality in severe acute henipavirus infection is high, ranging from about 50% in HeV infection and 40–70% in NiV infection. The higher mortality of acute NiV encephalitis in Bangladesh/India (70%) compared to Malaysia (40%) may partly be related to easier access to good hospital care in the latter (Parashar et al. 2000; Harit et al. 2006; Hossain et al. 2008). However, other factors, e.g. possible higher virulence of some NiV strains needs to be investigated. Brain stem involvement, presence of virus in the CSF and diabetes mellitus may be poor prognostic factors in acute NiV encephalitis (Goh et al. 2000; Chua et al. 2000; Chong et al. 2001).

Patients with severe infection may recover without serious sequelae (Goh et al. 2000). However, henipavirus infection may be complicated by relapsing encephalitis following initial recovery months or years later. A single case of relapsing HeV encephalitis and more than 20 cases of relapsing NiV encephalitis (probably in <10% of survivors) have been reported (O'Sullivan et al. 1997; Tan et al. 2002; Chong and Tan 2003; Sejvar et al. 2007). The case of relapsing HeV encephalitis occurred about 13 months after exposure, while an average of 8 months elapsed before relapsing NiV encephalitis occurred. Some cases of relapsing NiV encephalitis only had fever and headache during the acute phase, and have also been called

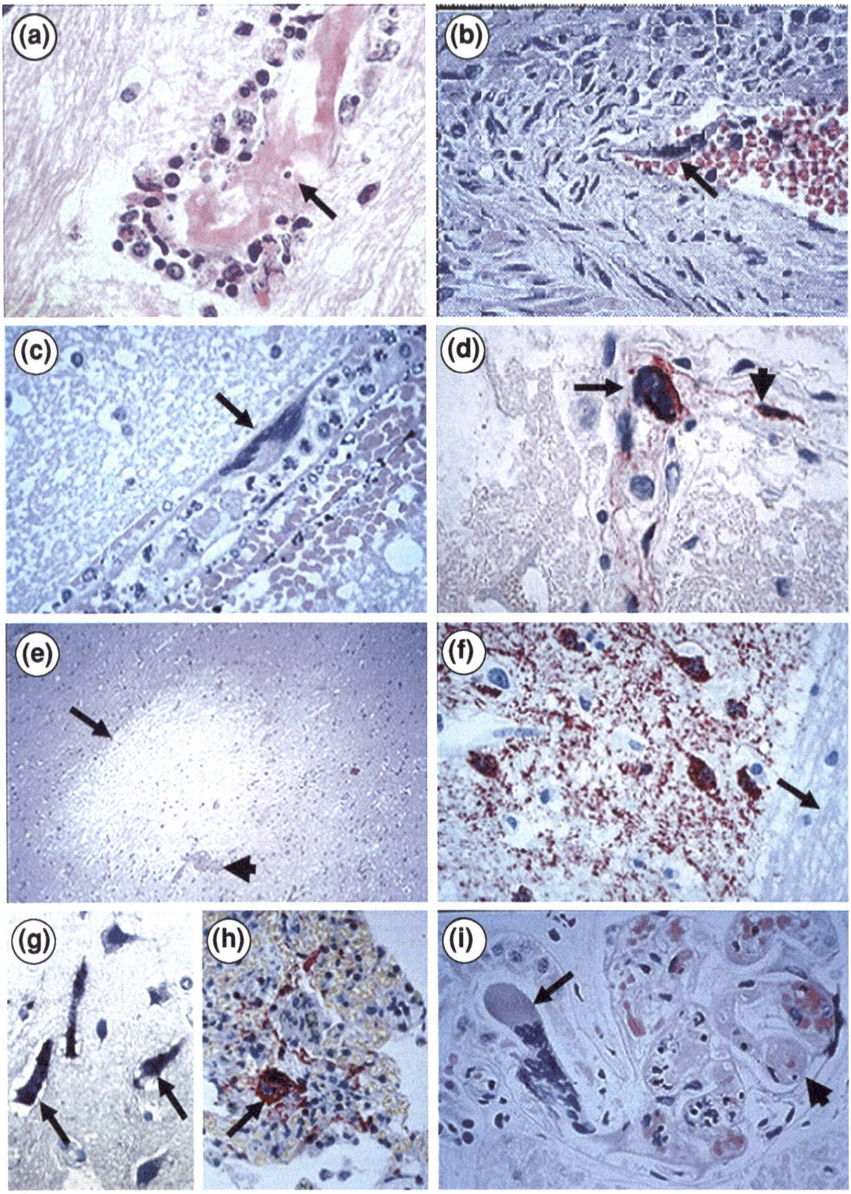

"late-onset" encephalitis. There was a patient who developed "late-onset" Nipah encephalitis 11 years after initial exposure (unpublished data, Tan CT).

Clinical, radiological and pathological findings suggest that relapsing NiV encephalitis is similar to "late-onset" encephalitis but is distinct from acute NiV encephalitis (Sarji et al. 2000; Tan et al. 2002; Wong et al. 2009; Wong 2010). The brain MR scans usually show confluent geographic lesions rather than

◀**Fig. 1** Pathology of human acute Nipah virus infection. **a** Vasculitis and intravascular thrombosis (*arrow*) in the brain. **b** Transmural inflammation of a pulmonary arteriole with endothelial multinucleated giant cell (*arrow*) **c** Multinucleated giant cell arising from the endothelial surface (*arrow*) of a cerebral blood vessel with perivascular haemorrhage. **d** Viral antigens immunolocalised to a multinucleated giant cell (*arrow*) and adjacent endothelium (*arrowhead*). **e** Necrotic plaque in cerebral parenchyma (*arrow*) adjacent to a thrombosed vessel (*arrowhead*). **f** Viral antigens found in numerous neuronal perikarya and processes in the putamen but not in the adjacent white matter (*arrow*). **g** Viral RNA in infected neurons (*arrows*). **h** Intra-alveolar cell in the lung with viral antigens (*arrow*). **i** Glomerulus with capillary thrombosis (*arrowhead*), necrosis and peripheral multinucleated giant cell formation (*arrow*). H&E stains (**a**, **b**, **c**, **e**, **i**), immunoperoxidase stains (**d**, **f**, **h**), in situ hybridisation (**g**). Magnification, objective ×4 (**c**); ×20 (**h**); ×40 (**a**–**d**, **f**, **g**, **i**). Figures **b**–**i** modified with permission from Elsevier

discrete lesions (Fig. 3b). There is good evidence that relapsing henipavirus encephalitis is probably the result of viral recrudescence from surviving viral foci in the CNS after the acute infection (Tan et al. 2002; Wong et al. 2009; Wong 2010). Further investigations are needed to fully understand the factors that could contribute to relapsing encephalitis. These may include mutations in the viral genotype and/or host immune response (Wong 2010). Interestingly, relapsing henipavirus encephalitis appears to be reminiscent of the rare subacute sclerosing panencephalitis due to mutant measles virus that affects the brain months and years after initial measles infection. It is not known whether acute henipavirus could evolve into the relapsing encephalitis phenotype without a significant symptom-free interval as suggested by brain MR scans of some "acute" henipavirus encephalitis cases that appeared to be show more confluent lesions. Currently, there is no useful laboratory biomarker for relapsing encephalitis.

3 Pathology of Acute Henipavirus Infection in Humans

Although published data on HeV infection consists of a single case and most of the information about human henipavirus infection is derived from NiV studies, both viruses probably cause a similar pathology. One characteristic feature is disseminated small vessel vasculopathy comprising true vasculitis, endothelial ulceration and intramural necrosis in the central nervous system (CNS), lung, kidney and many other major organs (Fig. 1a–b and Fig. 2a–b) (Wong et al. 2002a, b, 2009; Wong 2010). In NiV infection and possibly HeV as well, CNS involvement appears to be most severe compared to other organs. A careful search may reveal the occasional multinucleated giant cell or syncytia arising from the endothelial surface (Fig. 1b–d). Vascular occlusion by vasculitis-induced thrombosis (Fig. 1a, e, i) and perivascular haemorrhage possibly due to vascular wall destruction/weakening were observed (Fig. 1c). Viral antigens, RNA and nucleocapsids could be detected in vascular endothelium and smooth muscle, and multinucleated giant cells (Figs. 1d, 2c) (Wong et al. 2002a, b; Goldsmith et al. 2002).

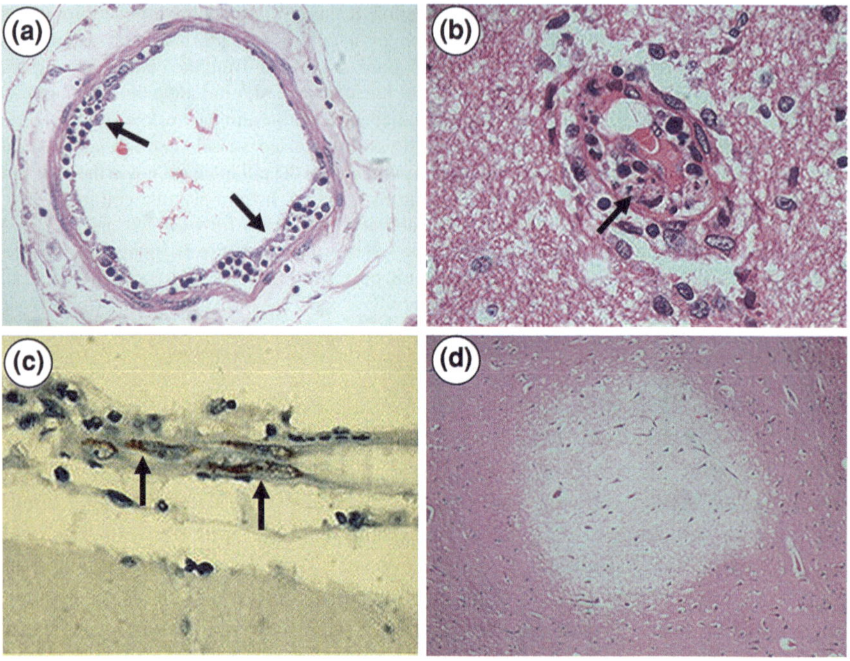

Fig. 2 Pathology of human acute hendra virus infection. Mild vasculitis or endotheliitis (*arrows*) in a meningeal blood vessel (**a**) and vessel in the cerebrum (**b**). Hendra viral antigens (**c**, *arrows*) in endothelial cells of a meningeal blood vessel. A necrotic plaque in the cerebral parenchyma (**d**). H&E stains (**a**, **b**, **d**) and immunoperoxidase stain (**c**). Magnification, objective ×10 (**d**); ×40 (**a–c**). Figures **a–c** modified with permission from John Wiley & Sons

Vasculopathy was often associated with discrete necrotic or more subtle vacuolar plaque-like lesions in both grey and white matter of the CNS (Figs. 1e, 2d and 3a) that probably corresponded to brain MR scan lesions. These lesions were characterised by varying degrees of necrosis, oedema and inflammation. In other organs, similar but less discrete parenchymal necrotic lesions were also described. Inflammatory cells were non-specific, consisting of macrophages, neutrophils, lymphocytes and plasma cells. Often viral inclusions, antigens and RNA could be demonstrated in adjacent peripheral parenchymal cells such as neuroglial cells, renal tubules, alveolar cells, etc. (Fig. 1f–h) (Wong et al. 2009). In the lung, kidney (Fig. 1i), lymph node and spleen, occasional multinucleated giant cells derived from parenchyma and possibly inflammatory cells may also be observed (Wong et al. 2002a, b, 2009).

The dual pathogenetic mechanisms of microinfarction associated with vasculopathy and direct parenchymal cell infection in multiple organs correlates well with the observed clinical manifestations, and appears to be rather unique in viral pathogenesis.

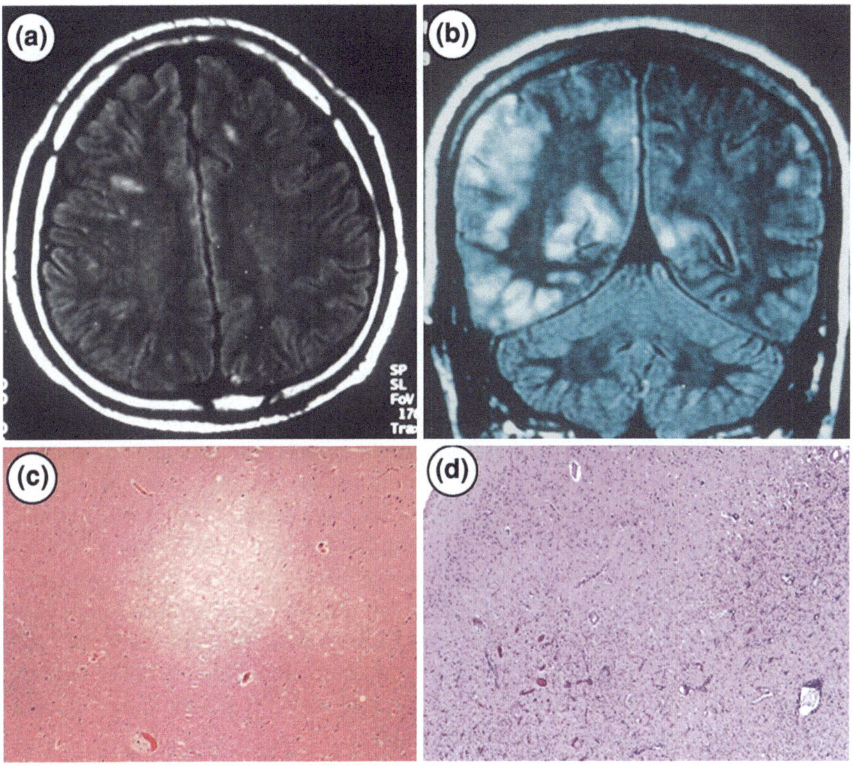

Fig. 3 Brain MR scans of acute Nipah (**a**) and relapsing Nipah encephalitis (**b**) showing hyperintense discrete and more confluent lesions, respectively, corresponding to discrete pathological lesions (**c**) and confluent lesions (**d**). H&E stains (**c**, **d**). Magnification, objective ×4

4 Pathology of Relapsing Henipavirus Encephalitis

The pathological features of relapsing henipavirus encephalitis were based on a few autopsies (Tan et al. 2002; Wong et al. 2009; Wong 2010) and again, appear to be similar for both HeV and NiV. In contrast to acute henipavirus infection in which there is systemic involvement, the pathology of relapsing encephalitis is confined only to the CNS. Non-CNS organs were essentially normal and there was no evidence of vasculopathy.

In affected neuronal areas and adjacent white matter, extensive confluent parenchymal necrosis, oedema and inflammation were seen (Fig. 3d). Inflammatory cells consisted of macrophages, lymphocytes and some plasma cells, with prominent perivascular cuffing. Reactive gliosis and prominent vascular proliferation in areas of severe neuronal loss was also observed (Wong 2010). Like acute henipavirus encephalitis, viral inclusions, antigens and RNA could be demonstrated mainly in neurons, ependyma and possibly in other glial or inflammatory cells as well. Severe meningitis was found in many areas. In contrast to acute encephalitis, vasculitis, thrombosis, endothelial syncytia, viral antigens and RNA were absent in blood vessels.

5 Comparative Pathology Between Human and Animal Infections

The ability of henipaviruses to successfully infect numerous mammalian species is well documented in in vitro infections of different cultured cell lines (Bossart et al. 2002), in vivo experimental infections of animal models and natural infections in some animals. These animals included bat, chick embryo, ferret, hamster, monkeys, cats, dog, pig, etc. (Weingartl et al. 2009; Wong and Ong 2011).

In general, the acute infectious disease pathology described in various animal species appears to be similar to that found in the human disease viz., disseminated vasculopathy and parenchymal lesions in multiple organs, and direct evidence of viral infection. However, there may be some significant or subtle differences between animals. In the pig and cat, respiratory tract inflammation, notably involving bronchial epithelium, is a prominent feature whereas in the guinea pig there is only mild lung inflammation. Neuronal infection may be more subtle in the pig and cat in contrast to human infections (Tan et al. 1999; Parashar et al. 2000; Hooper et al. 2001; Middleton et al. 2002; Tanimura et al. 2004; Mills et al. 2009; Li et al. 2010; Wong and Ong 2011). To date there is no published data on relapsing henipavirus encephalitis in animals.

It is not surprising that henipaviruses cause similar infectious disease pathology in both humans and animals as it has now been shown that the viruses share the same virus entry receptor. The main receptor has been identified as ephrin B2 (Negrete et al. 2005; Bonaparte et al. 2005) and the alternative receptor, ephrin B3 (Negrete et al. 2006). These receptors are ubiquitous on plasma membranes of many mammalian cells, particularly in the blood vessels and CNS, thus accounting for the prominent clinicopathological features of vasculitis and CNS involvement in humans and animals.

Acknowledgments The authors acknowledge the extensive assistance from colleagues, clinicians, students and fellow scientists in Malaysia and around the world in our investigations of henipavirus infections. The support of R&D grants from the Malaysian Government is also gratefully acknowledged.

References

Bonaparte MI et al (2005) Ephrin-B2 is a functional receptor for Hendra virus and Nipah virus. Proc Natl Acad Sci U S A 102:10652–10657

Bossart KN et al (2002) Membrane fusion tropism and heterotypic functional activities of the Nipah virus and Hendra virus envelope glycoproteins. J Virol 76:11186–11198

Chadha MS et al (2006) Nipah virus-associated encephalitis outbreak, Siliguri, India. Emerg Infect Dis 12:235–240

Chew NK et al (1999) Electroencephalography in acute Nipah encephalitis. Neurol J Southeast Asia 4:45–51

Chong HT, Tan CT (2003) Relapsed and late-onset Nipah encephalitis, a report of three cases. Neurol J Southeast Asia 8:109–112

Chong HT et al (2000) Nipah encephalitis outbreak in Malaysia, clinical features in patients from Seremban. Neurol J Southeast Asia 5:61–67

Chong HT et al (2001) Occupational exposure, age, diabetes mellitus and outcome of acute Nipah encephalitis. Neurol J Southeast Asia 6:7–11

Chua KB et al (1999) Fatal encephalitis due to Nipah virus among pig-farmers in Malaysia. Lancet 354:1257–1259

Chua KB et al (2000) High mortality in Nipah encephalitis is associated with presence of virus in cerebrospinal fluid. Ann Neurol 48:802–805

Field HE et al (2000) A fatal case of Hendra virus infection in a horse in north Queensland: clinical and epidemiological features. Aust Vet J 78:279–280

Goh KJ et al (2000) Clinical features of Nipah virus encephalitis among pig farmers in Malaysia. N Engl J Med 342:1229–1235

Goldsmith CS et al (2002) Elucidation of Nipah virus morphogenesis and replication using ultrastructural and molecular approaches. Virus Res 92:89–98

Hanna JN et al (2006) Hendra virus infection in a veterinarian. Med J Aust 185:562–564

Harit AK et al (2001) Nipah/Hendra virus outbreak in Siliguri, West Bengal, India in 2001. Indian J Med Res 123:553–560

Hooper P et al (2001) Comparative pathology of the diseases caused by Hendra and Nipah viruses. Microbes Infect 3:315–322

Hossain MJ et al (2008) Clinical presentation of Nipah virus infection in Bangladesh. Clin Infect Dis 46:977–984

Lee KE et al (1999) The neurological manifestations of Nipah virus encephalitis, a novel paramyxovirus. Ann Neurol 46:428–432

Li M, Embury-Hyatt C, Weingartl HM (2010) Experimental inoculation study indicates swine as a potential host for Hendra virus. Vet Res 41:33

Lim CC et al (2002) Nipah virus encephalitis: serial MR study of an emerging disease. Radiology 222:219–226

Middleton D et al (2002) Experimental Nipah virus infection in pigs and cats. J Comp Path 126:124–136

Mills JN et al (2009) Nipah virus infection in dogs, Malaysia, 1999. Emerg Infect Dis 15:950–952

Murray K et al (1995) A morbillivirus that caused fatal disease in horses and humans. Science 268:94–97

Negrete OA et al (2005) Ephrin B2 is the entry receptor for Nipah virus, an emergent deadly paramyxovirus. Nature 436:401–405

Negrete OA et al (2006) Two key residues in EphrinB3 are critical for its use as an alternative receptor for Nipah virus. Plos Pathog 2 (2):e7

O'Sullivan JD et al (1997) Fatal encephalitis due to novel paramyxovirus transmitted from horses. Lancet 349:93–95

Parashar UD et al (2000) Case-control study of risk factors for human infection with the new zoonotic paramyxovirus, Nipah virus, during a 1998–1999 outbreak of severe encephalitis in Malaysia. J Infect Dis 181:1755–1759

Paton NI et al (1999) Outbreak of Nipah virus infection among abattoir workers in Singapore. Lancet 354:1253–1256

Playford EG et al (2010) Human Hendra virus encephalitis associated with equine outbreak, Australia, 2008. Emerg Infect Dis 16:219–223

Quddus R et al (2004) A report of 4 patients with Nipah encephalitis from Rajbari district, Bangladesh in the January 2004 outbreak. Neurol Asia 9:33–37

Ramasundram V et al (2000) Kinetics of IgM and IgG seroconversion in Nipah virus infection. Neurol J Southeast Asia 5:23–28

Rogers RJ et al (1996) Investigation of a second focus of equine morbillivirus in coastal Queensland. Aust Vet J 74:243–244

Sarji SA et al (2000) Magnetic resonance imaging features of Nipah encephalitis. Am J Roentgenol 175:437–442

Sejvar JJ et al (2007) Long term neurological and functional outcome in Nipah virus infection. Ann Neurol 62:235–242

Selvey LA et al (1995) Infection of humans and horses by a newly described morbillivirus. Med J Aust 162:642–645

Siva SR, Chong HT, Tan CT (2009) Ten year clinical and serological outcomes of Nipah virus infection. Neurol Asia 14:53–58

Tan CT, Wong KT (2003) Nipah encephalitis outbreak in Malaysia. Ann Acad Med Singapore 32:112–117

Tan KS Tan CT, Goh KJ (1999) Epidemiological aspects of Nipah virus infection. Neurol J Southeast Asia 4:77–81

Tan CT et al (2002) Relapsed and late-onset Nipah encephalitis. Ann Neurol 51:703–708

Tanimura N et al (2004) Monoclonal antibody-based immunohistochemical diagnosis of Malaysian Nipah virus infection in pigs. J Comp Path 131:199–206

Wang LF et al (2000) The exceptionally large genome of Hendra virus: support for creation of a new genus within the family Paramyxoviridae. J Virol 74:9972–9979

Weingartl HM, Berhane Y Czub M (2009) Animal models of henipavirus infection: a review. Vet J 181:211–220

Williamson MM, Torres-Velez FJ (2010) Henipavirus: a review of laboratory animal pathology. Vet Pathol 47:871–880

Wong KT (2000) Emerging and re-emerging epidemic encephalitis: a tale of two viruses. Neuropathol Appl Neurobiol 26:313–318

Wong KT (2010) Emerging viral encephalitides with a special focus on henipaviruses. Acta Neuropathol 120:317–325

Wong KT (2010) Nipah and Hendra viruses: recent advances in pathogenesis. Future Virol 5:129–131

Wong KT, Ong KC (2011) Pathology of acute henipavirus infection in humans and animals. Pathol Res Int 2011:1–12. doi:10.4061/2011/567248

Wong KT et al (2002a) Nipah virus infection, an emerging paramyxoviral zoonosis. Springer Semin Immunopathol 24:215–228

Wong KT et al (2002b) Nipah virus infection: Pathology and pathogenesis of an emerging paramyxoviral zoonosis. Am J Pathol 161:2153–2167

Wong KT et al (2009) Human Hendra virus infection causes acute and relapsing encephalitis. Neuropathol Appl Neurobiol 35:296–305

Henipaviruses in Their Natural Animal Hosts

D. J. Middleton and H. M. Weingartl

Abstract Hendra virus (HeV) and Nipah virus (NiV) form a separate genus *Henipavirus* within the family *Paramyxoviridae*, and are classified as biosafety level 4 pathogens due to their high case fatality rate following human infection and because of the lack of effective vaccines or therapy. Both viruses emerged from their natural reservoir during the last decade of the twentieth century, causing severe disease in humans, horses and swine, and infecting a number of other mammalian species. The current review summarizes our up to date understanding of pathology and pathogenesis in the natural reservoir species, the *Pteropus* bat, and in the equine and porcine spill over species.

Contents

D. J. Middleton
Australian Animal Health Laboratory, CSIRO, Geelong, Australia

H. M. Weingartl (✉)
National Centre for Foreign Animal Disease,
Canadian Food Inspection Agency, Winnipeg, Canada
e-mail: hana.weingartl@inspection.gc.ca

H. M. Weingartl
Department of Medical Microbiology,
University of Manitoba, Winnipeg, Canada

Current Topics in Microbiology and Immunology (2012) 359: 105–121
DOI: 10.1007/82_2012_210

Published Online: 3 April 2012

1 Introduction

Several species of bats of the genus *Pteropus* are considered natural reservoir for Hendra (HeV) and Nipah (NiV) viruses (genus *Henipavirus*, family *Paramyxoviridae*). Systemic infection of spill over hosts by viruses in the *Henipavirus* genus may be associated with high morbidity and case fatality rates such as in HeV infection of horses (Murray et al. 1995) and NiV infections of people (Wong et al. 2002) or be characterized by higher transmission rates with lower morbidity and mortality such as NiV virus infection of pigs (Mohd Nor and Ong 2000). Clinical signs in henipavirus infected animals, when they occur, are respiratory and/or neurological. HeV was the first one to emerge from its natural reservoir, causing an outbreak of an acute respiratory syndrome that resulted in the death of one human and 13 horses in 1994 in Hendra, a suburb of Brisbane, Australia (Murray et al. 1995). A number of outbreaks in horses, often associated with human cases occurred in Australia since (Hess et al. 2011). For the first time in 2011, an infection of a dog with HeV was detected during the outbreak in horses (http://web.oie.int/wahis/public.php?page=single_report&pop=1&reported=10833).

NiV was isolated in 1999 from a human encephalitis patient during an epidemic associated with significant human mortality in 1998–1999 in Malaysia. During this outbreak, pigs were identified as the source of NiV for human infections (Chua 2003). In the more recent outbreaks in India and Bangladesh, only one cluster of cases was possibly associated with transmission from pigs (Luby et al. 2009). Although pigs are experimentally susceptible to HeV (Li et al. 2010), natural infections in pigs were observed only with NiV (Black et al. 2001). Rare natural NiV infections were reported also in horses, goats, cats and dogs (Mills et al. 2009; Hooper and Williamson 2000; Luby et al. 2009; Field et al. 2001; Hooper et al. 2001).

2 Henipaviruses in Bats

HeV has not been associated with any naturally occurring disease of Australian pteropid bats, although there is a greater than 40% prevalence of antibodies to HeV in these animals (Young et al. 1996; Field et al. 2001) and HeV was isolated from free-living colonies of *Pteropus alecto* (fetus) and *Pteropus poliocephalus* (fetus and uterine fluid) (Halpin et al. 2000). Similarly, experimental challenges of both pregnant and nonpregnant *Pt. poliocephalus and Pt. alecto* with HeV have not produced clinically detectable illness in these animals (Williamson et al. 1998, 1999; Halpin et al. 2011). On histological examination sporadic vasculitis was identified in some bats, involving vessels in the lung, spleen, meninges, kidney and gastrointestinal tract, with antigen identifiable in similar locations, as well as in the placenta of a pregnant *Pt. poliocephalus*. Viral antigen was primarily visualized within the tunica media as opposed to endothelial cells; relative sparing of endothelium from infection may contribute in part to the lack of an observed systemic disease response, but the more striking difference is the reduced level of antigen in tissues overall compared to other species with henipavirus associated disease. Similarly, in spite of rigorous sampling regimens few isolations of HeV have been made from experimental bats challenged with HeV: these comprise virus from urine of a single nonpregnant *Pt. alecto* on repeated occasions, and from a fetal spleen and kidney, maternal heart, blood, kidney and spleen of two pregnant *Pt. poliocephalus*. In addition, viral RNA has been detected in blood, urine, rectal and throat swabs of pregnant and nonpregnant *Pt. alecto* within the first week following experimental HeV challenge by a noninvasive route (Halpin et al. 2011). Interestingly, approximately 50% only of HeV challenged bats demonstrate a rise in antibody titer to HeV: these are low and sometimes of short duration (<3 weeks). It is also of interest that seroconversion is not necessarily detected in bats from which virus reisolation has been successful.

In Malaysia, there is a 25% prevalence of antibody to NiV in *Pteropus hypomelanus* and *Pteropus Vampyrus* (Johara et al. 2001), and NiV has been reisolated from chewed fruit and shed urine beneath free-living colonies of *Pt. hypomelanus* on Tioman Island (Chua et al. 2002). No naturally occurring disease has been attributed to NiV infection in these animals. Experimental challenge with NiV of HeV-free members of an Australian pteropid bat species (*Pt. poliocephalus*) did not result in a clinically detectable disease although virus was repeatedly reisolated from urine of one animal over a week after challenge (Middleton et al. 2007), and one animal had histological evidence of vasculitis involving the intestinal submucosa. Inconsistent and sometimes transient seroconversion to NiV was also noted in these animals.

Although several species of bats of the genus *Pteropus* are considered natural reservoirs of henipaviruses, and bats can be experimentally infected, they do not develop disease or gross pathology, and the seroconversion and virus reisolation from them is rather inconsistent.

3 Hendra Virus in Horses

Following a suggested incubation period of 4–16 days (Selvey et al. 1995; Murray et al. 1995; Baldock et al. 1996), HeV infected horses develop severe, acute, febrile respiratory disease sometimes accompanied by facial swelling, ataxia and, terminally, copious frothy nasal discharge. In some cases, including an outbreak involving several animals in one location, neurological signs predominate with

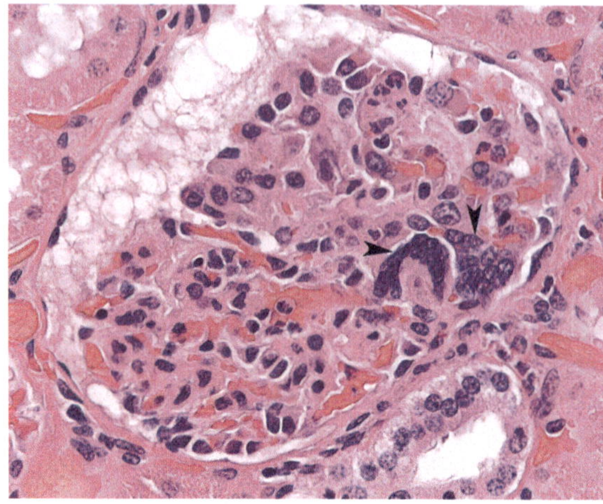

Fig. 1 Syncytial cells (*black arrows*) within renal glomerulus of horse infected with HeV (H&E). Magnification ×400

clinical features comprising hypersensitivity, ataxia, disorientation, facial paralysis, head tilt, circling, head pressing and stranguria (Field et al. 2010). Persistent infection of horses has not yet been confirmed: some convalescent animals have re-presented with neurological signs (myoclonic twitches) but did not undergo further laboratory assessment (Rogers et al. 1996). Recognition of seroconversion to HeV in asymptomatic horses has been confined to disease outbreaks suggesting that the virus is not maintained within the horse population.

In experimentally infected horses, HeV RNA may be recovered from nasal secretions from 48 h after exposure to virus and then throughout incubation period (Marsh et al. 2011), consistent with the proposal that systemic spread of virus may be preceded by local viral replication in the nasal cavity or nasopharynx (Weingartl et al. 2009). Increasing gene copy number recovered over time also suggests that the greater transmission risk is posed by animals in the immediate pre-symptomatic and symptomatic stages of infection, it which time routes of excretion may also include oral cavity secretions, urine, and feces as well as blood (Williamson et al. 1998; Marsh et al. 2011); this accords with field observations around human infection with HeV.

The predominant lesion in acute natural and experimental HeV infection of horses is vasculitis with fibrinoid degeneration and necrosis, most markedly affecting smaller vessels in major organs such as lung, brain (including meninges), spleen, lymph nodes, and kidney, but which is also detectable in nasal mucosa, liver, heart, stomach, intestine, uterus, and ovary. Pulmonary lesions are particularly extensive with severe necrotizing alveolitis with marked fibrinous alveolar exudates accounting for a commonly reported clinical presentation of the disease. Syncytial cells (Fig. 1) may also be identified in vascular endothelium and endothelium of lymphatic vessels as well as respiratory epithelial and other lymphoid cells. Viral antigen is readily demonstrable within the alveolar walls, syncytia, the media of small arterioles, and endothelium of affected blood vessels such those within meninges, lymph nodes, spleen, renal glomeruli and brain,

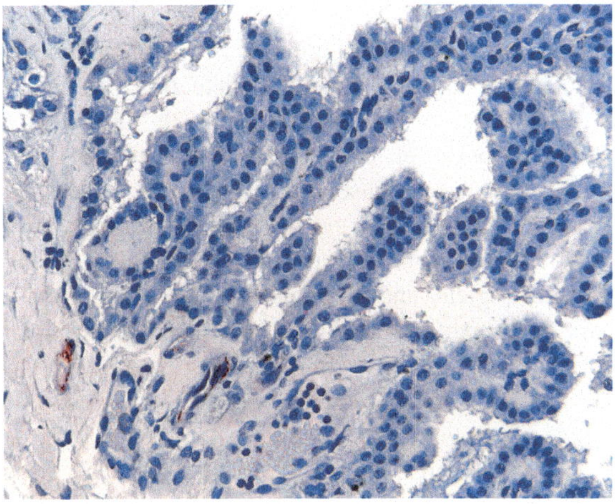

Fig. 2 HeV antigen (*red staining*) within endothelium of blood vessels supplying choroid plexus in an infected horse (rabbit polyclonal anti-Nipah N antibody). Magnification ×200

including choroid plexus (Fig. 2) (Hooper et al. 1997). Formation of endothelial syncytia and the presence of antigen within endothelium support direct infection of these cells by virus, and primary endothelial injury is known to play a crucial role in the pathogenesis of systemic vasculitis (Cid et al. 2004).

Horses that have seroconverted following acute HeV infection acquired in the field may have moderately severe focal nonsuppurative meningoencephalitis with gliosis and perivascular cuffing. In these cases, sparse amounts of HeV antigen may still be detectable in the brain several weeks after resolution of clinical signs, and HeV genome may be recovered from brain tissue—including olfactory pole—as well as bronchial and renal lymph nodes (unpublished observations). The pathogenesis and significance of these observations, particular with respect to the risk of recrudescence, is not understood.

4 Nipah Virus in Swine

4.1 Natural Infections

Nucleotide sequences of NiV isolates from pigs indicated a possibility that the virus was introduced into swine population on at least two occasions. Sequences from the secondary outbreak in the south of Malaysia were, in addition, identical with sequences from human isolates, confirming the pigs as a source of human infections with NiV (AbuBakar et al. 2004).

During the outbreak in swine, the infection rate was estimated to be 100% in affected farms with majority of the pigs not showing clinical disease, and mortality rate of 1–5% being only somewhat above normal. The acute febrile disease was initially called porcine respiratory and encephalitis syndrome, describing the main clinical characteristics in pigs. Direct and possibly also airborne exposure to secretions from infected animals is one of the presumed modes of transmission of NiV among pigs, supported by detection of NiV in the epithelium of upper and lower respiratory tract as well as in the lumen of the airways (Mohd Nor et al. 2000; Hooper et al. 2001; Tanimura et al. 2004).

Clinical disease observed in NiV infected pigs during the Malaysian outbreak varied in different age groups. Sows and boars sometimes died suddenly with no overt clinical signs. Neurological signs were more frequently observed than in younger piglets, and included head pressing, agitation (displayed as biting at the bars of the pen), tetanus-like spasms or seizures, an apparent pharyngeal muscle paralysis with an apparent inability to swallow, frothy salivation, and drooping of the tongue. Reported early abortions (first 3 months) were possibly due to NiV infections as well.

In younger pigs, the main clinical signs were respiratory. When present, clinical signs in weaners, growers and finishers included pyrexia (rectal temperatures of 40°C and higher), increased respiratory rate or forced respiration, harsh nonproductive cough. Lower number of pigs with clinical signs had accompanying neurological signs such as muscle spasm, rear leg weakness, and uncoordinated gate. Mortality rate in suckling piglets was occasionally quite high (40%); however, it was not clear whether inability of the sows to nurse was not the major reason.

The estimated incubation period in pigs during the outbreak was 7–14 days (Mohd Nor et al. 2000).

Gross pathological lesions observed in pigs during the NiV outbreak in Malaysia were not considered pathognomonic, and pathological changes due to other concurrent diseases were suspected (Hooper et al. 2001). Mild to severe pulmonary lesions with varying degrees of lobular consolidation and often distended interlobular septae were described. Bronchi and trachea of pigs were frequently filled with exudate or frothy fluid sometimes tinged with blood. In number of animals, congestion of brain or meninges were reported, and in some animals kidney congestion with petechial hemorrhages in the renal capsule and cortex were noted as well (Mohd Nor et al. 2000; Chua et al. 2000, Tanimura et al. 2004).

Due to frequent concurrent infections with other agents, only lesions with confirmed NiV antigen presence by immunohistochemistry were considered in histopathological evaluation of NiV pathology in swine. The most prominent histopathological changes in lungs were moderate to severe interstitial pneumonia, along with hyperplasia of columnar epithelium, tracheitis, peribronchial and peribronchiolar infiltration of lymphocytes, and exudates in the lumen of the airways. Increased numbers of alveolar macrophages within the alveolar, bronchiolar and bronchial lumen were noted as well. Alveolar macrophages, occasionally multinucleated, had frequent positive staining for NiV antigen, indicating that these cells are one of NiV targets. Formation of syncytial cells was reported in the endothelium of small blood and lymphatic vessels, although detectable vasculitis in lungs was rare. Syncytia formation in the respiratory tract, one of the characteristics of infection with henipaviruses, was most frequently observed in the epithelial cells (Hooper et al. 2001; Mohd Nor et al. 2000; Chua et al. 2000; Tanimura et al. 2004). Nonsuppurative lymphocytic meningitis and meningoencephalitis with gliosis were the main findings in the brain. Encephalitis, characterized by mild perivascular cuffing of lymphocytes and gliosis was considered to be rare. Vasculitis appeared to be somewhat more frequent than in the lungs. In the natural infections, NiV appeared to have a predilection for meningeal membranes, namely the arachnoid cells. The meningeal inflammatory infiltrates had viral antigen present in the inflammatory infiltrating cells as well.

Generalized vasculitis with fibrinoid necrosis, and infiltration of mononuclear cells sometimes associated with thrombosis was in addition to lung and brain observed in kidneys. One piglet in the report by Tanimura et al. (2004) had positive staining for NiV detected in number of other organs (spleen, larynx, heart, stomach and colon). It is, however, difficult to ascertain how typical that would be for other field cases.

The most frequently reported histopathological changes are summarized in Table 1 along with observations reported from experimental inoculations.

4.2 Experimental Infections

4.2.1 Clinical Disease

In general, the majority of young pigs (5–9 weeks old) experimentally infected with a human isolate of NiV passaged on Vero cells would develop subclinical infection or develop mild clinical signs, such as increased body temperature and mild respiratory signs (increased respiratory rate and a mild cough). Nasal and subcutaneous routes of inoculation, with virus dose in the range of 5×10^4–5×10^5 PFU/pig, lead to severe neurological signs requiring euthanasia in a proportion of pigs that represents a mortality rate higher than that reported in the field (Middleton et al. 2002; Weingartl et al. 2005). In a study by Weingartl and colleagues, out of 11 piglets inoculated with NiV by the oronasal and intraocular routes, two developed neurological signs and had to be euthanized, while nine remained clinically normal (Weingartl et al. 2005).

In a more recent experimental inoculation of two groups of piglets for development of positive control sera, 5-week-old piglets were inoculated nasally with virus re-isolated from lung of previously experimentally infected pig, and passaged once on Vero 76 cells. Out of 10 piglets, two had to be euthanized at 6 dpi likely due to NiV encephalitis, while one died and two more piglets had to be euthanized at 7 dpi with detected bacterial meningitis. In the first inoculation group, piglets remaining alive at 7 dpi were treated with Oxytetracyclin starting at 8 dpi, and survived until the end of the experiment at 28 dpi. Piglets in the second inoculation group were treated with the same antibiotic starting at day 3 post infection; however, the treatment this time did not prevent bacterial meningitis. Beside histopathological diagnosis of suppurative meningitis, the cerebrospinal fluid (CSF) was cloudy and bacterial growth was

Table 1 Histopathology in swine host in natural and experimental infections with NiV

Organ	Antigen detection	Microscopic pathological changes
Brain	endothelial cells and tunica media of small blood vessels in cerebral cortex and meninges neurons glial cells choroid plexus arachnoid cells granular cells olfactory bulb ependyma	generalized vasculitis with fibrinoid necrosis and mononuclear cells infiltration rare encephalitis (mild perivascular cuffs of lymphocytes; glial foci) suppurative or nonsuppurative meningoencephalitis with gliosis mononuclear infiltration of choroid plexus and meninges
Lung	choroid endothelial cells and tunica media of small blood and lymphatic vessels alveolar and bronchiolar repiratory epithelial cells (some cells form syncytia) exudative (syncytial) macrophages or mononuclear cells in alveoli and bronchioles infiltrating cells within the interstitium multinucleated giant cells	interstitial pneumonia (suppurative and nonsuppurative) alveolar macrophages (occasionally multinucleated) in the lumen of alveoli, with proteinaceous edema fluid endothelial syncytia of small blood and lymphatic vessels, as well as syncytia of bronchiolar respiratory epithelial cells focal or generalized vasculitis with fibrinoid necrosis and infiltration of mononuclear cells subpleural hemorrhages
Trachea, bronchi	cell debris in the lumen tracheal and bronchial epithelial cells multinucleated giant cells infiltrating cells within the submucosa	tracheitis bronchial pneumonia (suppurative and nonsuppurative) hyperplasia of the columnar epithelium with single cell necrosis of columnar cells peribronchial infiltration of lymphocytes lymphocytic inflamation of tracheal mucosa exudation of dead and live (macrophages) cells into the lumen

(continued)

Table 1 (continued)

Organ	Antigen detection	Microscopic pathological changes
Nasal turbinates (cavity)	osteoclasts respiratory epithelial cells endothelial cells of small blood vessels filae olfactoriae infiltrating lymphocytes and macrophages	lymphocytic inflamation of nasal mucosa
Lymph nodes	endothelial cells and tunica media of small blood and lymphatic vessels occasional lymphocytes folicular dendritic cells multinucleated giant cells	multiple foci of lymphocyte necrosis or depletion within the germinal centers and in cortical sinusoids increased number of dendritic cells nonvascular syncytia in paracortical regions syncytia formation in lymphatic endothelial cells
Spleen	see lymph nodes	focal necrosis extensive individual cell necrosis within periarteriolar lymphoid sheaths vascular lesions including perivascular accomodation of mononuclear cells nonvascular syncytia
Kidney	endothelial cells and tunica media of small blood vessels renal tubular epithelial cells epithelial cells of Bowman's capsule	generalized vasculitis with fibrinoid necrosis and infiltration of mononuclear cells focal renal tubular degeneration with interstitial accumulation of mononuclear cells syncytial cells and hyaline droplets formation within Bowman's space

Compiled from: Mohd Nor et al. (2000), Chua et al. (2000), Hooper et al. (2001), Middleton et al. (2002), Tanimura et al. (2004), Weingartl et al. (2005, 2006) and Berhane et al. (2008)

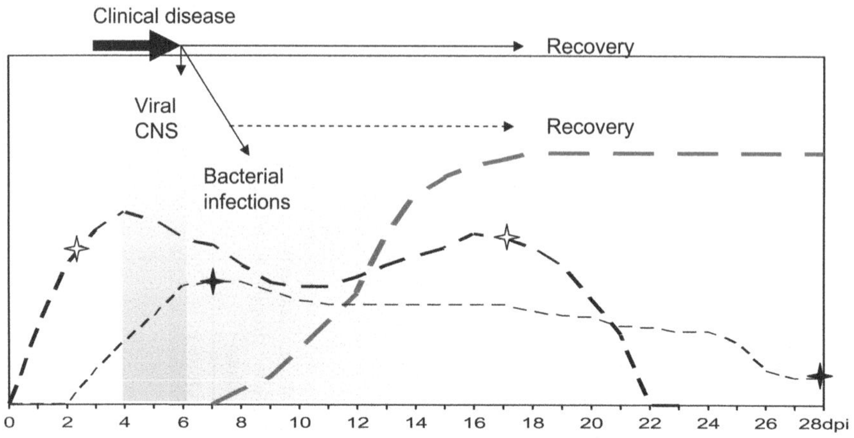

Fig. 3 NiV disease progression in experimentally infected swine. *Dashed medium black line* represents oro nasal shedding, *dashed thin black line* represents viremia, both based on copy number/ml as detected by real-time RT-PCR. *Stars* on the lines indicate first and last time point for virus isolation. *Thick dashed line* represents neutralizing antibodies as determined by plaque reduction neutralization assay (Middleton et al. 2002; Berhane et al. 2008). "Viral CNS" indicates period when only viral encephalitis was diagnosed. "Bacterial infection" represents mostly development of bacterial meningitis. *Bold black arrow* on top of the figure indicates development of primary clinical signs (increased rectal temperature, mild respiratory signs), which can be either cleared with recovery or the infection progresses to viral encephalitis and/or secondary bacterial infections, which can either be cleared or can be terminal (meningitis)

confirmed. Bacteria detected in CSF of piglets from the first "long term" study in 2007 were identified as *Streptococcus suis* and *Enterococcus faecalis*. This group was inoculated oronasally with human isolate passed in Vero 76 cells. One piglet had to be euthanized at 11 dpi due to bacterial meningitis, but four more had still bacteria in the CSF at 27–29 dpi (Berhane et al. 2008). This situation limits the in vivo pathogenesis studies to the early post inoculation phase, because at later points it is difficult to attribute the pathogenesis solely to the NiV infection itself.

Our current understanding of disease/infection progression during the experimental infections is summarized in Fig. 3. Briefly, shedding starts on day 2 post inoculation, prior to the onset of clinical signs, and lasts up to 3 weeks post infection, when the development of neutralizing antibodies reaches the peak. Although majority of experimentally infected piglets did not develop an apparent disease, and some experienced only mild respiratory signs, they all had transient increase in rectal temperatures from 3 to 6 dpi. About 20–40% of piglet inoculated nasally or subcutaneously developed CNS signs requiring euthanasia, suspected to be early post infection due only to viral encephalitis, and later due also to secondary bacterial meningitis (Middleton et al. 2002; Weingartl et al. 2005, 2006; Berhane et al. 2008; current report).

4.2.2 Pathology

Pathological findings in the experimentally infected pigs are consistent with the reports from field cases of NiV disease in pigs, with the main affected organs being brain, lung, spleen, lymph nodes, and kidneys (Table 1).

Association of lung lesions with individual lobules was confirmed in the experimental infections. This type of lesions is often observed in respiratory infections, such as influenza. Microscopic lesions in lungs included interstitial pneumonia with peribronchiolar, peribronchial and perivascular infiltration of mononuclear cells, and vasculitis often with fibrinoid necrosis. Viral antigen was detected in endothelial cells, smooth muscle cells of tunica media, as well as in macrophages, bronchiolar and less frequently alveolar epithelial cells (Hooper et al. 2001; Middleton et al. 2002; Weingartl et al. 2005; Berhane et al. 2008). In cases of neurological disease requiring euthanasia, nonsuppurative meningitis, meningoencephalitis or suppurative meningitis with whittish cloudy cerebrospinal fluid, indicative of possible secondary bacterial infection were more frequent then lymphocytic encephalitis. Prominent perivascular cuffing was observed in meninges and some also in the brain (Hooper et al. 2001; Middleton et al. 2002; Weingartl et al. 2005; Berhane et al. 2008). Meninges were often congested and edematous. Virus antigen was detected in the brain in neurons and glial cells, endothelial cells of blood vessels and smooth muscle cells of tunica media, and infiltrating mononuclear cells, as well as in ependyma, choroid plexus and meninges (Hooper et al. 2001; Middleton et al. 2002; Weingartl et al. 2005).

Experimental inoculations allowed some additional and more in detail observations, specifically concerning the lymphoid organs. Interestingly, no lesions in thymus were reported to date, while notable changes were observed in lymph nodes, with confirmed virus detection. Lymph nodes were often enlarged mottled, congested, and edematous. Lymphoid necrosis was reported both in natural and experimental infections early post inoculation (within first 2 weeks). Later post infection (around 28 dpi) lymphoid depletion in cortical sinusoids, and marked reduction in germinal centres was observed, but necrotic cells were no longer present. It appears that apoptosis (data not shown) may be one of the reasons for the observed cell death in lymph nodes of pigs. Cells within the lymph nodes containing antigen were identified by immunohistochemistry as endothelial cells of blood/lymph vessels, often forming syncytial cells, multinucleated giant cells and dendritic cells, with only occasional staining in the lymphocytes. In animals surviving NiV infection, the lesions appeared to be resolving around 3 weeks post infection (Hooper et al. 2001; Middleton et al. 2002; Weingartl et al. 2005, 2006; Berhane et al. 2008).

4.2.3 Virus Isolation

Virus shedding in experimentally infected piglets was detected from nasal and oral cavities, starting at 2 dpi and peaking during the first week post inoculation with titers reaching 10^4–10^5 PFU/ml of nasal wash or oral/nasal swab. The shedding commenced prior to clinical signs and lasted up to 3 weeks post inoculation, past the primary clinical signs.

NiV was isolated from the respiratory tract (nasal turbinates, nasopharynx, trachea, bronchi and lungs), lymphatic system (submandibular and bronchiolar lymph nodes, tonsil and spleen), central nervous system (all areas of brain, trigeminal ganglion and CSF) and blood/serum of both clinically affected, and clinically normal animals (Middleton et al. 2002; Weingartl et al. 2005, 2006; Berhane et al. 2008; current data). Quantitative virus isolation was not attempted in the first experimental infection study (Middleton et al. 2002); however in the later studies, same plaque assay format was used for determination of virus load, allowing the results of different experiments to be compared. The highest NiV titers were obtained from brain samples collected at 5 dpi, reaching 2×10^7 PFU/g, followed by submandibular lymph nodes at 4–6 dpi (titers around 10^6 PFU/g, and upper respiratory tract (trachea or nasal turbinates) at 3–5 dpi (titers around 10^6 PFU/g). Due to lobular involvement of lung, the virus isolation was dependent on sampling. NiV was cleared from the sampled tissues by 27–28 dpi, although very low virus titer was still isolated at 28 dpi from serum. (Weingartl et al. 2005, 2006; Berhane et al. 2008; current data).

Figure 4 summarizes the trends in virus isolation early post infection in piglets infected nasally (or via oro nasal route) with 5×10^4 to 5×10^5 PFU of NiV per pig. The values are based on samples from three pigs for days 3 and 4 post inoculation, two pigs for 5 dpi, and five pigs for days 6 and 7 post inoculation. Piglets at days 3–5 were euthanized based on experimental schedule, while piglets on days 6 and 7 post inoculation were either euthanized based on clinical sign, died, or were euthanized according to the experimental schedule. Values for days 6 and 7 may reflect the fact that some piglets no longer had clinical signs and started to recover, as some samples (such as brain) were negative on virus isolation, but still positive on real-time RT-PCR. Figure 4a illustrates a timeline when NiV is detected in brain and in CSF, supporting the hypothesis that the initial invasion of CNS in swine inoculated nasally is directly from nasal cavity via cranial nerves (Weingartl et al. 2005). Figure 4b summarizes the virus isolations from the respiratory tract. Virus detection in nasal turbinates is followed shortly thereafter by isolation of NiV from trachea, while virus presence in lungs is gradually increasing from 3 to 7 dpi. The virus load in lung associated lymph nodes reaches the peak at 6–7 dpi. This would be consistent with the suggestion, that the primary lung infection is via the airways, and the high virus load in the lung associated lymph node is due to the draining of the virus and infected cells from lungs. Figure 4c follows the infection of lymphoid system. Turbinates and submandibular lymph nodes are almost simultaneously invaded by NiV, reflecting the fact that the submandibular lymph nodes drain the nasal cavity. This may suggest a pivotal role of immune cells in viremia and dissemination of NiV throughout the porcine host via the lymphatic and cardiovascular systems.

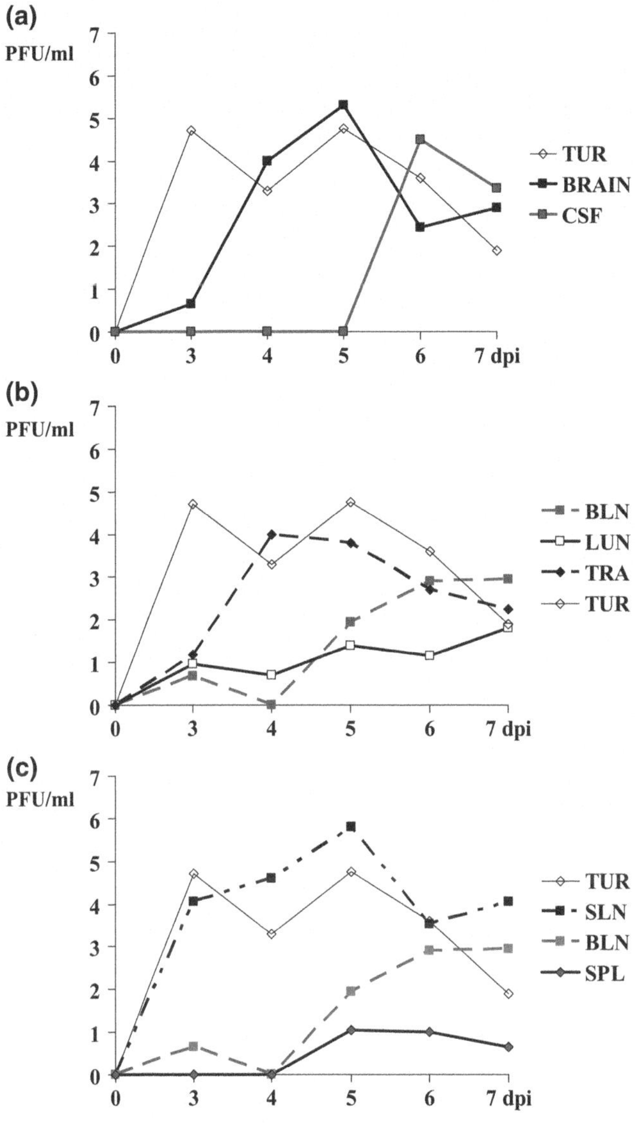

Fig. 4 Trends in NiV isolation early post infection. **a** Virus isolations from brain and CSF. *TUR* nasal turbinates, *CSF* cerebrospinal fluid. **b** Virus isolations from the respiratory tract. *BLN* lung associated (bronchial) lymph nodes, *LUN* lung, *TRA* trachea, *TUR* nasal turbinates. **c** NiV isolations from lymphoid system. *TUR* nasal turbinates, *SLN* submandibular lymph nodes, *BLN* lung associated lymph nodes, *SPL* spleen

4.2.4 Immune Response

Development of antibodies is associated in swine with protection against NiV (Weingartl et al. 2006) and also with clearance of the virus and recovery. First detectable neutralizing antibodies appeared around 7–10 dpi, and the experimentally infected pigs developed high neutralizing antibody titers of around 1,280 by 16 dpi, by which time nasal and oropharyngeal viral shedding had ceased. Despite the presence of neutralizing antibody, virus could still be isolated from the serum of these animals at 24 dpi, and viral RNA could be detected at 29 dpi (Berhane et al. 2008). In addition, development of neutralizing antibodies seemed to be somewhat delayed compared to infections with other viruses, such as influenza and also HeV (Li et al. 2010) where first neutralizing antibodies appear 5–6 dpi.

NiV infects specific porcine immune cells, and the antigen detection in alveolar macrophages and other infiltrating cells in naturally or experimentally infected pigs supports this observation. In porcine peripheral blood mononuclear cells population, the B lymphocytes, CD4−CD8−, as well as CD4+CD8− T lymphocytes are not permissive to NiV. The expansion of the CD4+CD8- cells (T helper cells) early post infection would be consistent with functional humoral response to NiV infection observed in swine. Interestingly, significant drop in the T helper cell frequency was observed in piglets which succumbed to the experimental infection, supporting the hypothesis that antibody development is the critical component of the protective immune response. Since productive viral replication was detected in monocytes, CD6+CD8+ T lymphocytes and NK cells (Stachowiak and Weingartl 2012), impact on the immune response has to be assessed. Infection of immune cells by NiV may interfere with immune cell signaling pathways, namely the ones employing STAT-1 (Shaw 2009) in the infected cells. NiV infected primary lymphocytes also failed to induce IFN-alpha (Stachowiak and Weingartl 2012). In addition, NiV appears to cause death of the infected but also of the "bystander" cells (e.g. in the lymph nodes) thus decreasing numbers of cells available to mount immune response. This may have an impact on the onset of immune response, as well as on development of frequently observed secondary bacterial infections in the experimental studies.

4.3 NiV Pathogenesis in the Swine Model

Pathogenesis studies with NiV in swine were so far limited to piglets between 4 and 10 weeks of age, and the models based on the experimental inoculations were summarized previously (Weingartl et al. 2009; Maisner et al. 2009). On oronasal inoculation, NiV infects number of cell types present in the oronasal mucosa: epithelial cells, cells of the immune system, and extracellular peripheral nerve endings of the cranial nerves leading in some animals to direct invasion of the brain (Weingartl et al. 2005). Endothelial cells of small blood and lymph vessels, a characteristic target for NiV are considered to be infected very early at the site of infection as well. Infection of the immune and the endothelial cells leads to viremia, and subsequent spread of NiV may be based on virus preference for endothelial cells of specific organs and tissues (Maisner et al. 2009). A current hypothesis of a mechanism of invasion of brain in swine includes the direct invasion via the cranial nerves followed by the invasion via the blood–brain barrier

(BBB). As the virus gains access to the brain directly from the nasal cavity by infecting naked nerve endings and proceeds into the brain, it infects not only the neurons, but also astrocytes (possibly participating in formation of BBB) and microglia, both being in direct physical contact with blood capillaries, potentially infecting the endothelial cells. Endothelial cells, astrocytes and microglia coordinate activation and regulation of local inflammatory reactions during infection, including release of cytokines and recruitment of monocytes and lymphocytes from circulation (Zachary 2007), some of them already infected with NiV, most importantly the CD6+CD8+ lymphocytes. From the recruited cells, the CD6+ cells would be the ones most strongly adhering to the microvascular endothelial cells due to a strong interaction with the activated leukocyte cell adhesion molecule ALCAM (CD166) highly expressed on the microvascular endothelial cell of the BBB. This molecule is also highly expressed on the endothelium forming the blood-air barrier in lung, and may explain NiV preferential tropism for small blood vessels of both, the lung and the brain (Stachowiak and Weingartl 2012). Infection of endothelial cells results in vasculitis, and permeability of the walls of blood and lymphatic vessels further increasing the virus load in the parenchyma of affected organs.

Originally proposed direct infection of the lung epithelial cells due to inhalation of the virus, or NiV spread along the respiratory epithelium from the upper respiratory tract is supported by the time lines for infection of the respiratory system by NiV and the gradual development of viremia (Figs. 3, 4). NiV invasion of the lower respiratory tract due to viremia, and the associated vascular damage and infiltration of infected mononuclear cells is also important, although occurring at the later point post infection.

Pathological changes were also observed in spleen and lymph nodes, consistent with the fact that the cells of the lymphoreticular system, including dendritic cells, are important target for NiV. Lymphocyte necrosis and depletion was observed especially in lymph nodes, and the specific staining for apoptosis was quite prominent (Berhane et al. 2008). Cell depletion may be due not only to the death of infected cells, but also due to a bystander death of noninfected cells caused by cell signaling. Infection of the immune system by NiV likely contributes to pathogenesis in swine, including transient immunomodulation.

5 Conclusion

Henipavirus studies in their natural animal hosts are restricted due to constraints of work with large animals or wildlife species, especially under biosafety level 4 conditions, as well as due to limited availability or lack of reagents, for example, for immune response studies. Consequently, our understanding of henipavirus infection and pathogenesis in their natural hosts is currently rather limited.

All of the above has an impact on veterinary vaccine development, efficacy testing, and ultimately availability of vaccines important not only for the veterinary field, but also for protection of human population by breaking the interspecies transmission cycle. In addition, understanding of mechanisms involved in protective immune response and pathogenesis (or lack of thereof) of henipavirus infections in bats, swine, or horses can highlight points of interest for development of intervention strategies in human infections.

References

AbuBakar S, Chang L-Y, Mohd Ali AR, Sharifah SH, Yusoff K, Zamrod Z (2004) EID 10: 2228–2230
Baldock FC, Douglas IC, Halpin K, Field H, Young PL, Black PF (1996) Vet J 20:57–61
Berhane Y, Weingartl HM, Lopez J, Neufeld J, Czub S, Embury-Hyatt C, Goolia M, Copps J, Czub M (2008) Transbound Emerg Dis 55:165–174
Black PF, Cronin JP, Morissy CJ, Westbury HA (2001) Aus Vet J 79:424–426
Chua KB (2003) J Clin Virol 26:265–275
Chua KB, Bellini J, Rota PA, Harcourt BH, Tamin A, Lam SK, Ksiazek TG, Rollin PE, Zaki SR, Shieh WJ, Goldsmith CS, Gubler DJ, Roehring JT, Eaton B, Gould AR, Olson J, Field H, Daniels P, Ling AE, Peters CJ, Anderson LJ, May BWJ (2000) Science 288:1432–1435
Chua KB, Koh CL, Hooi PS, Wee KF, Khong JH, Chua BH, Chan YP, Lim ME, Lam SK (2002) Microbes Infect 4:145–151
Cid MC, Segarra M, Garcia-Martinez A, Hernandez-Rodriguez J (2004) Curr Rheumatol Rep 6:184–194
Field H, Young P, Mohd Yob J, Mills J, Hall L, Mackenzie J (2001) Microbes Infect 3 :307–314
Field H, Schaff K, Kung N, Simon C, Waltisbuhl D, Hobert H, Moore F, Middleton D, Crook A, Smith G, Daniels P, Glanville R, Lovell D (2010) Emerg Infect Dis 16:338–340
Halpin K, Young PL, Field HE, Mackenzie JS (2000) J Gen Virol 81:1927–1932
Halpin K, Hyatt AD, Fogarty R, Middleton D, Bingham J, Epstein JH, Rahman SA, Hughes T, Smith C, Field HE, Daszak P, Henipavirus Ecology Research Group (2011) Am J Trop Med Hyg 85:946–951
Hess IMR, Massey PD, Walker B, Middleton DJ, Wright TM (2011) NSW public health bull. http://www.publish.csiro.au/index.cfm?paper=NB10077
Hooper PT, Williamson MM (2000) Vet Clin North Am Equine Pract 16:597–603, xi
Hooper PT, Ketterer PJ, Hyatt AD, Russell GM (1997) Vet Pathol 34:312–322
Hooper P, Zaki S, Daniels P, Middleton DA (2001) Microbes Infect 3:315–322
Johara MY, Field H, Rashdi AM, Morrissy C, van der Heide B, Rota P, bin Adzhar A, White J, Daniels P, Jamaluddin A, Ksiazek T (2001) Emerg Infect Dis 7:3
Li M, Embury-Hyatt C, Weingartl HM (2010) Vet Res 41:33. doi: 10.1051/vetres/2010005
Luby SP, Gurley ES, Hossain MJ (2009) Clin Infect Dis 49:1743–1748
Maisner A, Neufeld J, Weingartl H (2009) Thromb Haemost 102:1014–1023
Marsh GA, Haining J, Hancock TJ, Robinson R, Foord AJ, Barr JA, Riddell S, Heine HG, White JR, Crameri G, Field HE, Wang LF, Middleton DJ (2011) Emerg Infect Dis 17:2232–2238
Middleton DJ, Westbury HA, Morrissy CJ, van der Heide BM, Russel GM, Braun MA, Hyatt AD (2002) J Comp Path 126:124–136
Middleton DJ, Morrissey CJ, van der Heide BM, Russell GM, Braun MA, Westbury HA, Halpin K, Daniels PW (2007) J Comp Path 136(4):266–272
Mills JN, Alim ANM, Bunning ML, Lee OB, Wagoner KD, Amman BR, Stockton PC, Ksiazek TG (2009) Emerg Infect Dis 15:950–952

Mohd Nor MN, Ong BL (2000) In: Cargill C, McOrist S (ed) Proceedings of the 16th international pig veterinary society congress, Melbourne, Australia, pp 548–550
Mohd Nor MN, Gan CH, Ong BL (2000) Rev Sci Tech Off Int Epiz 19:160–165
Murray K, Selleck P, Hooper P, Hyatt A, Gould A, Gleeson L, Westbury H, Hiley L, Selvey L, Rodwell B, Ketterer P (1995) Science 268:94–97
Rogers RJ, Douglas IC, Baldock FC, Glanville RJ, Seppanen KT, Gleeson LJ, Selleck PW, Dunn KJ (1996) Aus Vet J 74:243–244
Selvey LA, Wells RM, McCormack JG, Ansford AJ, Murray K, Rogers RJ, Lavercombe PS, Selleck P, Sheridan JW (1995) MJA 162:642–645
Shaw ML (2009) Viruses 1:1190–1203. doi:10.3390/v1031190
Stachowiak B, Weingartl HM (2012) PLoS One 7(1):e30855 [Epub: 27 Jan, 2012]
Tanimura N, Imada T, Kashiwazaki Y, Shahirudin S, Sharifah SH, Aziz AJ (2004) J Comp Path 131:199–206
Weingartl H, Czub S, Copps J, Berhane Y, Middleton D, Marszal P, Gren J, Smith G, Ganske S, Manning L, Czub M (2005) J Virol 79:7528–7534
Weingartl HM, Berhane Y, Caswell JL, Loosmore S, Audonnet JC, Roth JA, Czub M (2006) J Virol 80:7929–7938
Weingartl HM, Berhane Y, Czub M (2009) Vet J 181:211–220
Williamson MM, Hooper PT, Selleck PW, Westbury HA, Slocombe RF (1999) J Comp Path 122:201–207
Williamson MM, Hooper PT, Selleck PW, Gleeson LJ, Daniels PW, Westbury HA, Murray PK (1998) Aus Vet J 76:813–818
Wong KT, Shieh WJ, Zaki SR, Tan CT (2002) Springer Semin Immunopathol 24:215–228
Young PL, Halpin K, Selleck PW, Field H, Gravel JL, Kelly MA, Mackenzie JS (1996) Emerg Infect Dis 2:3
Zachary JF (2007) In: Gavin MD, Zachary JF (ed) Pathologic basis of veterinary disease, 4th edn. Elsevier Inc., Missouri, pp 833–971

Nipah and Hendra Virus Interactions with the Innate Immune System

Christopher F. Basler

Abstract Nipah virus and Hendra virus are related, highly pathogenic paramyxoviruses with unusually broad host ranges. Henipaviruses encode several proteins that block innate immune responses, and these are likely to serve as virulence factors. Specfically, four virus-encoded proteins, the phosphoprotein (P), the V protein, the W protein, and the C protein have each been demonstrated to counteract aspects of the interferon (IFN)-α/β response, a key component of the innate immune response to virus infection. The available data indicate that V and W can inhibit the production of IFNα/β in response to various stimuli, while the P, V, and W proteins also block the ability of IFNs to signal and induce an antiviral state in cells. The C protein also inhibits the antiviral effects of IFNα/β by a poorly characterized mechanism. Reverse genetics systems, which allow the generation of recombinant viruses bearing specific mutations, have demonstrated the importance of the viral IFN-antagonists for replication. With these systems in hand, the field is now poised to define how specific viral IFN-antagonist functions influence viral pathogenesis.

Contents

C. F. Basler (✉)
Department of Microbiology, Mount Sinai School of Medicine,
Box 11241 Gustave L. Levy Place, New York, NY 10029, USA
e-mail: chris.basler@mssm.edu

Current Topics in Microbiology and Immunology (2012) 359: 123–152
DOI: 10.1007/82_2012_209

Published Online: 11 April 2012

1 Introduction

Nipah virus (NiV) and Hendra virus (HeV) are paramyxoviruses of the genus *Henipavirus* of the *Paramyxovirinae* subfamily within the family *Paramyxoviridae* (Lamb and Parks 2007). These viruses are notable for their ability to cause severe, often lethal disease in humans and for their broad host range (reviewed in (Ksiazek et al. 2011)). Both NiV and HeV are zoonotic pathogens, and bats of the *Pteropus* genus are thought to serve as reservoir hosts. Upon transmission to humans, these viruses can cause severe encephalitis or respiratory disease. In addition to infecting and causing disease in humans, these viruses can infect a broad range of mammals, including bats, horses, pigs, dogs, cats, and rodents (Ksiazek et al. 2011). Such a broad host range is unusual among paramyxoviruses. The abilities of a virus to effectively replicate in a host, to spread from individual to individual and to cause disease suggest that the virus can overcome or evade host innate antiviral defenses that might otherwise rapidly suppress virus replication. The broad host range of NiV and HeV suggests that these viruses have evolved mechanisms to target components of the innate immune response in Pteropus bats that are well conserved in other mammals. This review will focus on the mechanisms by which NiV and HeV evade innate immunity. As members of the paramyxovirus family, NiV and HeV employ immune evasion strategies similar to those used by other family members (Table 1). However, they also exhibit novel mechanisms of escape. While some similarities between henipaviruses and other paramyxoviruses are highlighted below, other reviews are available for more general information on paramyxovirus evasion of innate immunity [e.g. (Chambers and Takimoto 2009a, b; Goodbourn and Randall 2009; Ramachandran and Horvath 2009)].

Replication of NiV and HeV is thought to occur by mechanisms similar to those of other paramyxoviruses (Lamb and Parks 2007). The virions are enveloped and

Table 1 Select examples of innate immune evasion by P gene products from members of the *Paramyxovirinae* subfamily of the *Paramyxoviridae* virus family

Genus species *viral protein*	P gene products made by members of genus[1]	IFN signaling molecule or process targeted	Induction of innate responses, targeted molecule/process indicated	References[2]
Henipaviruses	P, V, W, C			
Nipah virus				
P		STAT1		Ciancanelli et al. (2009); Shaw et al. (2004)
V		STAT1 STAT2		Ciancanelli et al. (2009); Hagmaier et al. (2006); Ludlow et al. (2008); Park et al. (2003); Rodriguez et al. (2002)
			MDA5, RLR signaling	Andrejeva et al. (2004); Childs et al. (2007, 2009); Parisien et al. (2009); Virtue et al. (2011a, b)
W		STAT1		Ciancanelli et al. (2009); Park et al. (2003); Shaw et al. (2004)
			TLR signaling RLR signaling	Shaw et al. (2004)
C		IFN signaling		Park et al. (2003)
Hendra virus				
V		STAT1 STAT2		Rodriguez et al. (2003)
			MDA5	Andrejeva et al. (2004); Childs et al. (2007, 2009); Parisien et al. (2009); Virtue et al. (2011a, b)
Morbilliviruses				
Measles virus	P, V, W, C			
P		STAT1		Devaux et al. (2007)

(continued)

Table 1 (continued)

Genus species *viral protein*	P gene products made by members of genus[1]	IFN signaling molecule or process targeted	Induction of innate responses, targeted molecule/process indicated	References[2]
			TLR signaling	Yokota et al. (2008)
V		STAT1 and STAT2 Jak1		Caignard et al. (2007, 2009a); Ohno et al. (2004); Palosaari et al. (2003); Ramachandran et al. (2008); Fontana et al. (2008); Takeuchi et al. (2003); Yokota et al. (2003)
			MDA5 TLR7/9 NF-κB	Andrejeva et al. (2004); Childs et al. (2007, 2009); Pfaller and Conzelmann (2008); Schuhmann et al. (2011)
C		IFN-induced gene expression		Fontana et al. (2008); Shaffer et al. (2003)
			IFNβ transcription	Sparrer et al. (2012)
Rinderpest virus				
C			IFNα/β production	Boxer et al. (2009)
Respiroviruses	P, V, W or D, C', C, Y1, Y2			
Sendai virus				
V			MDA5, IFNα/β production	Andrejeva et al. (2004); Chambers and Takimoto (2009a); Childs et al. (2007, 2009); Komatsu et al. (2004)
C' C Y1 Y2		STAT1		Garcin et al. (1999, 2000, 2001, 2003); Gotoh et al. (1999, 2003); Kato et al. (2001); Komatsu et al. (2000, 2002); Saito et al. (2002); Takeuchi et al. (2001)

(continued)

Table 1 (continued)

Genus species *viral protein*	P gene products made by members of genus[1]	IFN signaling molecule or process targeted	Induction of innate responses, targeted molecule/process indicated	References[2]
			IFNβ production	Komatsu et al. (2004, 2007)
Human parainfluenza virus 1				
C protein(s)		IFN signaling		Bousse et al. (2006); Van Cleve et al. (2006)
			IFNα/β production	Van Cleve et al. (2006)
Human parainfluenza virus 3				
C protein(s)		STAT1		Malur et al. (2005); Caignard et al. (2009b)
Bovine parainfluenza virus 3				
V			IFNβ promoter activation	Komatsu et al. (2007)
C			IFNβ promoter activation	Komatsu et al. (2007)
Rubulaviruses	V, I, P			
Parainfluenza virus 5				
V		STAT1		Didcock et al. (1999); Parisien et al. (2002a, b); Precious et al. (2005a, b, 2007); Ulane and Horvath (2002); Ulane et al. (2005)
			MDA5, IFNβ production	Andrejeva et al. (2004); Childs et al. (2007, 2009); He et al. (2002)

(continued)

Table 1 (continued)

Genus species *viral protein*	P gene products made by members of genus[1]	IFN signaling molecule or process targeted	Induction of innate responses, targeted molecule/process indicated	References[2]
Mumps virus				
V		STAT1 and STAT3		Kubota et al. (2001, 2002); Nishio et al. (2002); Ulane et al. (2003); Yokosawa et al. (2002)
			MDA5, IFNβ production	Andrejeva et al. (2004); Childs et al. (2007)
I				
P				
Human parainfluenza virus 2				
V		STAT2		Nishio et al. (2005a); Parisien et al. (2001, 2002a); Precious et al. (2005b); Ulane and Horvath (2002)
I				
P				
Human parainfluenza virus 4				
V		Interacts with STAT1 and STAT2 but fails to inhibit IFN signaling		Nishio et al. (2005b)

(continued)

Table 1 (continued)

Genus species *viral protein*	P gene products made by members of genus[1]	IFN signaling molecule or process targeted	Induction of innate responses, targeted molecule/process indicated	References[2]
Mapuera virus				
V		STAT1 and STAT2		Hagmaier et al. (2007)
Avulaviruses	P, V, I			
Newcastle disease virus				
P				
V		STAT1		Huang et al. (2003); Park et al. (2003)
			MDA5, IFNβ production	Childs et al. (2007)

[1] For henipaviruses, morbilliviruses, respiroviruses, and avulaviruses the unedited P gene mRNA encodes the P protein, the +1G edited mRNA encodes the V protein and the +2G edited mRNA encodes a protein called W, D or I. For the rubulaviruses, the unedited P gene mRNA encodes the V protein, the +1G edited mRNA encodes the I protein and +2G edited mRNA encodes the P protein. The morbillivirus, respirovirus and henipavirus P gene mRNAs also contain alternate, overlapping open reading frames that encode C proteins. In the case of the respiroviruses, the use of alternate start codons results in production of multiple "C" proteins called C', C, Y1 and Y2

[2] This table does not address all examples of paramyxovirus IFN-antagonist proteins, for more information see (Chambers and Takimoto 2009a; Goodbourn and Randall 2009; Ramachandran and Horvath 2009)

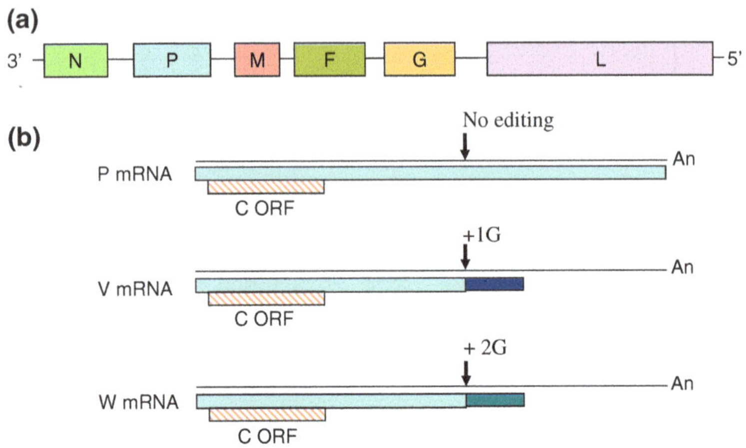

Fig. 1 The henipavirus genome and the mRNA products of the henipavirus P gene. **a** Schematic diagram showing the single-stranded, negative-sense NiV genome. The *line* indicates non-transcribed regions. The *boxes* indicate individual genes: N, the nucleoprotein gene; P, the phosphoprotein gene; M, the matrix protein gene; F, the fusion protein gene; G, the attachment glycoprotein gene; L, the Large (RNA-dependent RNA polymerase) gene. **b** The mRNAs and proteins produced from the P gene are illustrated. The *thin lines* indicate the polyadenylated (An) mRNAs, and arrows indicate the editing site. The *boxes* represent the proteins potentially translated from each species of mRNA. The P protein coding mRNA is colinear with the P gene. The V protein coding mRNA contains G insertions at the conserved P gene editing site such that translation is shifted to the +1 reading frame. The W protein coding mRNA contains G insertions at the conserved P gene editing site such that translation is shifted to the +2 reading frame. Each of the mRNAs contains coding sequences for the C protein which is expressed from an internal open reading frame

contain non-segmented negative-strand RNA genomes. The viral genomes possess, the N, P, M, F, G and L genes (Harcourt et al. 2000; Wang et al. 1998) (Fig. 1). N encodes the viral nucleocapsid (N) protein. The P gene encodes multiple protein products, including the phosphoprotein (P) and the C, V and W proteins, which are discussed in detail below and are summarized in Table 1. The M, F and G genes encode the viral matrix protein, fusion protein, and attachment glycoprotein, respectively. The L gene encodes the catalytic subunit of the viral polymerase known as the Large (L) protein. Viral replication takes place in the cell cytoplasm, although assembly requires nuclear translocation of the M protein (Wang et al. 2010). Viral RNA synthesis requires the N, P and L proteins (Halpin et al. 2004). These cytoplasmic RNA synthesis reactions likely generate molecules that trigger innate immune responses to infection.

2 Host Cell α/β Responses

IFNα/β refers to a family of related cytokines produced by most cell types that, upon release, signals through the IFNα/β receptor (IFNAR) to activate antiviral responses. These cytokines also modulate adaptive immune responses to infection (Biron 2001; Gonzalez-Navajas et al. 2010; Hoebe and Beutler 2004). Because of their antiviral activity, viruses have evolved mechanisms to evade or overcome IFN responses (Gale and Sen 2009).

2.1 Pattern Recognition Receptors that Trigger IFNα/β Gene Expression

IFNα/β gene expression can be triggered by signaling from several different pattern recognition receptors (PRRs) including select Toll-like receptors and RIG-I-like receptors (RLRs) [reviewed in (Jensen and Thomsen 2012)]. A schematic overview of these pathways and the steps targeted by NiV and HeV proteins is provided in Fig. 2. TLRs are type I transmembrane proteins that reside either at the cell surface or in endosomes. They possess extracellular/luminal domains possessing leucine-rich repeats and intracellular Toll/IL-1 receptor homology (TIR) domains. The latter associate with adaptor proteins to mediate signaling. The TLRs involved in triggering IFNα/β production in response to RNA virus infection include TLR3, TLR4, TLR7, and TLR8. Most relevant to this review, TLR3 is an endosomal TLR that responds to dsRNA to trigger IFNα/β gene expression. TLR3 signaling requires the adaptor protein TRIF, and over-expression of TRIF is sufficient to induce IFNβ promoter activation.

The RLRs, which include RIG-I, MDA-5 and LGP2, are cytoplasmic RNA helicases that signal in response to viral nucleic acids (reviewed in (Jensen and Thomsen 2012)). The RLRs likely serve as PRRs for detection of NiV and HeV, although the specific contributions of TLRs and RLRs toward innate immune responses to henipavirus infection remain to be experimentally defined. RIG-I signals in response to RNAs with 5′ triphosphates possessing double-stranded regions and to dsRNAs of < 1,000 base-pairs. In addition, studies have implicated other features, such as the AU content of an RNA, as important for inducing RIG-I signaling (Saito et al. 2008; Schlee et al. 2009; Schmidt et al. 2009). Notably, the genomes and antigenomes of many RNA viruses, including non-segmented negative-sense RNA viruses, such as NiV and HeV, possess 5′ triphosphates (Habjan et al. 2008). These viral genomic RNAs also exhibit inverse complementarity at their ends and can form base-paired panhandle structures that serve as promoters for the viral RNA-dependent RNA polymerase. These features would be relatively unique to viral RNAs, identifying them as "foreign" and mediate recognition by RIG-I. MDA5 is activated by long dsRNAs, by picornavirus infection and, at least in conventional dendritic cells by Sendai virus. LGP2 lacks the CARD domains required for RIG-I and MDA5 signaling and appears to modulate IFNα/β responses. As is discussed below, the V proteins of paramyxoviruses have been demonstrated to bind to and impair the functions of MDA5 and LGP2 (Andrejeva et al. 2004; Childs et al. 2007; Childs et al. 2009; Parisien et al. 2009; Ramachandran and Horvath 2009, 2010).

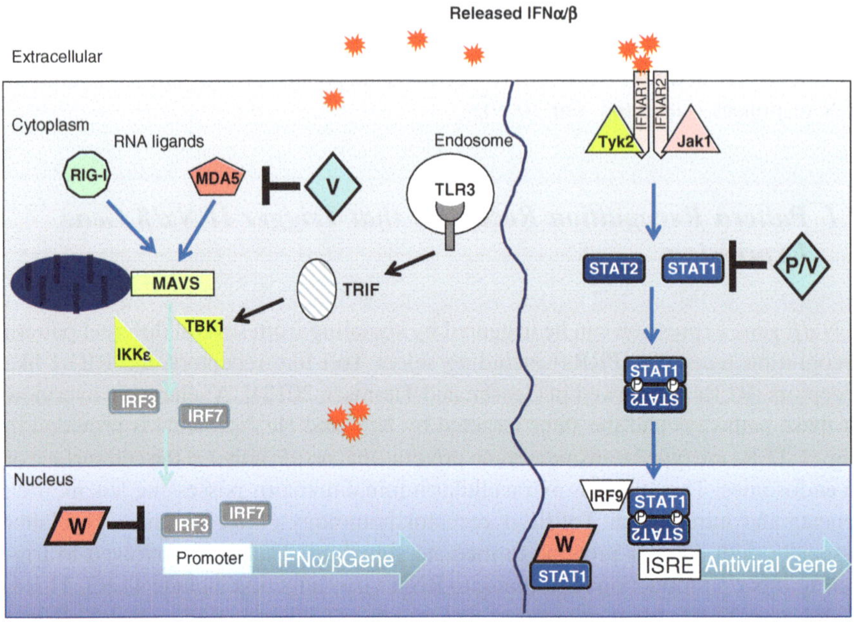

Fig. 2 Innate immune evasion mechanisms of the Nipah virus P, V and W proteins. *Left* The V and W proteins block IFNα/β expression by different mechanisms. Depicted are simplified versions of the RIG-I-like receptor (RLR) signaling and TLR3 signaling pathways leading to activation of interferon regulatory factors 3 and 7 (IRF-3 and IRF-7). RIG-I, MDA5 and TLR3 can sense RNAs produced during viral infection. Activation of RIG-I or MDA5 leads to their interaction with the mitochondria-associated adpater protein MAVS. This leads to activation of the kinases IKKε and/or TBK-1 which phosphorylate and activate IRF-3 and, when present, IRF-7. Activated IRF-3 and IRF-7 accumulate in the nucleus and contribute to IFNα/β gene transcription. TLR3 signals through adaptor protein TRIF to activate IKKε/TBK-1. The cytoplasmic NiV and HeV V proteins have been demonstrated to interact with and inhibit MDA5; however, NiV V did not inhibit TLR3 signaling. The nuclear W protein effectively blocks RLR and TLR3 activation of the IFNβ promoter by preventing the accumulation of phosphorylated IRF-3. This function requires the nuclear localization of W. *Right* The Nipah virus P, V and W proteins each block IFN signaling pathways. The P, V and W proteins can each inhibit the phosphorylation and activation of STAT1 and STAT2 in response to IFNα/β and block the phosphorylation and activation of STAT1 in response to IFNγ. The inhibition of the IFNα/β pathway is depicted. The cytoplasmic P and V proteins interact with and inhibit STAT1. The V protein has also been shown to interact with STAT2, in a STAT1 dependent manner. P and V prevent nuclear accumulation of STAT1 and STAT2 following IFNα/β addition to cells and cause the STAT proteins to accumulate in a high molecular weight complex. The nuclear W protein relocalizes non-phosphorylated STAT1 from its typically cytoplasmic localization to the nucleus. STAT1 relocalized to the nucleus by W remains unphosphorylated and does not activate transcription of genes

Activation results in structural rearrangements that allow RIG-I and MDA5 to interact with the CARD domains of the mitochondria-associated adaptor protein MAVS (also known as IPS-1, VISA and CARDIF) (Kawai et al. 2005; Meylan et al. 2005; Seth et al. 2005; Xu et al. 2005). MAVS signals via either of two IRF-3 kinases IKKε and TBK-1 to activate the transcription factor interferon regulatory factor (IRF)-3, or if present IRF-7. MAVS also activates NF-κB. The consequence of RLR signaling through MAVS is the induction of IFNα/β and cytokine gene expression (Jensen and Thomsen 2012).

The structure and mechanisms by which RLRs are activated have been reviewed elsewhere (O'Neill and Bowie 2011; Yoneyama et al. 2012). Briefly, RIG-I possesses two amino-terminal caspase recruitment domains (CARDs) at its amino-terminus; a central helicase domain and a carboxy-terminal repressor domain which also possesses RNA binding activity (Yoneyama et al. 2004). In its inactive state, the RIG-I CARD domains, which are crucial for signaling, are sequestered by portions of the helicase domain. The carboxy-terminal repressor domain is available to bind to dsRNA possessing 5′-triphosphate. This promotes binding by the helicase domain of the dsRNA and also of ATP. These binding events appear to trigger conformational changes that release the CARD domains from their inactive state such that they become available to interact with a ubiquitin ligase such as TRIM25. This promotes the ubiquitination of the CARD domains of RIG-I or the formation of K63-linked polyubiquitin chains that bind to the CARD domains, allowing interaction with MAVS and subsequent signaling. MDA5 possesses a similar structure as RIG-I, while LGP2 possesses helicase and carboxy-terminal domains analogous to those of RIG-I but lacks the amino-terminal CARD domains characteristic of RIG-I and MDA5. LGP2 cannot, therefore, signal through MAVs. The left side of Fig. 2 illustrates how NiV and HeV proteins impair RLR and TLR3 signaling pathways.

2.2 IFNα/β-Induced Signaling

IFNα and IFNβ are released from cells and bind to the IFNAR, activating a JAK/STAT signaling pathway [reviewed in (Platanias 2005)]. The tyrosine kinases Jak1 and Tyk2 become activated and phosphorylate tyrosines on the cytoplasmic tails of the receptor subunits. This creates docking sites for STAT1 and STAT2 via their SH2 domains. The STAT proteins are tyrosine phosphorylated, and then heterodimerize via phosphotyrosine-SH2 domain interactions. This dimeric complex further complexes with interferon-regulatory factor 9 (IRF9), forming a transcription factor complex, ISGF-3. The dimerized STATs reveal a nuclear localization signal on STAT1 that promotes nuclear accumulation. The ISGF3 complex activates IFN-α/β responsive genes and induces in cells an antiviral state. IFNγ is distinct from IFNα/β and binds to its own receptor, the IFNγ receptor. IFNγ is produced primarily by T cells and natural killer (NK) cells and is relevant to this review because it also can exert antiviral effects, through a Jak-STAT signaling cascade. However, IFNγ activates Jak1 and Jak2 tyrosine kinases and leads primarily to the tyrosine phosphorylation of STAT1 with consequent formation of STAT1 homodimers. The right side of Fig. 2 illustrates how NiV and HeV proteins impair the IFN signaling pathways.

3 NiV and HeV P Gene Products Counter IFNα/β Responses

Paramyxoviruses encode a number of mechanisms to counter host innate immunity, particularly the IFNα/β response (reviewed in (Goodbourn and Randall 2009; Ramachandran and Horvath 2009)). Viral evasion strategies identified to date include mechanisms to limit the production of immunostimulatory RNAs, to prevent the function of PRRs and to block IFN-induced signaling pathways (Goodbourn and Randall 2009; Ramachandran and Horvath 2009). Many, although not all, paramyxoviruses use products of the viral P gene to carry out these innate immune evasion functions (Table 1). This prompted the investigation of the NiV and HeV P gene products as IFN-antagonists.

3.1 Henipaviruses Encode Accessory Proteins from their P Genes

NiV and HeV encode multiple proteins from their P genes (Harcourt et al. 2000; Kulkarni et al. 2009; Lo et al. 2009; Wang et al. 1998; Yoneda et al. 2010). As is discussed in detail below, each of these multifunctional proteins has been shown to antagonize one or more components of the host IFNα/β antiviral response. That inhibition of IFNα/β production is important for the outcome of infection in vivo is supported by the observation that daily treatment with the IFN-inducer Poly(I)-poly($C_{12}U$) can prevent death in a majority of NiV-infected hamsters (Georges-Courbot et al. 2006).

Faithful transcription of the P gene by the viral polymerase, based on predicted gene start and gene end sequences, would yield an mRNA encoding P, an essential, non-enzymatic component of the viral RNA-dependent RNA polymerase complex (Harcourt et al. 2000; Wang et al. 1998). Within the P gene is a single editing site, homologous to editing sites of other paramyxoviruses, at which non-templated G residues are added during transcription by the viral polymerase (Kulkarni et al. 2009; Lo et al. 2009). Depending on the number of additional residues inserted, translation of edited transcripts results in a protein with an amino-terminus identical to the P protein, up to where editing has occurred. After this point, a frameshift into an alternate reading frame would yield proteins with distinct carboxy-termini. The protein with the +1 frameshift is known as the V protein, and the protein with the +2 frameshift is known in NiV and HeV as the W protein. Editing has been demonstrated to occur with a relatively high frequency in NiV P gene transcripts as compared to several other paramyxoviruses (Kulkarni et al. 2009) Editing frequency also varies over time. In the case of NiV, early in infection, little editing was detected. However, a higher percentage of transcripts are edited at later time points with insertions of one to eleven residues (Kulkarni et al. 2009). This high frequency of editing was confirmed to occur for Malaysian and Bangladesh NiV isolates as well as for HeV (Lo et al. 2009).

3.2 The P Gene Encoded Products are Produced in Infected Cells

Using specific antibodies, the C, P, V and W proteins have been detected in infected cells and have also been demonstrated to be present in sucrose gradient-purified virus particles (Kulkarni et al. 2009; Lo et al. 2009). Consistent with studies on tranfected cells (discussed below), C, P and V are cytoplasmic; while W is nuclear in infected Vero cells (Lo et al. 2009).

3.3 P Gene Products Inhibit Innate Antiviral Responses

Each of the NiV P gene products has been demonstrated to counteract aspects of the host IFNα/β response. Transfection of expression plasmids into chicken embryo fibroblasts resulted in IFNα/β production that was sufficient to suppress replication of a Newcastle disease virus expressing green fluorescent protein (NDV-GFP), as evidenced by lack of GFP fluorescence. However, when the plasmids expressed any of the NiV C, P, V or W proteins, NDV-GFP replication was restored. Expression of the common amino-terminal domain shared by P, V and W was sufficient for this effect. Expression plasmids encoding NiV C also rescued NDV-GFP replication, albeit with lower efficiency than P, V or W (Park et al. 2003). Rescue of NDV-GFP replication correlated with the ability of P, V, the amino-terminus of V and W to each inhibit IFNβ-induced gene expression, whereas C did not inhibit IFN signaling, as measured by this assay (Park et al. 2003).

4 NiV and HeV Inhibit RLR and TLR Signaling

The specific PRRs that recognize and respond to NiV and HeV in either infected cells or animals have not yet been defined. Nonetheless, based on the fact that they are negative-sense RNA viruses, the henipaviruses have the potential to signal through RIG-I (Yoneyama and Fujita 2009). In support of this, RNA extracted from NiV particles could stimulate IFNβ promoter transcription upon transfection into 293T cells, and this activation was RIG-I-dependent (Habjan et al. 2008). However, such a study does not exclude a role for MDA5, LGP2, or TLRs in the innate response to henipavirus infection. As discussed below, NiV proteins have been demonstrated to block RLR and TLR-mediated induction of IFNα/β responses (Andrejeva et al. 2004; Childs et al. 2007, 2009; Parisien et al. 2009).

4.1 The NiV V and W Proteins Inhibit IFNα/β Production

The Henipavirus V and W proteins are able to inhibit signaling pathways that trigger activation of IRF-3 and lead to IFNβ gene expression (Andrejeva et al.

2004; Shaw et al. 2005). Expression of either V or W in 293T cells was found to inhibit the Sendai virus-triggered activation of the ISG54 promoter. The ISG54 promoter can be activated either by IRF-3 or by the IFN-induced transcription factor ISGF3. In this experimental system, the promoter activation occurs via Sendai virus activation of RLR signaling (Baum et al. 2010; Shaw et al. 2005; Strahle et al. 2006; Yount et al. 2008). The ability of V and W to counteract IFNβ-inducing signals is not equivalent, however. W proved more effective than V at blocking activation of the ISG54 promoter via the TLR3 signaling pathway (Shaw et al. 2005). This was demonstrated by transfecting NiV expression plasmids into 239T cells expressing TLR3 and adding dsRNA to the culture medium to activate TLR3 signaling. Alternatively, the TLR3 adaptor protein, TRIF, was over-expressed in the absence or presence of V or W. In either case, W proved a more effective inhibitor than V, with the inhibitory activity correlating with a reduction of phospho-IRF-3 levels in W expressing cells (Shaw et al. 2005). That NiV V does not effectively block TLR3 signaling was confirmed in a separate study that looked at the ability of several paramyxovirus V proteins to inhibit TLR3 signaling. In this study, the V proteins of the rubulaviruses human parainfluenza virus 2, parainfluenza virus 5, and mumps virus were found to effectively block TLR3 signaling by acting as alternative substrates for IKKε and TBK-1 while, once again, NiV V was not an effective inhibitor of TLR3 signaling (Lu et al. 2008).

The NiV W protein exhibits a distinct nuclear localization and possesses a basic nuclear localization signal (NLS) within its unique carboxy-terminal domain (Shaw et al. 2005). The nuclear localization of W appears to be critical for its impairment the TLR3/TRIF pathway to IFNβ production. Mutation of the NLS such that W becomes mainly cytoplasmic impairs its ability to block TRIF-mediated activation of an IRF-3-responsive promoter. Interestingly, relocalization of V to the nucleus by addition of a heterologous NLS allows V to block the pathway, suggesting that the shared amino-terminus may interact with host factors in the nucleus or relocate host factors from the cytoplasm to the nucleus to mediate the inhibitory effect (Shaw et al. 2005). Interestingly, V was found to effectively inhibit activation of the ISG54 promoter by over-expressed IKKε but not TBK-1. In contrast, W could inhibit promoter activation induced by over-expression of either kinase (Shaw et al. 2005). Therefore, if TLR3 signaling proceeds primarily via TBK-1, this might explain why W but not V is an effective inhibitor of this pathway. The molecular basis by which W (or V) can selectively impair TLR3 signaling from the nucleus remains, however, to be fully defined.

4.2 *Henipavirus V Proteins Inhibit MDA5*

The V proteins of a number of paramyxoviruses have been shown to block IFNβ promoter activation [e.g. (He et al. 2002; Andrejeva et al. 2004; Goodbourn and Randall 2009; Lu et al. 2008; Poole et al. 2002; Shaw et al. 2005)]. The SV5 (now known as PIV5) V protein was first demonstrated to coprecipitate with MDA5, and additional experiments demonstrated that the V proteins of other paramyxoviruses, including HeV, also interact with MDA5 (Andrejeva et al. 2004). A characteristic feature of V proteins is a conserved carboxy-terminal domain that corresponds to the region of the protein encoded by sequences after the editing site. This region of V proteins contains a number of conserved residues including histidine and cysteine residues that mediate binding to two zinc atoms (Li et al. 2006; Liston and Briedis 1994; Paterson et al. 1995). MDA5 signals in response to dsRNA to promote IFNβ promoter activation, and it is the conserved cysteine-rich carboxy-terminus of the V proteins that mediates the binding and inhibition of MDA5 (Andrejeva et al. 2004). Subsequently, other V proteins, including the NiV V, were found to interact with MDA5 through interaction of the V carboxy-terminus with the helicase domain of MDA5 (Childs et al. 2007). Importantly for zoonotic pathogens with broad host ranges such at NiV and HeV, the V proteins could inhibit MDA-5 function in human, murine, bovine and avian cells (Childs et al. 2007). Given the fact that NiV and HeV apparently do not cause disease in experimentally-infected mice (Wong et al. 2003), this latter observation suggests that the ability of the virus to inhibit MDA5 is not, on its own, sufficient for viral disease. Whether IFNα/β responses or other aspects of innate immunity serve as determinants of henipavirus host range remains to be determined.

The V binding site on MDA5 has been mapped to the helicase domain. This domain was demonstrated to be involved in dsRNA binding and oligomerization of activated MDA5. V proteins, including the HeV V, or the conserved cysteine-rich carboxyterminal domains of V proteins are sufficient to mediate this MDA5 inhibitory activity (Childs et al. 2009). The MDA5-V interaction has been mapped to a 130 amino acid region of the MDA5 helicase domain, corresponding to the second RecA-like domain. An analogous domain is present in both RIG-I and LGP2. V proteins also interact with this region of LGP2 but not that of RIG-I (Parisien et al. 2009). RLR ATPase activity is stimulated by activating RNA ligands, but V proteins, including the NiV V, inhibit both MDA5 and LGP2 ATPase activity, providing further mechanistic insight into how Vs antagonist RLR signaling (Parisien et al. 2009). V interaction with and inhibition of MDA5 has been shown to involve a subset of conserved residues involved in zinc binding as well as select other residues (Ramachandran and Horvath 2010). This information, in addition to providing insight into how V proteins interact with MDA5, identifies mutations that might be built into recombinant henipaviruses to define the contribution of MDA5 antagonism to virus replication in cell culture and in vivo.

4.3 Variable Ability of NiV and HeV to Block IFN Production in Different Cell Types

The ability of henipaviruses to prevent virus induced IFNα/β production may not be uniform in all cell types. When NiV replication was compared in a human neuroblastoma cell line, BE(2)-M17 (M17) and in several types of human endothelial cells, replication was generally more robust in the neuronal cells. Efficient replication correlated inversely with the production of IFNβ by the infected cells, with little IFN response in the neuronal cells but more IFNβ production in the endothelial cells (Lo et al. 2010). Interestingly, the cells with more pronounced IFNβ production exhibited a cytoplasmic localization of the W protein. In contrast, W was nuclear in the neuronal cells, a localization similar to what has been seen in transfected cells (Lo et al. 2010; Shaw et al. 2005). Given that nuclear localization of W was important for its ability to suppress IFNβ production in transfection experiments, it is possible that this altered W localization might lead to the enhanced IFNβ production in endothelial cells. What the cell type specific determinants of W localization are remains to be uncovered and how the W and V proteins cooperate to regulate IFNα/β synthesis in infected cells must also be defined.

5 Inhibition of IFN Signaling by the P, V and W Proteins

Inhibition of IFN signaling pathways is a common immune evasion strategy employed by paramyxoviruses (Young et al. 2000). Although the strategies employed to block signaling are diverse, products of the P gene are frequently involved [reviewed in (Ramachandran and Horvath 2009)]. In the case of rubulaviruses, such as parainfluenza virus 5, parainfluenza virus 2 or mumps virus, the V proteins target STAT1, STAT2 or STAT1 and STAT3, respectively, for proteasome-mediated degradation (Didcock et al. 1999; Kubota et al. 2001; Parisien et al. 2001, 2002a, b; Yokosawa et al. 2002). In contrast, the V protein of measles virus, a morbillivirus, also interferes with IFN signaling but without promoting STAT degradation (Ramachandran and Horvath 2009). Measles virus V can interact with STAT2 via the conserved V cysteine-rich carboxy-terminal domain and with STAT1 via sequences near the amino-terminus of V and prevent the IFN-induced nuclear accumulation of the STAT proteins (Caignard et al. 2009a, b; Palosaari et al. 2003; Ramachandran et al. 2008). The STAT1 binding domain in measles virus V lies in the amino-terminal domain shared with P, and the measles virus P protein has also been demonstrated to contribute to inhibition of STAT1 phosphorylation and function (Devaux et al. 2007; Ramachandran et al. 2008). These latter mechanisms are similar to the mechanisms employed by NiV and HeV to impair IFN signaling, where P gene products have been demonstrated to bind and inhibit STAT1 and STAT2 but not to promote STAT protein degradation.

5.1 The NiV and HeV V Proteins Interact with and Inhibit STAT1 and STAT2

Rodriguez et al. characterized the predicted NiV and HeV V proteins in terms of their ability to block IFN signaling pathways (Rodriguez et al. 2002; Rodriguez et al. 2003). The NiV V was expressed with an HA-tag and demonstrated to inhibit the IFNα or the IFNγ-induced expression of IFN-responsive reporter genes. NiV V did not promote the degradation of either STAT1 or STAT2, whereas a control PIV-5 V did, as expected, promote STAT1 degradation. NiV V was further demonstrated to co-immunoprecipitate with STAT1 and STAT2, and expression of V shifted STAT1 and STAT2 into a high molecular weight complex when cell lysates were assayed by gel filtration chromatography. NiV V was also noted to be strongly cytoplasmic, and in untreated cells, STAT1 and STAT2 were exclusively localized to the cytoplasm. Addition of IFNα/β or IFNγ normally causes a rapid localization of STAT1 and, in the case of IFNα/β, also STAT2 to the nucleus. However, STAT1 and STAT2 remained cytoplasmic in NiV V-expressing cells after IFNα or IFNγ addition. The inhibition of STAT1 relocalization was found to be STAT2 independent following addition of IFNγ. Finally, it was also demonstrated that the relocalization correlated with a block to the IFNα or IFNγ-induced phosphorylation of STAT1 (Rodriguez et al. 2002). Similar observations were made for the HeV V protein which also inhibited IFNα and IFNγ induced reporter gene expression; prevented STAT1 and STAT2 nuclear accumulation in response to IFNγ and IFNα; co-immunoprecipitated with STAT1 and STAT2 but not STAT3 and shifted STAT1 and STAT2 into high molecular weight complexes (Rodriguez et al. 2003).

For some paramyxoviruses, V proteins block Jak-STAT signaling in a host species-specific manner (e.g. (Parisien et al. 2002a; Young et al. 2001)). As zoonotic pathogens, the ability of NiV and HeV to antagonize innate immune responses in the cells of different host species is of particular interest. Consistent with its ability to infect a wide range of mammals and consistent with a model where NiV has evolved to use bats as reservoir hosts, the NiV V protein was demonstrated to inhibit IFN signaling in a broad range of mammalian species. Using a reporter gene assay, NiV could inhibit IFN signaling in bat TB1 (*Tadarida brasilensis*) cells, Vero (African green monkey) cells as well as in pig, dog, rabbit, horse cells. Interestingly, a point mutant which exhibited impaired inhibition of IFN signaling in other cell types, retained residual activity in the bat cells, suggesting that NiV V may interact differently with bat IFN signaling pathway components as compared to other species (Hagmaier et al. 2006).

5.2 A STAT1 Binding and Inhibitory Domain Resides in the Amino-Terminal Domain Common to the P, V and W Proteins

The STAT1 binding domain on V was mapped to amino acids 100–160, while binding to STAT2 also requires additional V sequences carbboxy-terminal to residue 160 (Rodriguez et al. 2004). Interestingly, STAT2 binding requires that STAT1 be present. Using STAT1-STAT3 chimeras, a V binding site was mapped to the STAT1 SH2 and linker domains (Rodriguez et al. 2004). The STAT1 binding domain lies well within the amino-terminal domain common to the P, V and W proteins. This would suggest that P and W should also bind to STAT1, and this was confirmed by co-immunoprecipitation studies (Shaw et al. 2004). P and W were also demonstrated to inhibit IFN-induced expression of an IFN-responsive promoter, although P seemed somewhat less potent as compared with V or W. The common amino-terminal 50–150 amino acids were sufficient to co-precipitate STAT1, to inhibit IFNα/β-induced gene expression and to overcome the antiviral effects of IFNβ (Shaw et al. 2004). Interestingly, W was found to redistribute an over-expressed STAT1 almost completely to the nucleus, as assessed by indirect immunofluorescence, whereas P or V resulted in near exclusive cytoplasmic STAT1. W and P, like V, each prevented IFNα-induced STAT1 tyrosine phosphorylation (Shaw et al. 2004). Therefore, P, V and W each are inhibitors of IFN signaling, although they exert these effects from different subcellular compartments and with somewhat different efficiencies.

5.3 *The STAT1 Binding Domain Mediates Additional Protein–Protein Interactions*

As noted above, the cysteine-rich carboxy-terminus of NiV V mediates its interaction with MDA5 and suppresses IFNβ promoter activation by this PRR (Andrejeva et al. 2004). In addition, the conserved carboxy-termini of V proteins have oligomerization activity (Ulane et al. 2005). The NiV and HeV V proteins were demonstrated to mediate self interaction and interactions with other paramyxovirus V proteins. Purified PIV5 V protein was also demonstrated to oligomerize as assessed by gel filtration assay and to form ring structures that could be visualized by electron microscopy. It was proposed that such oligomeric structures could serve as platforms for V interactions with their multiple binding partners (Ulane et al. 2005). Given that inhibition of STAT1 and STAT2 activation by NiV V do not require the carboxy-terminal region, oligomerization via the V carboxy-terminus also appears unnecessary for STAT1 inhibition. Nonetheless, this does not rule out the possibility that oligomerized Vs interact with STATs and that oligomerization might contribute to the inhibition of STAT1 and STAT2 in the context of NiV infection.

In addition to its STAT1 and STAT2 binding activity, other functions have been assigned to specific regions of V. Although V was strongly cytoplasmic, leptomycin B (LMB)-treatment of V-expressing cells could relocalize the protein to the nucleus, suggesting that V shuttles between the nucleus and cytoplasm (Rodriguez et al. 2002). Using V truncation mutants, a nuclear export signal (NES) was mapped to amino acids 100–220 (Rodriguez et al. 2004). Further mapping, employing an assay in which heterologous sequences were screened for the ability to functionally substitute for a mutated HIV-1 Rev protein NES, identified the

STAT1 binding region (residues 160–220) as possessing NES activity. When a Leu-rich region in this area was either deleted or mutated at specific Leu residues, V was partially relocalized V to the nucleus. Despite this, the mutant Vs could still inhibit STAT1 nuclear accumulation following IFN addition (Rodriguez et al. 2004). Therefore, the functional significance of V nuclear-cytoplasmic shuttling remains to be defined.

The NiV and HeV V proteins have also been demonstrated to interact with the cellular kinase Polo-like kinase 1 (PLK1) (Ludlow et al. 2008). PLK1 interacts with henipavirus V proteins by co-immunoprecipitation as well as with P and W proteins. The interaction is specific toward the henipavirus proteins, because measles virus V did not interact with PLK1. The PLK1 interacting domain mapped to amino acids 100–160, the same region that binds STAT1, and requires a $_{129}$-T/S SSP-$_{132}$ motif. Mutation of V residue S130 disrupted interaction of V with both PLK1 and STAT1. However, sequential co-immunoprecipitation assays indicated that independent interactions occur between V and either STAT1 or PLK1. Although PLK1 has been reported to phosphorylate the P protein of PIV5, affecting viral gene expression, when S130 was mutated in the NiV P protein, no effect was seen on NiV P function in a reconstituted viral polymerase assay (Ludlow et al. 2008; Sun et al. 2009). Therefore, the importance for NiV replication of PLK1-P, V or W interaction still must be elucidated.

5.4 The P Gene Products Disrupt STAT1 Trafficking and Activation in NiV-Infected Cells

The shared amino-terminal domain of the NiV P, V and W proteins not only mediates interactions with host factors but is also required for P to function as a component of the viral polymerase complex. This was illustrated by testing P mutants in a minigenome assay. In this assay, the viral polymerase complex is reconstituted by co-expression of the viral N, P and L proteins. This complex transcribes and replicates a co-expressed model viral genomic RNA that encodes only a reporter gene (Sleeman et al. 2008). The polymerase was inactive when P protein deletion mutants lacking the first 50, 100, or 150 amino acids were tested (Ciancanelli et al. 2009). A series of mutants across the amino-terminal 150 amino acids of P, in which 10 amino acids were deleted at a time, identified a region spanning residues 81–120 where mutations negatively-impacted function in the minigenome assay. How residues 81–120 influence viral RNA synthesis remains to be clarified (Ciancanelli et al. 2009).

Interestingly mutants in the 51–110 and 141–150 regions retained the ability to bind to STAT1 and to inhibit IFN signaling. However, deletion mutants between residues 111 and 140 failed to detectably bind STAT1 and to inhibit IFN-induced gene expression (Ciancanelli et al. 2009). The identification of residues 111–140 as critical was consistent with previously published mapping data and with a study which identified a single amino acid change, G125E, as sufficient to disrupt V-STAT1 and STAT2 interactions (Hagmaier et al. 2006). Based on the latter study, several glycine codons in this region, G120, G121, G125, G127, and G135, were individually mutated to code for glutamic acid. Each mutant P was functional in minigenome assays, but mutations at positions 121, 125 and 127 disrupted inhibition of IFN-induced reporter gene expression. The G135E mutant was partially impaired in this assay. STAT1 interaction was impaired when these mutants were tested in the context of NiV V or W (Ciancanelli et al. 2009).

To assess the impact of P, V, and W STAT1 binding in the context of NiV replication, a recombinant virus was made possessing the G121E mutation. Because the mutation would also alter the coding sequence of the overlapping C open reading frame (ORF), the G121E mutation was built into a background where expression of C was disrupted by mutation of the C ORF start codon and by insertion of a stop codon early in the C reading frame. In Vero cells, the C knockout mutation attenuated growth relative to a wild-type virus, reducing maximal virus titers by 10–100-fold. However, addition of the G121E mutation did not further attenuate replication as compared to the C knockout virus, perhaps because Vero cells cannot produce IFNα/β.

Examination of STAT1 revealed that NiV does, in fact, modulate STAT1 trafficking during the course of its replication cycle. Viruses encoding wild-type P, V, and W proteins relocalized inactive STAT1 to the nucleus. In cells infected with the G121E virus, STAT1 was cytoplasmic if the cells were not treated with IFN. When IFNβ was added to infected cells, tyrosine phosphorylated STAT1 was detected by immunofluorescence in 91% of G121E-infected cells and only 33% of the C knockout (wild-type P/V/W)-infected cells. Also, in those cells infected with virus encoding wild-type P, V, and W proteins, the signal for phospho-STAT1 was generally less intense than in the G121E mutant infected cells (Ciancanelli et al. 2009). These studies clearly established that the P gene products alter STAT1 trafficking and activation in NiV-infected cells. The relocalization of inactive STAT1 to the nucleus in untreated but infected cells also suggests that the nuclear W protein is the dominant determinant of STAT1 localization in infected Vero cells.

The impact of the henipavirus-encoded STAT1 inhibitory functions in the context of virus infection remains a work in progress. When 293T cells were first transfected with IFNβ-promoter-responsive reporter genes and then infected at a multiplicity of infection of 10 with HeV or NiVs, little induction of reporter gene expression was detected (Virtue et al. 2011b). This indicates that henipavirus infection induces little IFN response and is consistent with transfection experiments that implicated V and W proteins as inhibitors of IFNα/β gene expression. When similarly transfected cells were infected at a multiplicity of infection of 1, resulting in about 90% infection, addition of IFNα/β led to the upregulation of the reporter gene and to the expression of endogenous, IFN-inducible ISG54 and ISG56 genes. This suggested that, at least under the conditions examined, the viruses did not effectively block IFN-induced Jak-STAT signaling (Virtue et al. 2011b).

The observation that IFN signaling is not entirely blocked by infection with wild-type viruses is not inconsistent with the previously described data from Ciancanelli et al. 2009, where mutant viruses were examined. In this study, some phospho-STAT1 was detected in Vero cells infected with viruses expressing wild-type P, V and W proteins, although STAT1 transcriptional activity was not assessed (Ciancanelli et al. 2009). W, which was hypothesized to be the dominant determinant of STAT1 nuclear localization in infected cells, was reported, as noted above, to localize to the cytoplasm in human endothelial cells (Lo et al. 2010). Although W expressed by transient transfection is nuclear in 293T cells (unpublished observation), its localization in infected 293T cells was not determined. It would be of interest to determine whether the ability of infection to block IFN signaling in different cell types correlates with W localization and to identify host factors that influence W localization. It will also be critical to determine to what extent inhibition of IFN signaling contributes to virus replication in IFN-competent systems and how this influences virus pathogenesis.

Although the status of IFN responses in henipavirus infected cells has been reported to vary in various human cell types, the viruses do appear to effectively block innate immune signaling pathways in cells of Pteropus bats. Specifically, *Pteropus alecto* cell lines and primary cells, including SV40 large T antigen-transformed lung cells, fetal cells and primary kidney cells were infected with HeV or Malaysian or Bangladesh NiV isolates (Crameri et al. 2009; Virtue et al. 2011a). Using mainly RT-PCR to monitor IFN and IFN-responsive genes, it was demonstrated that henipavirus infection blocks both IFN production and IFN signaling in these cells (Virtue et al. 2011a). As was noted earlier, the molecular determinant of henipavirus virulence in different mammalian species remains to be defined. The fact that these viruses can block bat IFN signaling while failing to cause disease in experimentally infected bats (Middleton et al. 2007) suggests that the ability of the virus to suppress IFN signaling is not a sole determinant of virulence. It is reasonable to hypothesize, however, that henipaviruses unable to block host innate antiviral responses, will be unable to sustain their replication and fail to cause disease. It will be of interest to determine whether these viruses inhibit IFN responses to different degrees in different host species and whether such differences will correlate with the capacity of the virus to cause disease.

6 How do NiV-Encoded IFN-Antagonists Influence Replication?

The requirements for BSL4 containment provide significant constraints upon researchers and hamper studies to determine how viral immune antagonists influence viral replication and pathogenesis. The development of methods to model how individual viral components can cooperate to influence host responses is one potential remedy to this difficulty. Seto et al. explored a computational modeling approach in order to clarify how the P, V, and W proteins will cooperate to influence antiviral responses to NiV infection (Seto et al. 2010). Recombinant

Newcastle disease viruses expressing individual P, V, and W proteins (NDV-P, NDV-V, and NDV-W, respectively) were constructed. These were used to infect primary human monocyte-derived dendritic cells, and cellular responses to infection were monitored. A parental NDV is a potent inducer of IFNα/β antiviral responses, inflammatory cytokine responses and a trigger of class I major histocompatibility complex (MHCI) expression on the cell surface. Expression of P, V, or W inhibited MHCI upregulation, suppressed IFNα/β responses and decreased cytokine release. Inhibition of cytokine responses was most potent for W, slightly less potent for V and less so for P. Using the data obtained in this system and incorporating data on P, V, and W expression obtained from NiV-infected cells, a mathematical model of antiviral signaling in NiV-infected cells was developed. This model suggests that the infection may suppress innate antiviral responses (i.e. IFNα/β responses) and effectively inhibit induction of adaptive immunity by blocking MHCI upregulation. At the same time, some degree of inflammatory cytokine responses would occur, particularly early in infection. It was proposed that the residual cytokine response might promote vascular permeability and virus dissemination (Seto et al. 2010). This model is consistent with studies demonstrating that NiV infection induces cytokine production by human endothelial cells (Lo et al. 2010). Because there have been only very limited studies characterizing virus relication and kinetics of gene expression in general and because the kinetics of NIV replication in human dendritic cells has not been explored in detail, this model was built on a number of assumptions. It nonetheless provides a framework to guide future studies.

The availability of genetically engineered viruses allows the contributions of henipavirus-encoded IFN-antagonists to be assessed in vivo as well as in cell culture (Ciancanelli et al. 2009; Yoneda et al. 2009). Toward this end, NiVs engineered to prevent the specific expression of the C, V, and W proteins were constructed (Yoneda et al. 2010). The C-deficient virus was generated by introduction of a single nucleotide change that introduced a stop codon into the C ORF at the sixth codon while leaving the overlapping P/V/W amino acid sequence unchanged. A V-deficient virus and a W-deficient virus were each generated by introducing a stop codon in the appropriate reading frame immediately after the P gene editing site. This approach eliminates the expression of viral proteins possessing the unique carboxy-termini of the V or W proteins. However, in the V- and W-deficient viruses, edited P gene-derived mRNAs would still direct translation of a protein corresponding to the common P, V, W amino-terminus. Such a protein would be unique to the mutant virus and would retain the STAT1 binding domain as well as sequences which, in the context of a V protein, can inhibit a reconstituted viral polymerase (Sleeman et al. 2008). Vero cells were transfected with an IFN-responsive reporter gene, infected for 48 h and then treated with IFNα (1,000 units/ml). All of the viruses completely suppressed reporter activation, indicating that the viruses effectively block IFN signaling and that loss of individual gene products did not abrogate this effect. Nonetheless, growth of the C-deficient and V-deficient viruses was attenuated in Vero cells, suggesting that C and V may contribute to replication (Yoneda et al. 2010). This is consistent with the growth impairment in Vero cells of C-deficient viruses previously seen by others (Ciancanelli et al. 2009). The molecular basis by which these viruses are attenuated has not been defined but should reflect functions other than inhibition of IFNα/β responses, because Vero cells lack IFNα/β genes. Interestingly, the C-deficient virus was more attenuated in 293 cells than the other viruses, but the basis for this also remains unclear (Yoneda et al. 2010). The viruses were also assessed for virulence in Syrian golden hamsters. The W-deficient virus killed hamsters, as has been reported for wild-type virus. However, neither the V-deficient nor the C-deficient viruses killed hamsters at any dose tested, indicating that these proteins influence pathogenesis. The molecular basis for the in vivo attenuation remains to be defined (Yoneda et al. 2010). In this respect, it is worth noting that expression of the C, V. and W proteins can each, when individually expressed, inhibit NiV polymerase activity as assessed by minigenome assay (Sleeman et al. 2008). The capacity of these proteins to regulate viral RNA synthesis could serve as a means to regulate or limit production of immunostimulatory RNAs that might otherwise trigger host innate immune responses. Elimination of C, V, or W expression may impair virus replication either through activation of cellular antiviral responses or simply by perturbing the normal rate of RNA synthesis. Additional studies performed using recombinant viruses with mutations that disrupt specific molecular functions, such as STAT1 binding or MDA5 interaction, may clarify how immune evasion functions regulate replication and pathogenesis.

7 Epilogue

As is evident from this review, a significant effort has defined several strategies by which NiV and HeV can block host IFNα/β responses. The proteins most heavily studied in this regard, the C, P, V, and W proteins, were candidates for these functions based on analogy to other paramxoviruses which also use P gene products to regulate host innate immunity. The studies described above demonstrate IFN-antagonist function for the C, P, V, and W proteins, but they do not exclude roles for other viral gene products in immune evasion nor do they exclude the possibility that P gene products will have additional functions. The requirement for BSL4 containment has limited the characterization of henipavirus-innate immune system interactions. The development of efficient reverse genetics systems should now allow more detailed assessments regarding the contribution of specific IFN-antagonist functions to virus replication. For example, it will be important to dissect out the contribution to pathogenesis of antagonism of STAT1 and STAT2 versus inhibition of IFNα/β-induction. Studies should examine how these functions influence replication in biologically-relevant cell types and in animal models. At the same time, additional studies to characterize the kinetics of expression and the determinants of subcellular localization of these gene products are clearly warranted. Basic characterization of the host cell pathways that detect and respond to both wild-type and mutant viruses and integration of this information into models of viral pathogenesis should clarify the role of innate immune response in the outcome of these deadly infections. Finally, because viral IFN-antagonist functions are likely to be required for virulence, the immune evasion factors described represent potential targets for therapeutics.

References

Andrejeva J, Childs KS, Young DF, Carlos TS, Stock N, Goodbourn S, Randall RE (2004) The V proteins of paramyxoviruses bind the IFN-inducible RNA helicase, MDA-5, and inhibit its activation of the IFN-beta promoter. Proc Natl Acad Sci U S A 101:17264–17269

Baum A, Sachidanandam R, Garcia-Sastre A (2010) Preference of RIG-I for short viral RNA molecules in infected cells revealed by next-generation sequencing. Proc Natl Acad Sci U S A 107:16303–16308

Biron CA (2001) Interferons alpha and beta as immune regulators—a new look. Immunity 14:661–664

Bousse T, Chambers RL, Scroggs RA, Portner A, Takimoto T (2006) Human parainfluenza virus type 1 but not Sendai virus replicates in human respiratory cells despite IFN treatment. Virus Res 121:23–32

Boxer EL, Nanda SK, Baron MD (2009) The rinderpest virus non-structural C protein blocks the induction of type 1 interferon. Virology 385:134–142

Caignard G, Guerbois M, Labernardiere JL, Jacob Y, Jones LM, Wild F, Tangy F, Vidalain PO (2007) Measles virus V protein blocks Jak1-mediated phosphorylation of STAT1 to escape IFN-alpha/beta signaling. Virology 368:351–362

Caignard G, Bourai M, Jacob Y, Tangy F, Vidalain PO (2009a) Inhibition of IFN-alpha/beta signaling by two discrete peptides within measles virus V protein that specifically bind STAT1 and STAT2. Virology 383:112–120

Caignard G, Komarova AV, Bourai M, Mourez T, Jacob Y, Jones LM, Rozenberg F, Vabret A, Freymuth F, Tangy F, Vidalain PO (2009b) Differential regulation of type I interferon and epidermal growth factor pathways by a human Respirovirus virulence factor. PLoS Pathog 5:e1000587

Chambers R, Takimoto T (2009a) Host specificity of the anti-interferon and anti-apoptosis activities of parainfluenza virus P/C gene products. J Gen Virol 90:1906–1915

Chambers R, Takimoto T (2009b) Antagonism of innate immunity by paramyxovirus accessory proteins. Viruses 1:574–593

Childs K, Stock N, Ross C, Andrejeva J, Hilton L, Skinner M, Randall R, Goodbourn S (2007) mda-5, but not RIG-I, is a common target for paramyxovirus V proteins. Virology 359:190–200

Childs KS, Andrejeva J, Randall RE, Goodbourn S (2009) Mechanism of mda-5 Inhibition by paramyxovirus V proteins. J Virol 83:1465–1473

Ciancanelli MJ, Volchkova VA, Shaw ML, Volchkov VE, Basler CF (2009) Nipah virus sequesters inactive STAT1 in the nucleus via a P gene-encoded mechanism. J Virol

Crameri G, Todd S, Grimley S, McEachern JA, Marsh GA, Smith C, Tachedjian M, De Jong C, Virtue ER, Yu M, Bulach D, Liu JP, Michalski WP, Middleton D, Field HE, Wang LF (2009) Establishment, immortalisation and characterisation of pteropid bat cell lines. PLoS ONE 4:e8266

Devaux P, von Messling V, Songsungthong W, Springfeld C, Cattaneo R (2007) Tyrosine 110 in the measles virus phosphoprotein is required to block STAT1 phosphorylation. Virology 360:72–83

Didcock L, Young DF, Goodbourn S, Randall RE (1999) The V protein of simian virus 5 inhibits interferon signalling by targeting STAT1 for proteasome-mediated degradation. J Virol 73:9928–9933

Fontana JM, Bankamp B, Bellini WJ, Rota PA (2008) Regulation of interferon signaling by the C and V proteins from attenuated and wild-type strains of measles virus. Virology 374: 71–81
Gale M Jr, Sen GC (2009) Viral evasion of the interferon system. J Interferon Cytokine Res 29:475–476
Garcin D, Latorre P, Kolakofsky D (1999) Sendai virus C proteins counteract the interferon-mediated induction of an antiviral state. J Virol 73:6559–6565
Garcin D, Curran J, Kolakofsky D (2000) Sendai virus C proteins must interact directly with cellular components to interfere with interferon action. J Virol 74:8823–8830
Garcin D, Curran J, Itoh M, Kolakofsky D (2001) Longer and shorter forms of Sendai virus C proteins play different roles in modulating the cellular antiviral response. J Virol 75:6800–6807
Garcin D, Marq JB, Goodbourn S, Kolakofsky D (2003) The amino-terminal extensions of the longer Sendai virus C proteins modulate pY701-Stat1 and bulk Stat1 levels independently of interferon signaling. J Virol 77:2321–2329
Georges-Courbot MC, Contamin H, Faure C, Loth P, Baize S, Leyssen P, Neyts J, Deubel V (2006) Poly(I)-poly(C12U) but not ribavirin prevents death in a hamster model of Nipah virus infection. Antimicrob Agents Chemother 50:1768–1772
Gonzalez-Navajas JM, Lee J, David M, Raz E (2010) Immunomodulatory functions of type I interferons. Nat Rev Immunol
Goodbourn S, Randall RE (2009) The regulation of type I interferon production by paramyxoviruses. J Interferon Cytokine Res 29:539–547
Gotoh B, Takeuchi K, Komatsu T, Yokoo J, Kimura Y, Kurotani A, Kato A, Nagai Y (1999) Knockout of the Sendai virus C gene eliminates the viral ability to prevent the interferon-alpha/beta-mediated responses. FEBS Lett 459:205–210
Gotoh B, Takeuchi K, Komatsu T, Yokoo J (2003) The STAT2 activation process is a crucial target of Sendai virus C protein for the blockade of alpha interferon signaling. J Virol 77:3360–3370
Habjan M, Andersson I, Klingstrom J, Schumann M, Martin A, Zimmermann P, Wagner V, Pichlmair A, Schneider U, Muhlberger E, Mirazimi A, Weber F (2008) Processing of genome 5' termini as a strategy of negative-strand RNA viruses to avoid RIG-I-dependent interferon induction. PLoS ONE 3:e2032
Hagmaier K, Stock N, Goodbourn S, Wang LF, Randall R (2006) A single amino acid substitution in the V protein of Nipah virus alters its ability to block interferon signalling in cells from different species. J Gen Virol 87:3649–3653
Hagmaier K, Stock N, Precious B, Childs K, Wang LF, Goodbourn S, Randall RE (2007) Mapuera virus, a rubulavirus that inhibits interferon signalling in a wide variety of mammalian cells without degrading STATs. J Gen Virol 88:956–966
Halpin K, Bankamp B, Harcourt BH, Bellini WJ, Rota PA (2004) Nipah virus conforms to the rule of six in a minigenome replication assay. J Gen Virol 85:701–707
Harcourt BH, Tamin A, Ksiazek TG, Rollin PE, Anderson LJ, Bellini WJ, Rota PA (2000) Molecular characterization of Nipah virus, a newly emergent paramyxovirus. Virology 271:334–349
He B, Paterson RG, Stock N, Durbin JE, Durbin RK, Goodbourn S, Randall RE, Lamb RA (2002) Recovery of paramyxovirus simian virus 5 with a V protein lacking the conserved cysteine-rich domain: the multifunctional V protein blocks both interferon-beta induction and interferon signaling. Virology 303:15–32
Hoebe K, Beutler B (2004) LPS, dsRNA and the interferon bridge to adaptive immune responses: Trif, Tram, and other TIR adaptor proteins. J Endotoxin Res 10:130–136
Huang Z, Krishnamurthy S, Panda A, Samal SK (2003) Newcastle disease virus V protein is associated with viral pathogenesis and functions as an alpha interferon antagonist. J Virol 77:8676–8685
Jensen S, Thomsen AR (2012) Sensing of RNA viruses-A review on innate immune receptors invovled in recognizing RNA virus invasion. J Virol 86:2900–2910

Kato A, Ohnishi Y, Kohase M, Saito S, Tashiro M, Nagai Y (2001) Y2, the smallest of the Sendai virus C proteins, is fully capable of both counteracting the antiviral action of interferons and inhibiting viral RNA synthesis. J Virol 75:3802–3810

Kawai T, Takahashi K, Sato S, Coban C, Kumar H, Kato H, Ishii KJ, Takeuchi O, Akira S (2005) IPS-1, an adaptor triggering RIG-I- and Mda5-mediated type I interferon induction. Nat Immunol 6:981–988

Komatsu T, Takeuchi K, Yokoo J, Tanaka Y, Gotoh B (2000) Sendai virus blocks alpha interferon signaling to signal transducers and activators of transcription [in process citation]. J Virol 74:2477–2480

Komatsu T, Takeuchi K, Yokoo J, Gotoh B (2002) Sendai virus C protein impairs both phosphorylation and dephosphorylation processes of Stat1. FEBS Lett 511:139–144

Komatsu T, Takeuchi K, Yokoo J, Gotoh B (2004) C and V proteins of Sendai virus target signaling pathways leading to IRF-3 activation for the negative regulation of interferon-beta production. Virology 325:137–148

Komatsu T, Takeuchi K, Gotoh B (2007) Bovine parainfluenza virus type 3 accessory proteins that suppress beta interferon production. Microbes Infect 9:954–962

Ksiazek TG, Rota PA, Rollin PE (2011) A review of Nipah and Hendra viruses with an historical aside. Virus Res 162:173–183

Kubota T, Yokosawa N, Yokota S, Fujii N (2001) C terminal CYS-RICH region of mumps virus structural V protein correlates with block of interferon alpha and gamma signal transduction pathway through decrease of STAT 1-alpha. Biochem Biophys Res Commun 283:255–259

Kubota T, Yokosawa N, Yokota S, Fujii N (2002) Association of mumps virus V protein with RACK1 results in dissociation of STAT-1 from the alpha interferon receptor complex. J Virol 76:12676–12682

Kulkarni S, Volchkova V, Basler CF, Palese P, Volchkov VE, Shaw ML (2009) Nipah virus edits its P gene at high frequency to express the V and W proteins. J Virol 83:3982–3987

Lamb RA, Parks GD (2007) Paramyxoviridae: the viruses and their replication. In: Knipe DM, Howley PM (eds) Fields Virology, 5th edn, vol 1. Lippincott Williams and Wilkins, Philadelphia

Li T, Chen X, Garbutt KC, Zhou P, Zheng N (2006) Structure of DDB1 in complex with a paramyxovirus V protein: viral hijack of a propeller cluster in ubiquitin ligase. Cell 124:105–117

Liston P, Briedis DJ (1994) Measles virus V protein binds zinc. Virology 198:399–404

Lo MK, Harcourt BH, Mungall BA, Tamin A, Peeples ME, Bellini WJ, Rota PA (2009) Determination of the henipavirus phosphoprotein gene mRNA editing frequencies and detection of the C, V and W proteins of Nipah virus in virus-infected cells. J Gen Virol 90:398–404

Lo MK, Miller D, Aljofan M, Mungall BA, Rollin PE, Bellini WJ, Rota PA (2010) Characterization of the antiviral and inflammatory responses against Nipah virus in endothelial cells and neurons. Virology 404:78–88

Lu LL, Puri M, Horvath CM, Sen GC (2008) Select paramyxoviral V proteins inhibit IRF3 activation by acting as alternative substrates for inhibitor of kappaB kinase epsilon (IKKe)/TBK1. J Biol Chem 283:14269–14276

Ludlow LE, Lo MK, Rodriguez JJ, Rota PA, Horvath CM (2008) Henipavirus V protein association with Polo-like kinase reveals functional overlap with STAT1 binding and interferon evasion. J Virol 82:6259–6271

Malur AG, Chattopadhyay S, Maitra RK, Banerjee AK (2005) Inhibition of STAT 1 phosphorylation by human parainfluenza virus type 3 C protein. J Virol 79:7877–7882

Meylan E, Curran J, Hofmann K, Moradpour D, Binder M, Bartenschlager R, Tschopp J (2005) Cardif is an adaptor protein in the RIG-I antiviral pathway and is targeted by hepatitis C virus. Nature 437:1167–1172

Middleton DJ, Morrissy CJ, van der Heide BM, Russell GM, Braun MA, Westbury HA, Halpin K, Daniels PW (2007) Experimental Nipah virus infection in pteropid bats (Pteropus poliocephalus). J Comp Pathol 136:266–272

Nishio M, Garcin D, Simonet V, Kolakofsky D (2002) The carboxyl segment of the mumps virus V protein associates with Stat proteins in vitro via a tryptophan-rich motif. Virology 300:92–99

Nishio M, Tsurudome M, Ito M, Garcin D, Kolakofsky D, Ito Y (2005a) Identification of paramyxovirus V protein residues essential for STAT protein degradation and promotion of virus replication. J Virol 79:8591–8601

Nishio M, Tsurudome M, Ito M, Ito Y (2005b) Human parainfluenza virus type 4 is incapable of evading the interferon-induced antiviral effect. J Virol 79:14756–14768

Ohno S, Ono N, Takeda M, Takeuchi K, Yanagi Y (2004) Dissection of measles virus V protein in relation to its ability to block alpha/beta interferon signal transduction. J Gen Virol 85:2991–2999

O'Neill LA, Bowie AG (2011) The powerstroke and camshaft of the RIG-I antiviral RNA detection machine. Cell 147:259–261

Palosaari H, Parisien JP, Rodriguez JJ, Ulane CM, Horvath CM (2003) STAT protein interference and suppression of cytokine signal transduction by measles virus V protein. J Virol 77:7635–7644

Parisien JP, Lau JF, Rodriguez JJ, Sullivan BM, Moscona A, Parks GD, Lamb RA, Horvath CM (2001) The V protein of human parainfluenza virus 2 antagonizes type I interferon responses by destabilizing signal transducer and activator of transcription 2. Virology 283:230–239

Parisien JP, Lau JF, Horvath CM (2002a) STAT2 Acts as a Host Range Determinant for Species-Specific Paramyxovirus Interferon Antagonism and Simian Virus 5 Replication. J Virol 76:6435–6441

Parisien JP, Lau JF, Rodriguez JJ, Ulane CM, Horvath CM (2002b) Selective STAT protein degradation induced by paramyxoviruses requires both STAT1 and STAT2 but is independent of alpha/beta interferon signal transduction. J Virol 76:4190–4198

Parisien JP, Bamming D, Komuro A, Ramachandran A, Rodriguez JJ, Barber G, Wojahn RD, Horvath CM (2009) A shared interface mediates paramyxovirus interference with antiviral RNA helicases MDA5 and LGP2. J Virol 83:7252–7260

Park M-S, Shaw ML, Munoz-Jordan J, Cros JF, Nakaya T, Bouvier N, Palese P, Garcia-Sastre A, Basler CF (2003) Newcastle Disease Virus (NDV)-Based Assay Demonstrates Interferon-Antagonist Activity for the NDV V Protein and the Nipah Virus V, W, and C Proteins. J. Virol. 77:1501–1511

Paterson RG, Leser GP, Shaughnessy MA, Lamb RA (1995) The paramyxovirus SV5 V protein binds two atoms of zinc and is a structural component of virions. Virology 208:121–131

Pfaller CK, Conzelmann KK (2008) Measles virus V protein is a decoy substrate for IkappaB kinase alpha and prevents Toll-like receptor 7/9-mediated interferon induction. J Virol 82:12365–12373

Platanias LC (2005) Mechanisms of type-I- and type-II-interferon-mediated signalling. Nat Rev Immunol 5:375–386

Poole E, He B, Lamb RA, Randall RE, Goodbourn S (2002) The V proteins of simian virus 5 and other paramyxoviruses inhibit induction of interferon-beta. Virology 303:33–46

Precious B, Childs K, Fitzpatrick-Swallow V, Goodbourn S, Randall RE (2005a) Simian virus 5 V protein acts as an adaptor, linking DDB1 to STAT2, to facilitate the ubiquitination of STAT1. J Virol 79:13434–13441

Precious B, Young DF, Andrejeva L, Goodbourn S, Randall RE (2005b) In vitro and in vivo specificity of ubiquitination and degradation of STAT1 and STAT2 by the V proteins of the paramyxoviruses simian virus 5 and human parainfluenza virus type 2. J Gen Virol 86:151–158

Precious BL, Carlos TS, Goodbourn S, Randall RE (2007) Catalytic turnover of STAT1 allows PIV5 to dismantle the interferon-induced anti-viral state of cells. Virology 368:114–121

Ramachandran A, Horvath CM (2009) Paramyxovirus disruption of interferon signal transduction: STATus report. J Interferon Cytokine Res 29:531–537

Ramachandran A, Horvath CM (2010) Dissociation of paramyxovirus interferon evasion activities: universal and virus-specific requirements for conserved V protein amino acids in MDA5 interference. J Virol 84:11152–11163

Ramachandran A, Parisien JP, Horvath CM (2008) STAT2 is a primary target for measles virus V protein-mediated alpha/beta interferon signaling inhibition. J Virol 82:8330–8338

Rodriguez JJ, Parisien JP, Horvath CM (2002) Nipah virus V protein evades alpha and gamma interferons by preventing STAT1 and STAT2 activation and nuclear accumulation. J Virol 76:11476–11483

Rodriguez JJ, Wang LF, Horvath CM (2003) Hendra virus V protein inhibits interferon signaling by preventing STAT1 and STAT2 nuclear accumulation. J Virol 77:11842–11845

Rodriguez JJ, Cruz CD, Horvath CM (2004) Identification of the nuclear export signal and STAT-binding domains of the Nipah virus V protein reveals mechanisms underlying interferon evasion. J Virol 78:5358–5367

Saito S, Ogino T, Miyajima N, Kato A, Kohase M (2002) Dephosphorylation failure of tyrosine-phosphorylated STAT1 in IFN- stimulated Sendai virus C protein-expressing cells. Virology 293:205–209

Saito T, Owen DM, Jiang F, Marcotrigiano J, Gale M Jr (2008) Innate immunity induced by composition-dependent RIG-I recognition of hepatitis C virus RNA. Nature 454:523–527

Schlee M, Roth A, Hornung V, Hagmann CA, Wimmenauer V, Barchet W, Coch C, Janke M, Mihailovic A, Wardle G, Juranek S, Kato H, Kawai T, Poeck H, Fitzgerald KA, Takeuchi O, Akira S, Tuschl T, Latz E, Ludwig J, Hartmann G (2009) Recognition of 5' triphosphate by RIG-I helicase requires short blunt double-stranded RNA as contained in panhandle of negative-strand virus. Immunity 31:25–34

Schmidt A, Schwerd T, Hamm W, Hellmuth JC, Cui S, Wenzel M, Hoffmann FS, Michallet MC, Besch R, Hopfner KP, Endres S, Rothenfusser S (2009) 5′-triphosphate RNA requires base-paired structures to activate antiviral signaling via RIG-I. Proc Natl Acad Sci U S A 106:12067–12072

Schuhmann KM, Pfaller CK, Conzelmann KK (2011) The measles virus V protein binds to p65 (RelA) to suppress NF-kappaB activity. J Virol 85: 3162–3171

Seth RB, Sun L, Ea CK, Chen ZJ (2005) Identification and characterization of MAVS, a mitochondrial antiviral signaling protein that activates NF-kappaB and IRF 3. Cell 122:669–682

Seto J, Qiao L, Guenzel CA, Xiao S, Shaw ML, Hayot F, Sealfon SC (2010) Novel Nipah virus immune-antagonism strategy revealed by experimental and computational study. J Virol 84:10965–10973

Shaffer JA, Bellini WJ, Rota PA (2003) The C protein of measles virus inhibits the type I interferon response. Virology 315:389–397

Shaw ML, Garcia-Sastre A, Palese P, Basler CF (2004) Nipah virus V and W proteins have a common STAT1-binding domain yet inhibit STAT1 activation from the cytoplasmic and nuclear compartments, respectively. J Virol 78:5633–5641

Shaw ML, Cardenas WB, Zamarin D, Palese P, Basler CF (2005) Nuclear localization of the Nipah virus W protein allows for inhibition of both virus- and toll-like receptor 3-triggered signaling pathways. J Virol 79:6078–6088

Sleeman K, Bankamp B, Hummel KB, Lo MK, Bellini WJ, Rota PA (2008) The C, V and W proteins of Nipah virus inhibit minigenome replication. J Gen Virol 89:1300–1308

Sparrer KM, Pfaller CK, Conzelmann KK (2012) Measles virus C protein interferes with Beta interferon transcription in the nucleus. J Virol 86: 796–805

Strahle L, Garcin D, Kolakofsky D (2006) Sendai virus defective-interfering genomes and the activation of interferon-beta. Virology 351:101–111

Sun D, Luthra P, Li Z, He B (2009) PLK1 down-regulates parainfluenza virus 5 gene expression. PLoS Pathog 5:e1000525

Takeuchi K, Komatsu T, Yokoo J, Kato A, Shioda T, Nagai Y, Gotoh B (2001) Sendai virus C protein physically associates with Stat1. Genes Cells 6:545–557

Takeuchi K, Kadota SI, Takeda M, Miyajima N, Nagata K (2003) Measles virus V protein blocks interferon (IFN)-alpha/beta but not IFN-gamma signaling by inhibiting STAT1 and STAT2 phosphorylation. FEBS Lett 545:177–182

Ulane CM, Horvath CM (2002) Paramyxoviruses SV5 and HPIV2 assemble STAT protein ubiquitin ligase complexes from cellular components. Virology 304:160–166

Ulane CM, Rodriguez JJ, Parisien JP, Horvath CM (2003) STAT3 ubiquitylation and degradation by mumps virus suppress cytokine and oncogene signaling. J Virol 77:6385–6393

Ulane CM, Kentsis A, Cruz CD, Parisien JP, Schneider KL, Horvath CM (2005) Composition and assembly of STAT-targeting ubiquitin ligase complexes: paramyxovirus V protein carboxyl terminus is an oligomerization domain. J Virol 79:10180–10189

Van Cleve W, Amaro-Carambot E, Surman SR, Bekisz J, Collins PL, Zoon KC, Murphy BR, Skiadopoulos MH, Bartlett EJ (2006) Attenuating mutations in the P/C gene of human parainfluenza virus type 1 (HPIV1) vaccine candidates abrogate the inhibition of both induction and signaling of type I interferon (IFN) by wild-type HPIV1. Virology 352:61 73

Virtue ER, Marsh GA, Baker ML, Wang LF (2011a) Interferon production and signaling pathways are antagonized during henipavirus infection of fruit bat cell lines. PLoS One 6:e22488

Virtue ER, Marsh GA, Wang LF (2011b) Interferon signaling remains functional during henipavirus infection of human cell lines. J Virol 85:4031–4034

Wang LF, Michalski WP, Yu M, Pritchard LI, Crameri G, Shiell B, Eaton BT (1998) A novel P/V/C gene in a new member of the Paramyxoviridae family, which causes lethal infection in humans, horses, and other animals. J Virol 72:1482–1490

Wang YE, Park A, Lake M, Pentecost M, Torres B, Yun TE, Wolf MC, Holbrook MR, Freiberg AN, Lee B (2010) Ubiquitin-regulated nuclear-cytoplasmic trafficking of the Nipah virus matrix protein is important for viral budding. PLoS Pathog 6:e1001186

Wong KT, Grosjean I, Brisson C, Blanquier B, Fevre-Montange M, Bernard A, Loth P, Georges-Courbot MC, Chevallier M, Akaoka H, Marianneau P, Lam SK, Wild TF, Deubel V (2003) A golden hamster model for human acute Nipah virus infection. Am J Pathol 163:2127–2137

Xu LG, Wang YY, Han KJ, Li LY, Zhai Z, Shu HB (2005) VISA is an adapter protein required for virus-triggered IFN-beta signaling. Mol Cell 19:727–740

Yokosawa N, Yokota S, Kubota T, Fujii N (2002) C-terminal region of STAT-1alpha is not necessary for its ubiquitination and degradation caused by mumps virus V protein. J Virol 76:12683–12690

Yokota S, Saito H, Kubota T, Yokosawa N, Amano K, Fujii N (2003) Measles virus suppresses interferon-alpha signaling pathway: suppression of Jak1 phosphorylation and association of viral accessory proteins, C and V, with interferon-alpha receptor complex. Virology 306:135–146

Yokota S, Okabayashi T, Yokosawa N, Fujii N (2008) Measles virus P protein suppresses Toll-like receptor signal through up-regulation of ubiquitin-modifying enzyme A20. FASEB J 22:74–83

Yoneda M, Fujita K, Sato H, Kai C (2009) Reverse genetics of Nipah virus to probe viral pathogenicity. Methods Mol Biol 515:329–337

Yoneda M, Guillaume V, Sato H, Fujita K, Georges-Courbot MC, Ikeda F, Omi M, Muto-Terao Y, Wild TF, Kai C (2010) The nonstructural proteins of Nipah virus play a key role in pathogenicity in experimentally infected animals. PLoS ONE 5:e12709

Yoneyama M, Fujita T (2009) RNA recognition and signal transduction by RIG-I-like receptors. Immunol Rev 227:54–65

Yoneyama M, Fujita T, Wang R (2012) Cytoplasmic sensing of viral double-stranded RNA and activation of innate immunity by RIG-I-like receptors innate immune regulation and cancer imunotherapy. Springer, New York, pp 51–60

Yoneyama M, Kikuchi M, Natsukawa T, Shinobu N, Imaizumi T, Miyagishi M, Taira K, Akira S, Fujita T (2004) The RNA helicase RIG-I has an essential function in double-stranded RNA-induced innate antiviral responses. Nat Immunol 5:730–737

Young DF, Didcock L, Goodbourn S, Randall RE (2000) Paramyxoviridae use distinct virus-specific mechanisms to circumvent the interferon response. Virology 269:383–390

Young DF, Chatziandreou N, He B, Goodbourn S, Lamb RA, Randall RE (2001) Single amino acid substitution in the V protein of simian virus 5 differentiates its ability to block interferon signaling in human and murine cells. J Virol 75:3363–3370
Yount JS, Gitlin L, Moran TM, Lopez CB (2008) MDA5 participates in the detection of paramyxovirus infection and is essential for the early activation of dendritic cells in response to Sendai virus defective interfering particles. J Immunol 180:4910–4918

Animal Challenge Models of Henipavirus Infection and Pathogenesis

Thomas W. Geisbert, Heinz Feldmann and Christopher C. Broder

Abstract The henipaviruses, Hendra virus (HeV), and Nipah virus (NiV), are enigmatic emerging pathogens that causes severe and often fatal neurologic and/or respiratory disease in both animals and humans. Amongst people, case fatality rates range between 40 and 75% and there are no vaccines or treatments approved for human use. A number of species of animals including guinea pigs, hamsters, cats, ferrets, pigs, and African green monkeys have been employed as animal models of human henipavirus infection. Here, we review the development of animal models for henipavirus infection, discuss the pathology and pathogenesis of these models, and assess the utility of each model to recapitulate important aspects of henipavirus-mediated disease seen in humans.

Contents

T. W. Geisbert (✉)
Galveston National Laboratory, Department of Microbiology and Immunology,
University of Texas Medical Branch, Galveston, TX, USA
e-mail: tom.geisbert@utmb.edu

H. Feldmann
Laboratory of Virology, Rocky Mountain Laboratories,
Division of Intramural Research,
National Institute of Allergy and Infectious Diseases,
National Institutes of Health, Hamilton, MT, USA

C. C. Broder
Department of Microbiology and Immunology,
Uniformed Services University, Bethesda, MD, USA

Current Topics in Microbiology and Immunology (2012) 359: 153–177

DOI: 10.1007/82_2012_208

Published Online: 4 April 2012

1 Introduction

Hendra virus (HeV) and Nipah virus (NiV) are members of the genus *Henipavirus* (family *Paramyxoviridae*) that can cause severe respiratory illness and/or encephalitis in a wide variety of mammals, including humans (Selvey et al. 1995; Chua et al. 1999). HeV was identified as the causative agent of an acute respiratory disease in horses in 1994 in Queensland, Australia (Selvey et al. 1995) and to date there have now been 32 outbreaks in Australia since, with at least one or more occurrences per year since 2006. Every outbreak of HeV has involved horses as the initial infected host and there have been a total of seven human cases arising from exposure to infected horses. Four human fatalities have occurred (Playford et al. 2010) with the most recent occurring in August 2009 (Anonymous 2009). NiV was first identified during an outbreak of severe encephalitis in Malaysia and Singapore in 1998–1999 with at least 265 human cases and 105 deaths, with pigs serving as the intermediate amplifying host (Chua et al. 1999; Paton et al. 1999; Goh et al. 2000; Chua 2003). Since 1998 there have been more than a dozen recognized occurrences of human NiV infection, primarily in Bangladesh and India (Chadha et al. 2006; Harit et al. 2006; Luby et al. 2009b; Arankalle et al. 2011). In the majority of subsequent spillover events, the mortality rate among humans has been higher ($\sim$ 75%) along with evidence of multiple rounds of person-to-person transmission (Gurley et al. 2007; Homaira et al. 2010).

Several species of fruit bats of the *Pteropus* genus (flying foxes) appear to be the principle natural reservoirs of both NiV and HeV [reviewed in Bishop and Broder (2008)] but serological evidence of NiV or Nipah-like virus infection has recently been reported in several additional frugivorous and insectivorous bats (Hayman et al. 2008; Li et al. 2008). NiV has been isolated from bat urine and partially eaten fruit (Chua et al. 2002; Reynes et al. 2005) and direct transmission of NiV from flying foxes to humans from contaminated food sources has been suggested (Harit et al. 2006; Luby et al. 2006).

The development and characterization of suitable animal models for henipavirus infection is essential for studying features of the virus such as shedding and transmission and in the case of henipaviruses the underlying mechanisms of pathogenesis and recrudescence. In addition, the availability of well characterized animal models is essential for fulfilling the critical needs for the in vivo evaluation of potential prophylactic and antiviral modalities for human use. Of particular importance, the U.S. Food and Drug Administration (FDA) implemented the Animal Efficacy Rule which came into effect in 2002. This rule specifically applies to the development of medical countermeasures when human efficacy studies are not possible or ethical, such as is often the case with highly virulent emerging pathogens like the henipaviruses. Essentially, this rule allows for the evaluation of vaccines or therapeutics using data derived from studies carried out in two animal models. Here, we provide a literature review of small and large animal models of HeV and NiV infection.

2 Naturally Occurring Henipavirus Infections

2.1 Henipavirus Infection in Humans

Both HeV and NiV are zoonotic viral infections of humans but each can also infect various mammalian species in spillover events from their natural pteropid bat reservoir hosts [reviewed in Eaton et al. (2006)]. Among the known susceptible animal hosts, including humans, each can cause a systemic infection and the central features of acute henipavirus infection and pathogenesis in humans are widespread vasculitis with endothelial cell tropism with syncytia, thrombosis, ischemia and necrosis, along with parenchymal cell infection and pathogenesis of many major organs but is particularly prominent in the brain, lung, heart, and kidney (Chua et al. 1999; Wong et al. 2002, 2009). Reflecting the major involvement of both the lung and brain, severe henipavirus disease in humans can manifest as acute interstitial pneumonitis, encephalitis, or a combination of both.

To date there have been only seven (four fatal) confirmed human cases of HeV, all in Australia, since 1994 (Murray et al. 1995a; Selvey et al. 1995; Hooper et al. 1996; Rogers et al. 1996; Hanna et al. 2006; Anonymous 2009; Playford et al. 2010). Of these seven human cases (three in 1994 and one in 2004) four had initially presented with influenza-like illness and two survived infection, whereas the two linked cases in 2008 experienced initial influenza-like illness but later, following clinical resolution including abatement of fever, progressed to encephalitic disease with widespread cortical, subcortical, and deep white matter involvement similar to NiV encephalitis with one patient succumbing to infection (Playford et al. 2010). The seventh and fatal case of HeV infection presented with encephalitis $\sim$3 weeks following exposure to a horse that died of HeV infection (Anonymous 2009).

In contrast, there have been many hundreds of cases of human NiV infection since the initial Malaysian outbreak together with more than a dozen outbreaks in Bangladesh and India [reviewed in Bishop and Broder (2008); Luby et al. (2009b)]. NiV human infection reveals central nervous system (CNS) vascular pathology as more severe in comparison to other organ systems, with necrosis, edema and inflammation often associated with viral antigen (Wong et al. 2002). Systemic vasculitis, extensive thrombosis, and parenchymal necrosis, particularly within the CNS, along with endothelial cell damage, evident syncytia and widespread presence of viral antigen in endothelial and smooth muscle cells of blood vessels. Viral antigen was also present in various parenchymal cells, notably in neurons, and NiV infection of endothelial cells and neurons along with vasculitis and thrombosis seem to be critical to the pathogenesis of NiV infection in humans (Wong et al. 2002; Wong and Ong 2011).

In addition, human henipavirus infection outcomes can take a prolonged course following an initial infection, a process first observed in the second fatal HeV case which occurred in an individual who died from encephalitis 13 months after an aseptic meningitic illness that was only retrospectively determined to have been caused by HeV (O'Sullivan et al. 1997). The majority of human cases of NiV infection in Malaysia presented with acute encephalitis with $\sim$25% of cases also showing respiratory signs (Chua 2003); however, it was noted that neurological disease could present later (>10 weeks) following a nonencephalitic, asymptomatic infection, or patients who recovered from acute encephalitis. Relapsed-encephalitis presented from several months to as late as 4 years after infection (Wong et al. 2001; Tan et al. 2002; Chong and Tan 2003; Tan and Wong 2003). A recent comparative analysis of the first two fatal human cases of HeV infection (acute respiratory and relapsed encephalitis) has been reported (Wong et al. 2009). Immuno- and histopathological findings showed that HeV was neurotropic in both cases and could cause CNS infections that can result in either acute encephalitis (in the absence of apparent clinical encephalitis) or relapsed encephalitis with pathological features similar to acute and relapsed NiV encephalitis. How these viruses escape immunological clearance for such an extended period is unknown, and relapsed or late onset encephalitis are considered to be caused by recrudescence and replication of virus within the CNS (Tan et al. 2002; Wong et al. 2009). To our knowledge, attempts to recreate this recrudescence process of HeV or NiV infection in an animal model has not been attempted although several groups have proposed it.

2.2 Henipavirus Infections in Animals

The natural reservoirs for HeV and NiV are several species of *Pteropus* fruit bats. Early, serological surveys revealed that all four Australian-mainland species of *Pteropus* fruit bats can be naturally infected with HeV (Young et al. 1996) and HeV has been isolated from two of these bat species from uterine fluids or fetal tissues (Halpin et al. 2000, 2011) and more recently from pooled urine samples (Smith et al. 2011). NiV neutralizing antibodies were identified in both Island flying foxes (*P. hypomelanus*) and Malayan flying foxes (*P. vampyrus*) (Yob et al. 2001). Subsequently, NiV was isolated from pooled urine samples as well as from a swab sample taken from a partially eaten fruit from Island flying foxes (Chua et al. 2002). NiV has also been isolated from the urine of flying foxes in Cambodia (Reynes et al. 2005). No occurrence of clinical disease caused by HeV (Halpin et al. 2000; Field et al. 2001) or NiV (Rahman et al. 2010; Sohayati et al. 2011) infection has been reported in naturally infected fruit bats.

The first appearance and recognition of HeV with associated disease occurred in Australia in 1994 among horses whereas the initial occurrence of NiV was associated with an outbreak in farmed pigs in Malaysia in 1998 [reviewed in Eaton et al. (2006)]. HeV occurs only in Australia, and in total, there have now been 32 separate spillover events of HeV along that country's coast from Queensland to New South Wales, with the last 18 episodes occurring between June and October of 2011 (Anonymous 2011b; Smith et al. 2011). Every appearance of HeV infection in animals with associated disease has been in domestic horses. The only other known natural HeV infection in another animal species was reported in 2011 with serological evidence observed in a healthy dog from a HeV-infected property (Anonymous 2011a).

In addition to pigs, there were other naturally occurring NiV infections recorded in cats, dogs, and horses in the initial Malaysian outbreak (Hooper et al. 2001). Serological evidence of natural NiV infection among other animals revealed that dogs in those areas associated with the farms in the Malaysian outbreak were susceptible to NiV infection (Field et al. 2001). Diseased dogs however were uncommon, and only two animals (one found dead and another sick) were examined (Hooper et al. 2001; Wong and Ong 2011). A more recent assessment of natural NiV infection in dogs by serological analysis has confirmed that natural infection is possible, but the presence of NiV does not appear to be maintained within the population in an absence of infected pigs (Mills et al. 2009). Subsequent NiV outbreaks have occurred nearly every year in Bangladesh and India since its discovery in Malaysia in 1999 (Harit et al. 2006; Luby et al. 2009b; Arankalle et al. 2011); however, these appearances of the virus have all been associated with severe disease and high fatality rates in humans, with more than half of these cases attributed to person-to-person transmission (Luby et al. 2009b; Homaira et al. 2010). The initial introduction of the virus into humans was primarily associated with the consumption of raw date palm sap contaminated with NiV from bats (Luby et al. 2009a; Rahman et al. 2011); however, a few cases were associated with contact with sick animals including cows (Hsu et al. 2004), pigs, and diseased goats (Luby et al. 2009a). Experimental infection of a variety of animals with HeV, NiV, or both has since been carried out by a number of laboratories with most of the species that were noted to be susceptible to natural infection, including fruit bats, and these will be discussed below.

3 Animal Modeling of Henipavirus Infection

3.1 Initial Experimental Infections of Animals

The first animal infection experiments with HeV (then called equine morbillivirus) were carried out in horses as a means of reproducing the syndrome observed in those animals during the first recorded outbreak (Murray et al. 1995a, b). The initial HeV infection experiments in horses produced primarily a respiratory disease yielding pulmonary edema and congestion and hemorrhage. Follow-up experiments have provided more detail and will be discussed below.

Additional HeV challenge experiments in horses have detailed the presence of viral antigen noted in endothelial cells in a wide variety of organs including lungs, lymph nodes, kidneys, spleen, bladder, and meninges. A series of other animal infection experiments were conducted with HeV soon after its discovery in 1994 at the Australian Animal Health Laboratories (AAHL). A challenge experiment was carried out in mice, guinea pigs, rats, chickens, rabbits, cats, and dogs, each of which were challenged with a 5,000 $TCID_{50}$ dose of HeV by subcutaneous (s.c.) inoculation (Westbury et al. 1995). Mice, rats, rabbits, chickens, and dogs did not develop any signs of clinical disease, and only cats and guinea pigs developed HeV disease. Two of two cats developed respiratory distress and inappetance by the 5th and 6th day and both succumbed to infection a day later, while four of five guinea pigs developed similar clinical signs between day 7 and 12 each died a day following disease onset. The remaining guinea pig remained well and did not develop HeV neutralizing antibody. At necropsy, both cats and guinea pigs exhibited gross pathological lesions of pneumonia with cats having a more severe presentation, and histologically these lesions were similar to those reported from horses that died of HeV disease (Murray et al. 1995a, b). Virus could also be isolated from a variety of tissue samples from infected cats and guinea pigs. No lesions were observed in any of the other animal species tested; however, rats, rabbits, and one dog seroconverted, while mice and chickens did not. A limited number of experimental horse inoculations with HeV were also carried out which essentially reproduced the clinical and pathological findings reported from equine field cases (Hooper et al. 1997a), characterized as a systemic vascular disease and parenchymal infection in a wide variety of organs with severe lesions in the lungs with hemorrhage, edema, and necrosis and syncytia in the endothelium of pulmonary vessels. Vascular and parenchymal lesions were also seen in the heart, kidney, brain, spleen, lymph node, and stomach. Follow-up HeV infection experiments carried out at the AAHL focused on the cat and guinea pig models (Westbury et al. 1996; Hooper et al. 1997b; Williamson et al. 1998, 2000, 2001).

The isolation and characterization of NiV as a newly discovered agent of encephalitic disease in humans and respiratory disease among farmed pigs in Malaysia came at the close of the 1990s (Anonymous 1999; Chua et al. 1999, 2000). The first animal infection experiments with NiV were also conducted at the AAHL as early as 1999, and these included pigs, cats, and bats (Hooper et al. 2001;

Middleton et al. 2002). Both nervous and respiratory disease with fever was reproduced in pigs, between day 7 and 9, by s.c. inoculation with a 50,000 $TCID_{50}$ dose of NiV (a nonplaque purified, low passaged isolate obtained from the CNS of a fatal human case in Malaysia), while pigs exposed orally or by in-contact with infected pigs resulted in active infection with virus shedding in respiratory secretions, particularly from tonsillar swabs, but in the absence of any significant clinical signs or pathological findings. Diseased pigs exhibited histological evidence of systemic vasculitis, alveolitis, and meningitis with some endothelial syncytia, along with detectable viral antigen (Middleton et al. 2002). In this same study, two cats that were inoculated oronasally with the same dose of virus were more severely diseased than pigs and became febrile and depressed with increased respiratory rates by day 6. One animal progressed to severe breathing difficulty and was euthanized and second began to recover by day 10 (Middleton et al. 2002). The clinical and pathological findings were comparable to those observed in HeV-infected cats (Westbury et al. 1996) but infection of the respiratory tract was more severe. Virus could be recovered from urine, tonsils, and blood. The pathological findings by histological and immunohistochemical analysis also revealed the systemic vasculitis, alveolitis, and meningitis in a field a case of NiV infection in a cat with most severely affected organs being the brain, kidney, liver, and lung (Hooper et al. 2001; Middleton et al. 2002).

3.2 *Henipavirus Infection Experiments in Their Pteropid Bat Hosts*

A limited number of studies have also examined experimental henipavirus infection in fruit bats, including HeV infection in *P. poliocephalus* and *P. alecto* (Williamson et al. 1998, 2000; Halpin et al. 2011) and NiV infection in *P. poliocephalus* and *P. vampyrus* (Middleton et al. 2007; Halpin et al. 2011). All of these studies have been carried out at the AAHL, and in all experiments, no clinical disease or gross pathological findings has ever been observed in fruit bats experimentally infected with either HeV or NiV, even with high doses (50,000 $TCID_{50}$) of virus by s.c. inoculation, but most bats do seroconvert. Only occasional and varied histopathological findings were noted in some bats and most of these observations came from animals inoculated s.c. with HeV (Williamson et al. 1998, 2000; Middleton et al. 2007). Vasculitis with positive immunostaining, associated with the spleen and kidney was observed in some *P. poliocephalus* bats infected with HeV, and this study also confirmed transplacental transmission of HeV in bats showing positive immunostaining of two placentas and virus isolation from a fetus (Williamson et al. 2000). However, fetal tissues showed no evidence of pathology or HeV by immunostaining.

Notably, in the two largest studies of henipavirus infection in pteropid bats, all tissues were found negative by immunohistochemical staining for viral antigen, including tissues which showed histopathological findings (Middleton et al. 2007; Halpin et al. 2011). These studies also revealed that virus shedding from experimentally infected bats does occur, although recovery of virus was rare and only from urine (Middleton et al. 2007; Halpin et al. 2011). HeV genome could be detected in urine, blood, and rectal and throat swabs, as well as some tissues *P. alecto*, whereas NiV genome was rarely detected in *P. vampyrus* and only in mucosal swabs and not in tissues (Halpin et al. 2011).

4 Well-Established Animal Models of Henipavirus Infection

The principal clinical and pathological findings of HeV and NiV infection in animal models and their potential limitations in comparison to human infection are shown in Table 1.

4.1 Guinea Pig Model

Guinea pigs were first examined by experimental infection with HeV but the pathology seen in guinea pigs differed significantly in several respects as compared to human cases as well as both naturally and experimentally infected horses such as little or no pulmonary edema being noted (Hooper et al. 1997b; Williamson et al. 2001). The clinical response to either HeV or NiV infection in guinea pigs has been reported to be often mild and variable, ranging from an inapparent infection following challenge to sudden death. Further, in several reports, only a proportion of animals presented with clinical signs, such as transient weight loss, depression, ataxia, lethargy, and twitching (Williamson et al. 2001; Wong et al. 2003). One study demonstrated that intraperitoneal (i.p.) inoculation with high doses of NiV (10^7 PFU) resulted only a transient fever with minor weight loss after 5–7 days and later recovered (Wong et al. 2003). Another study conducted a 50,000 $TCID_{50}$ challenge with NiV by i.p. administration, and here also only three of eight animals exhibited any signs of clinical disease which were only ruffled fur, mild behavioral changes, and ataxia (Middleton et al. 2007). The vascular tropism of HeV infection in guinea pigs is evident in many organs and when severe disease occurred death from widespread vascular disease was apparently the cause (Hooper et al. 1997b; Torres-Velez et al. 2008). In a model of HeV encephalitis in the guinea pig, it was reported that only a proportion of infected animals would develop encephalitis with virus observed in blood vessels and neurons (Williamson et al. 2001). The general consensus of the guinea pig is that it is not a suitable animal model for challenge and protection studies.

4.2 Pig Model

NiV infection of pigs is frequently asymptomatic, this was especially noted following natural infections but also later during experimental challenge of pigs with NiV by the ocular and oronasal route (Mohd Nor et al. 2000; Middleton et al. 2002;

Table 1 Comparison of disease in animal models of henipavirus infection with human infection

Animal model	Principal clinical and pathological findings in animal models in comparison to human infections	
	Hendra Virus[a]	Nipah Virus[b]
Guinea pig[c] HeV: Subcutaneous; 5×10^3; 3×10^4–5×10^4; or 5×10^4 $TCID_{50}$ NiV: Intranasal; 6×10^5 PFU, Intraperitoneal; 1×10^7 PFU Intraperitoneal; 5×10^4 $TCID_{50}$ Intraperitoneal; 2.5×10^5 PFU	Systemic vascular disease Endothelial syncytia evident; gross signs of pneumonia Virus in kidney and urine; female reproductive tract; placenta with fetal transmission Encephalitis with neuronal infection *Limitations* Clinical response frequently mild and variable- from unapparent to sudden death. Evidence of infection in only a proportion of animals Virus targets larger vessels in preference to capillaries. Viral antigen located preferentially in tunica media Lesser pulmonary involvement. Only a proportion of animals develop encephalitis Virus isolate, route of inoculum and dose effect outcome. Intranasal infection often not productive	Systemic vascular disease Vasculitis with endothelial syncytia Viral antigen in endothelium and vessels *Limitations* Clinical response frequently mild. Evidence of infection in only a small proportion of challenged animals Transient fever and weight loss with full recovery Encephalitis not observed No lesions in the lung Virus isolate, route of inoculum, and dose effect outcome Intranasal infection not productive
Pig[d,e] HeV: Oronasal; 6.6×10^6 PFU, Landrace pigs Intranasal; 2×10^7 PFU, Gottingen minipigs NiV: Subcutaneous; 5×10^4 $TCID_{50}$ Oral; 5×10^4 $TCID_{50}$	Respiratory and neurological disease signs (minipig); respiratory disease (Landrace pig), with fever Interstitial pneumonia Syncytia frequent Virus shed in nasal, oral, recta and ocular secretions *Limitations* Respiratory system is primary organ target with involvement of surrounding lymphoid tissue Rare or no CNS involvement No virus in urine	Respiratory and neurological disease signs Interstitial pneumonia Meningoencephalitis Virus shed in nasal, oral, pharyngeal, rectal, and ocular secretions *Limitations* Respiratory system is primary organ target Although evident, CNS involvement was rarer Syncytia rare No virus in urine

(continued)

Table 1 (continued)

Animal model	Principal clinical and pathological findings in animal models in comparison to human infections	
	Hendra Virus[a]	Nipah Virus[b]
Hamster[f,g,h] HeV: Intranasal; 1–10^5 PFU Intranasal; 1–10^5 $TCID_{50}$ Intraperitoneal; 1–10^5 $TCID_{50}$ NiV: Intraperitoneal; 10–10^6 PFU Intranasal; 1–10^4 PFU	Systemic vascular disease and parenchymal infection in a wide variety of organs Lung and brain severely affected with vascular and parenchymal lesions. Endothelial syncytia evident Neurological signs evident Virus shed in urine Route of inoculum and dose shown to effect outcome, intranasal high dose yielding greater acute respiratory disease while low dose yielding greater neurological disease with respiratory disease	Systemic vascular disease and parenchymal infection in a wide variety of organs Lung and brain severely affected with vascular and parenchymal lesions. Endothelial syncytia evident Neurological signs evident Virus shed in urine Route of inoculum and dose shown effect outcome, intranasal high dose yielding greater acute respiratory disease while low dose yielding greater neurological disease with respiratory disease
Cat[i] HeV: Subcutaneous; 5×10^3 $TCID_{50}$ Intranasally; 5×10^3 $TCID_{50}$ Oral; 5×10^3 $TCID_{50}$ NiV: Subcutaneous; 500–5×10^3 $TCID_{50}$ Oronasal; 5×10^4 $TCID_{50}$	Systemic vascular disease and parenchymal infection in a wide variety of organs. Fever and respiratory distress Severe effects in the lungs, interstitial pneumonia Endothelial syncytia evident Extensive inflammation in the respiratory epithelium associated with viral antigen Virus in kidney and in urine *Limitations* Virus detected in brain but rare and no evidence of encephalitis	Systemic vascular disease and parenchymal infection in a wide variety of organs. Fever and respiratory distress Severe effects in the lungs with edema and hemorrhage. Endothelial syncytia evident Extensive inflammation in the respiratory epithelium associated with viral antigen Virus shed in urine Virus in female reproductive tract; placenta with fetal transmission *Limitations* Virus detected in brain but encephalitis uncommon

(continued)

Table 1 (continued)

Animal model	Principal clinical and pathological findings in animal models in comparison to human infections	
	Hendra Virus[a]	Nipah Virus[b]
Ferret[j,g] HeV: Oronasal; 50–5 × 10^4 $TCID_{50}$ NiV: Oronasal; 500–5 × 10^4 $TCID_{50}$	Systemic vascular disease and parenchymal infection in a wide variety of organs. Fever and respiratory distress Lung and brain severely affected with vascular and parenchymal lesions Neurological signs evident Virus shed in pharyngeal and rectal secretions Syncytia prominent	Systemic vascular disease and parenchymal infection in a wide variety of organs. Fever and respiratory distress Lung and brain severely affected with vascular and parenchymal lesions; vascular fibrinoid necrosis Neurological signs evident Virus shed in pharyngeal and rectal secretions. Syncytia prominent
African Green Monkey[k,g] HeV: Intratracheal; 4 × 10^5 $TCID_{50}$ NiV: Intratracheal; oral; 8.1 × 10^4–1.3 x 10^6 PFU Intratracheal; 2.5 × 10^3–6.5 x 10^4 PFU	Systemic vascular disease and parenchymal infection in a wide variety of organs Lung and brain severely affected with vascular and parenchymal lesions, with edema and hemorrhage, with fibrinoid necrosis in the lung Neurological signs evident Endothelial syncytia prominent Potential rectal, throat and oronasal shedding	Systemic vascular disease and parenchymal infection in a wide variety of organs Lung and brain severely affected with vascular and parenchymal lesions, with edema and hemorrhage, with fibrinoid necrosis in the lung Neurological signs evident Endothelial syncytia prominent Potential rectal, throat and oronasal shedding
Squirrel monkey[l] NiV: Intravenous; 1 × 10^3 or 1 × 10^7 PFU Intranasal; 1 × 10^3 or 1 × 10^7 PFU	Experimental infections not reported	Systemic vascular disease and parenchymal infection in a variety of organs Respiratory and neurological signs evident *Limitations* Route of inoculum and dose effect outcome. Intranasal infection less productive, with subject recovery Clinical response can be mild even with intravenous virus challenge. Evidence of infection and clinical signs in only a proportion of animals Vasculitis and brain abnormalities were less evident No evidence of virus in urine

(continued)

Table 1 (continued)

Animal model	Principal clinical and pathological findings in animal models in comparison to human infections	
	Hendra Virus[a]	Nipah Virus[b]
Horse[m,n] HeV: Intravenous + intranasal-aerosol; infected spleen and lung tissue homogenate intravenous; 1×10^8 $TCID_{50}$ + intranasal-aerosol; 2×10^8 $TCID_{50}$ oronasal; 2×10^6 $TCID_{50}$	Systemic vascular disease and parenchymal infection in a wide variety of organs. Endothelial syncytia prominent Elevated heart rate, fever Respiratory disease, interstitial pneumonia Neurological signs, CNS vasculitis including meninges Virus shed in nasal, oral, pharyngeal and rectal secretions, and urine	Experimental infections not reported

Summary of clinical and pathological outcomes of experimental animal infection models of Hendra virus and Nipah virus; and noted potential model limitations in comparison to natural cases of human infection. *PFU* plaque forming unit, *$TCID_{50}$* 50% tissue culture infectious dose

[a] Human infection and pathogenesis (Hanna et al., 2006; O'Sullivan et al., 1997; Playford et al., 2010; Selvey et al., 1995; Wong et al., 2009)

[b] Human infection and pathogenesis (Chadha et al., 2006; Goh et al., 2000; Hsu et al., 2004; Wong et al., 2002)

[c] HeV-infected guinea pig (Hooper et al., 1997b; Torres-Velez et al., 2008; Westbury et al., 1995; Williamson et al., 2000, 2001); NiV-infected guinea pig (Middleton et al., 2007; Wong et al., 2003)

[d] HeV-infected pig (Li et al., 2010); NiV-infected Pig (Berhane et al., 2008; Middleton et al., 2002; Weingartl et al., 2005, 2006)

[e] Model developed for porcine and/or livestock vaccine development

[f] HeV-infected hamster (Guillaume et al., 2009; Rockx et al., 2011); NiV-infected hamster (Georges-Courbot et al., 2006; Guillaume et al., 2004; Guillaume et al., 2006; Rockx et al., 2011; Wong et al., 2003)

[g] Pulmonary and neurological pathology is similar to human pathology

[h] LD50 for NiV and HeV in hamsters is 270 PFU (Wong et al., 2003) vs 12 PFU (Guillaume et al., 2009), respectively

[i] HeV-infected cat (Hooper et al., 1997b; Middleton et al., 2002; Westbury et al., 1996); NiV-infected cat (McEachern et al., 2008; Mungall et al., 2006; Mungall et al., 2007)

[j] HeV-infected ferret (Pallister et al., 2011); NiV-infected ferret (Bossart et al., 2009; Pallister et al., 2009)

[k] HeV-infected African green monkey (Bossart et al., 2011; Rockx et al., 2010); NiV-infected African green monkey (Geisbert et al., 2010)

[l] NiV-infected squirrel monkey (Marianneau et al., 2010)

[m] HeV-infected horse (Hooper et al., 1997a; Marsh et al., 2011; Murray et al., 1995a, 1995b; Williamson et al., 1998)

[n] Model developed for equine and/or livestock vaccine development

Weingartl et al. 2006; Berhane et al. 2008; also refer to the chapter on Henipaviruses in their natural host (Middleton and Weingartl, this issue). When symptoms were apparent, they varied in relation to the age of the pig, with older animals predominantly exhibiting a neurological disease syndrome in contrast to younger pigs which primarily presented with a respiratory syndrome. The virus manifests respiratory and neurologic tropisms in both asymptomatic and clinical infections (Middleton et al. 2002; Weingartl et al. 2005). In pigs, neurological disease manifested as trembling, twitches, muscle spasms, and uncoordinated gait (Mohd Nor et al. 2000). NiV challenge of younger pigs by the ocular and oronasal routes revealed that virus replication occurs in the oropharnyx and then spreads sequentially to the upper respiratory tract and submandibular lymph nodes, followed by replication in the lower respiratory tract, and additional lymphoid tissues (Weingartl et al. 2005). Virus replication and presence of detectable viral antigen can be widespread and accompanied by frequent syncytial cell formation, which is particularly evident in clinically ill animals, resulting from virus tropism and replication in endothelial and smooth muscle cells of medium to large veins and arteries within the respiratory system, CNS, and lymphoid tissues (Hooper et al. 2001; Middleton et al. 2002; Weingartl et al. 2005). NiV has also been reported to invade the CNS by route of the cranial nerves and crossing the blood–brain barrier (Weingartl et al. 2005). Infectious NiV could also be recovered from a wide variety of tissues in the infected pig, including tonsil, nasal, and throat swabs and lung, but only infrequently recovered from the urine of shedding animals (Daniels et al. 2001; Middleton et al. 2002; Weingartl et al. 2005).

There has been one report of HeV infection of Landrace and Gottingen minipig breeds. Oronasal or nasal inoculations were both demonstrated to cause infection (Li et al. 2010) with clinical signs mainly respiratory, but with possible neurological involvement seen only in the Gottingen minipig. Virus was detected mainly in tissues from respiratory and lymphoid systems in both Landrace and Gottingen pigs and histological and immunohistological analysis revealed frequent syncytia. 5-week-old Landrace pigs appeared to produce a more severe clinical disease in comparison to NiV infection (Weingartl et al. 2005, 2006), but in these experiments the inoculating dose of HeV was significantly higher compared to NiV ($\sim 10^7$ compared to 10^5). Importantly, virus could be isolated from nasal, oral, and rectal swabs, indicating the possible routes for virus shedding and risk of transmission.

4.3 Horse Model

Experimental HeV infection of horses by either intravenous (i.v.) or intranasal (i.n.) routes is almost uniformly fatal with death or euthanasia occurring $\sim$5–10 days following virus challenge. Experimental NiV infection of horses has not been carried out. Horses have been identified to have been naturally infected, and some do survive and some can experience asymptomatic infection with seroconversion (Murray et al. 1995a, b; Williamson et al. 1998). HeV infection in horses yields a predominantly respiratory syndrome initially with elevated heart rate and fever which can rapidly progress to more severe respiratory disease. In field cases, the airways are often filled with a blood-tinged frothy exudate but this has not been experimentally reproduced, and could be a result of the infecting dose of virus. In addition, signs of neurological disease can occur but appear less frequent and have been reported in both terminally ill horses as well as in horses that recovered from HeV infection (Rogers et al. 1996; Williamson et al. 1998).

The first experimental reproductions of HeV disease were carried out using a mixture of virus-infected lung and spleen homogenate that was prepared from two naturally infected horse field cases and used to inoculate horses i.v. and i.n. In addition, other horses were inoculated i.v. with a high dose of cell culture derived virus, (1×10^7 $TCID_{50}$ + i.n.-aerosol; 2×10^7 $TCID_{50}$) (Murray et al. 1995a, b). HeV infection was characterized by pulmonary edema and congestion, hemorrhage, thrombosis of capillaries, necrosis, and syncytial cells in the endothelium of pulmonary vessels. Additional HeV challenge experiments in horses have detailed the presence of viral antigen in endothelial cells in a wide variety of organs including lungs, lymph nodes, kidneys, spleen, bladder, and meninges. Virus can be recovered from a number of internal organs, including lung, and from saliva and urine (Murray et al. 1995a, b; Hooper et al. 1997a; Williamson et al. 1998; Marsh et al. 2011). The development of an experimental model of HeV infection in horses has been important for understanding the pathogenesis of the virus and the associated risk of transmission to humans, and will also be critical for the testing of vaccines against HeV infection of horses.

4.4 Cat Model

Following either NiV or HeV infection in the cat by either s.c. or oronasal inoculation, the first clinical signs are typically observed between 4 and 8 days and include depression, fever, and elevated rates of respiration (Westbury et al. 1995, 1996; Hooper et al. 1997b; Mungall et al. 2006). Most infected cats die or require euthanasia within 24 h after the appearance of respiratory distress. HeV or NiV infection and the resulting disease in cats closely resembles the respiratory disease presentation that is seen in naturally occurring HeV-infected horses, often with copious frothy sanguineous fluid in the bronchi and hemorrhage or congestion of the tracheal epithelium (Hooper et al. 1997b). Virus replication and resulting systemic vasculitis is present in both arteries and arterioles and syncytia are observed in the endothelia especially in the lungs, but also in gastrointestinal, spleen, and lymphoid organs.

Cats inoculated with HeV also shed virus in urine and infected cats in similar cages were able to transmit infection to one of three contact horses. A notable difference between NiV and HeV infection of cats is the extensive degree to which NiV, but not HeV, infects the respiratory epithelium (Middleton et al. 2002).

In addition, NiV can transplacentally infect and replicate in fetal tissues with high levels of recoverable virus from the placenta and uterine fluid (Mungall et al. 2007). Taken together, despite some limitations, particularly that henipavirus encephalitis in the cat has not been experimentally reproduced these studies indicated that the cat was a viable and reproducible animal model for lethal henipavirus infection and disease. The cat model has been successfully used in challenge and protection experiments in the evaluation of a subunit vaccine for NiV (McEachern et al. 2008).

4.5 Golden Syrian Hamster Model

Studies in hamsters have yielded different results following NiV infection. One study showed that pathological lesions were most severe and extensive in the hamster brain (Wong et al. 2003). In this study hamsters succumbed 5–9 days after i.p. administration of 100–10,000 PFU and 24 h after the appearance of tremors and limb paralysis. In this same study, hamsters inoculated i.n. with doses as high as 10^3–10^6 PFU succumbed between 9 and 15 days after NiV challenge, displaying progressive deterioration with limb paralysis, lethargy, limb twitching, and breathing difficulties. Vascular pathology was observed in a range of organs, including brain, lung, liver, kidney, and heart, and viral antigen and genome were found in endothelial cells. The vascular and parenchyma lesions were consistent with CNS-mediated clinical signs. A more recent study reported that hamsters inoculated i.n. or i.p. with 10^5 $TCID_{50}$ succumbed within 5 days after exposure to NiV while hamsters inoculated i.n. or i.p. with 10^2 $TCID_{50}$ died by day 12 after NiV challenge (Rockx et al. 2011). This study also reported that disease in hamsters exposed to higher doses of NiV resulted in acute respiratory distress (ARDS) whereas infection of hamsters with lower doses of NiV resulted in the development of neurological signs and more systemic spread of the virus through involvement of the endothelium. It is unclear why one study using doses of 10^3 to 10^6 PFU caused a disease dominated by neurologic signs (Wong et al. 2003) where a similar study using a similar dose (10^5 $TCID_{50}$) caused a disease dominated by respiratory signs (Rockx et al. 2011). The NiV hamster model has been successfully used in experiments to evaluate several vaccines and postexposure treatments (Guillaume et al. 2004, 2006; Georges-Courbot et al. 2006; Freiberg et al. 2010).

HeV infection of hamsters appears to more closely resemble the pathology seen in acute human cases, including both respiratory and brain pathology (Guillaume et al. 2009). HeV-induced pathology in the hamster consisted of endothelial infection and vasculitis with thrombosis and microinfarction, with evidence of direct parenchymal cell infection, notably in the CNS. As was reported for NiV disease in hamsters exposed to higher doses NiV (Rockx et al. 2011), higher doses of HeV (10^5 $TCID_{50}$) resulted in ARDS whereas infection of hamsters with lower doses of HeV (10^2 $TCID_{50}$) resulted in the development of neurological signs (Rockx et al. 2011). Hamsters have been used in a few studies to evaluate the potential of antiviral therapies against HeV (Guillaume et al. 2009; Freiberg et al. 2010).

Of additional importance regarding the use of hamsters as models for henipavirus infection are findings suggesting that the sensitivity to HeV infection may decrease with the increasing age of the animals (Guillaume et al. 2009).

4.6 Ferret Model

Ferret models of NiV and HeV infection and pathogenesis have been recently developed and characterized (Bossart et al. 2009; Pallister et al. 2009, 2011). The henipavirus infected ferret reveals both severe respiratory and neurological disease, along with generalized vasculitis following oral-nasal challenge with NiV with low doses (500 $TCID_{50}$) within 6–10 days post-infection (Bossart et al. 2009; Pallister et al. 2009). Disease signs in NiV-infected ferrets are various combinations of severe depression, cough, serous nasal discharge, dyspnea, s.c edema of the head, cutaneous ecchymoses, and obtundation along with tremor and hind limb paresis depending on the challenge dose. Pathological findings in the ferret included vascular fibrinoid necrosis in multiple organs, necrotizing alveolitis, and syncytia of endothelium and alveolar epithelium. Histopathological lesions included severe focal necrotizing alveolitis, vasculitis, degeneration of glomerular tufts, and focal necrosis in a wide-range of other tissues. Significant amounts of viral antigen are present in blood vessel walls and endothelial syncytia are frequently present. Viral antigen was also present in the brain parenchyma including neurons and infectious NiV was isolated from multiple organs including the brain. In total, the NiV-mediated disease process observed in the ferret accurately reflects the disease manifestation observed in NiV-infected humans.

Ferrets infected by an oral-nasal challenge with doses of HeV ranging from 50 to 50,000 $TCID_{50}$ succumbed 6–9 days after exposure (Pallister et al. 2011). Clinical signs in HeV-infected ferrets are essentially identical to those reported for NiV-infected ferrets and included severe depression and generalized tremors. Histological examination showed systemic vasculitis, splenitis, and bronchiolalveolitis with syncytial cell formation.

4.7 Nonhuman Primate Model

Nonhuman primate models of henipavirus infection have been developed using the African green monkey (AGM) which yields a consistent lethal infection and disease with low dose challenge; $\sim 2 \times 10^4$ PFU (NiV) (Geisbert et al. 2010) or 4×10^5 $TCID_{50}$ of HeV (Rockx et al. 2010). Virus challenge by intratracheal (i.t.) inoculation with either NiV or HeV of AGMs, results in a rapid spread of virus within 3–4 days and the establishment of infection in multiple organ systems. Both NiV- and HeV-infected monkeys developed a severe ARDS-like disease, associated with copious amounts of sanguinous fluid and froth. The lungs are

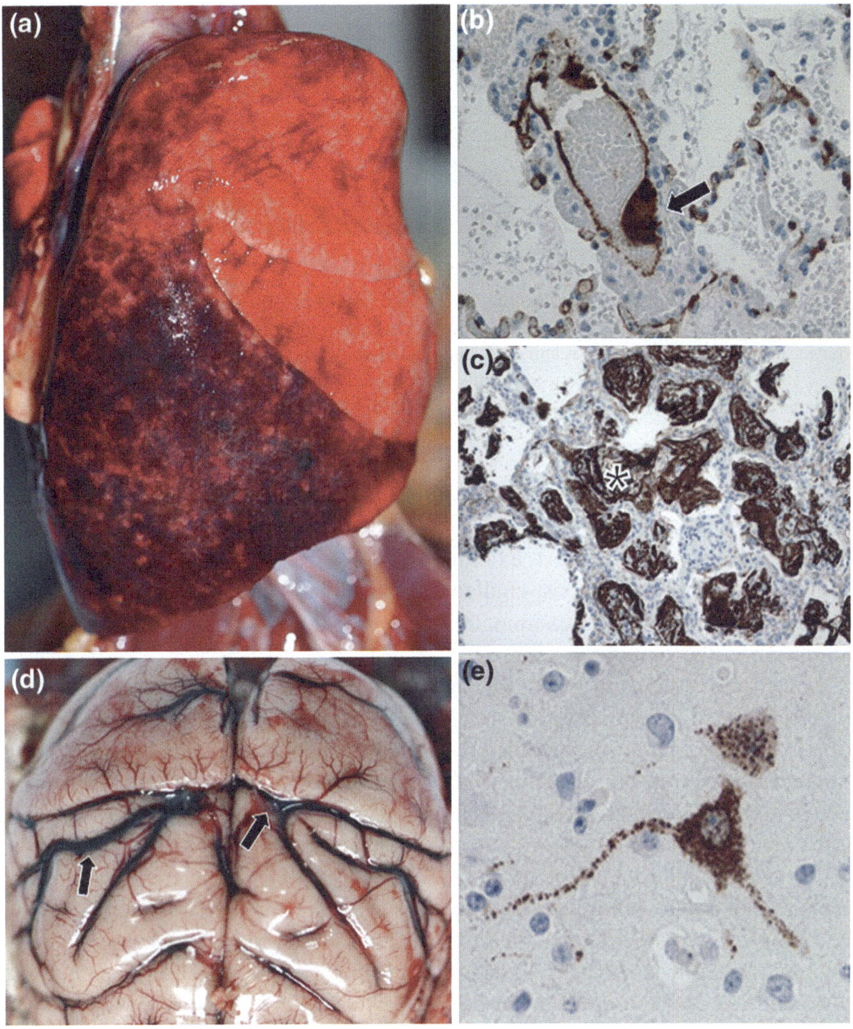

Fig. 1 Henipavirus pathology in the African green monkey. **a** Severely enlarged lungs; multifocal areas of congestion and hemorrhage of NiV-infected animal; **b** localization of NiV by immunohistochemical stain within a lung blood vessel with endothelial syncytia (*arrow*) and scattered foci of immunopositive cells (*brown*) abundant in alveolar septae (original magnification, 40×); **c** right diaphragmatic lobe of lung of a NiV-infected animal by immunohistochemical stain showing abundance of polymerized fibrin (*asterisk*) in and around alveolar spaces (original magnification, 40×); **d** congestion of the brain of a HeV-infected animal (*black arrows*); **e** detection of HeV antigen in neuron cell body and axon (original magnification, 40×). Panels **b** and **c** from (Geisbert et al. 2010); panel **d** from (Rockx et al. 2010)

consistently enlarged with multifocal areas of congestion and hemorrhage (Fig. 1a) and immunohistochemical and histopathological examination revealed significant amounts of NiV antigen (Fig. 1b) and polymerized fibrin (Fig. 1c).

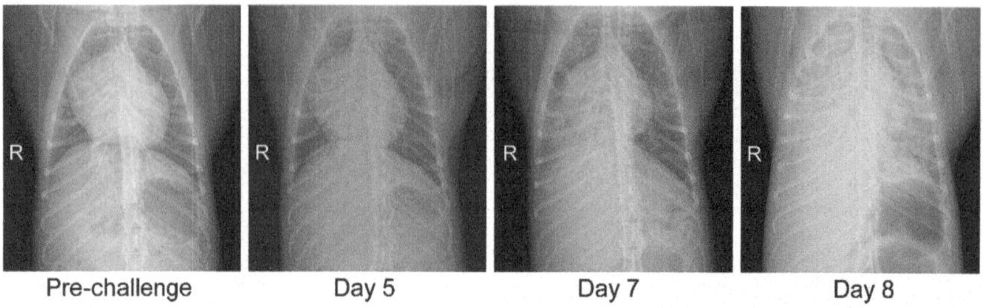

Fig. 2 Radiological and gross pathological changes in lungs of HeV-infected African green monkeys. Radiological progression of respiratory disease in an animal euthanized on day 8 due after i.t. exposure due to severe respiratory distress. First evidence of congestion was observed at day 7 post-infection and infection rapidly progressed to diffuse interstitial infiltrates and pulmonary consolidation by day 8 post-infection. R, right side. Figure from (Rockx et al. 2010)

Evidence of endothelial syncytia are prominent in most of the tissues and vasculitis was systemic. Viral antigen was also present in endothelial and arterial smooth muscle cells in most examined tissues. Respiratory disease typically manifested within 7 days post-challenge with either NiV or HeV following i.t. inoculation by radiological examination, and progressed to severe congestion and infiltration in the lung fields (Fig. 2).

At the time of death in the AGM, virus could be found in virtually every organ system sampled. Immunohistochemical analysis revealed the presence of viral antigen, predominantly in endothelial cells and smooth muscle cells, along with associated pathology. In addition, and of importance to the similarity of disease seen in humans, most infected AGMs have shown evidence of henipavirus-induced neurologic disease, with severe congestion (Fig. 1d) and evidence of meningeal hemorrhaging and edema. In the CNS, viral antigen was detected in endothelial cells in brain along with infection of neurons (Fig. 1e) and typically widespread in the brainstem. As was shown in the hamster models for henipavirus infection challenge dose appears to have some influence on the disease course and pathology. In general, AGM exposed to lower doses of NiV ($<10^5$ PFU) survive longer than animals exposed to higher doses of NiV ($>10^5$ PFU) with animals that succumb earlier showing more respiratory signs of disease and animals surviving longer showing more neurologic signs of disease (Geisbert and Broder Unpublished).

The AGM HeV model has been used in two challenge and protection experiments: (1) an evaluation of ribavirin as a prophylactic and postexposure treatment (Rockx et al. 2010) and (2) an evaluation of the recombinant human monoclonal antibody m102.4 as a postexposure treatment (Bossart et al. 2011).

NiV infection of squirrel monkeys has also been examined (Marianneau et al. 2010); however, only some animals demonstrated limited similarities to NiV pathogenesis in humans and only half of the number of challenged animals exhibited any clinical signs of disease with most challenged animals remaining well, even after i.n. or i.v. delivery of doses as high as 10^7 PFU of NiV.

In contrast to the guinea pig, feline, hamster, and squirrel monkey models of henipavirus infection, both neurological disease and severe respiratory pathology along with generalized vasculitis all occur in AGMs infected with either NiV or HeV, and the AGM model provides an accurate reflection of what is observed in henipavirus-infected humans.

5 Routes of Experimental Exposure and Dose

Early studies to develop animal models evaluated a number of challenge routes including intradermal, s.c., i.n., oral, i.t., i.v., and i.p., and in some cases various combinations of these routes. As studies have progressed, most animal models now focus on either the presumed natural oronasal routes or explore oronasal versus i.p. or s.c. exposures in order to examine pathogenic outcomes and disease processes (Rockx et al. 2011).

In addition, there has also been a limited number of animal transmission or in-contact studies in horses, bats, pigs, cats, and hamsters (Westbury et al. 1996; Williamson et al. 1998; Middleton et al. 2002; Wong et al. 2003; de Wit et al. 2012). Horse–horse, bat–horse, bat–bat, and cat–horse transmission studies have nearly all been unsuccessful under the limited experimental conditions and attempts to date. Notably, even horse-to-horse transmission of HeV has not been convincingly demonstrated (Williamson et al. 1998); however, good evidence of cat–cat transmission of HeV has been show between infected and in-contact cats within the same confines (Westbury et al. 1996). The conclusions from these studies indicate that HeV is not highly infectious, even between horses, and indeed HeV is rarely found in the bronchi or bronchioles of infected horses, suggesting that aerosol transmission to either humans or horses maybe less likely (Hooper et al. 1997a).

In regards to natural routes of virus transmission, the importance of pregnancy and fetal materials in disease spread was first hypothesized after the discovery that the index case of the 1994 HeV outbreak was a pregnant mare (Murray et al. 1995a, b) and vertical (in utero) transmission of HeV was later experimentally confirmed in guinea pigs and bats (Williamson et al. 2000). More recently, in utero transmission of NiV in cats has also been reported with evidence of a high level of virus replication in many tissues of a pregnant adult cat and in fetal tissues suggesting both vertical and horizontal transmission of this virus is possible (Mungall et al. 2007).

Guinea pigs and pigs are somewhat resistant to infection and disease caused by HeV and NiV, respectively, when the viruses are administered by the oronasal compared with the s.c. route (Williamson et al. 2001; Middleton et al. 2002). HeV has only been used to challenge pigs oronasally (Landrace pigs; 6.6×10^7 PFU) or i.n. (Gottingen minipigs; 2.0×10^7 PFU) (Li et al. 2010). In contrast, HeV and NiV appear to be equally infectious for cats following either parenteral (5,000 $TCID_{50}$ virus) or oronasal (50,000 $TCID_{50}$ virus) administration (Westbury et al. 1996; Hooper et al. 1997b; Middleton et al. 2002; Mungall et al. 2006;

McEachern et al. 2008). In general, higher doses in a number of animal species, particularly monkeys (Geisbert and Broder, unpublished) and hamsters (Rockx et al. 2011) has shown more rapid onset of fatal disease.

6 Epilogue

In this chapter, we have outlined what is known about the pathology and pathogenesis of human henipavirus infection and reviewed studies describing the development and use of animal models. Neither HeV or NiV productively infects or causes disease in typical small animal models, including mice or rabbits (reviewed in Bossart et al. (2007) and Bishop and Broder (2008)). Among other smaller animal models explored, the response to either HeV or NiV in guinea pigs was reported to be highly variable. Although a generalized vascular disease was observed with HeV in guinea pigs there was little or no pulmonary edema (Hooper et al. 1997b; Williamson et al. 2001). Infection of guinea pigs with high doses of NiV (10^7 PFU) showed only transient clinical signs followed by recovery (Wong et al. 2003) and another study using a high dose of virus by intraperitoneal administration produced disease in less than half of challenged animals (Middleton et al. 2007). There have also been some potential limitations noted in two other well-characterized animal models, the cat and hamster (Westbury et al. 1996; Hooper et al. 1997b; Wong et al. 2003; Mungall et al. 2006, 2007), particularly no overt CNS pathogenesis or respiratory disease, respectively. Of the small animal models evaluated to date, only the recently described ferret models of HeV and NiV infection exhibited both severe respiratory and neurological disease and generalized vasculitis in which the underlying pathology closely resembled henipavirus-mediated disease seen in humans (Bossart et al. 2009). Not surprisingly, the animal model which appears to most faithfully reproduce human henipavirus infection is the AGM (Geisbert et al. 2010; Rockx et al. 2010; Bossart et al. 2011).

References

Anonymous (1999) Update: outbreak of Nipah virus—Malaysia and Singapore. MMWR Morb Mortal Wkly Rep 48:335–337

Anonymous (2009) Hendra virus, human, equine—Australia (04): (Queensland) Fatal. ProMed Mail Archive Number: 20090903.3098

Anonymous (2011a) Hendra virus, equine—Australia (21): (Queensland) Canine. ProMed Mail Archive Number: 20110802.2324

Anonymous (2011b) Hendra virus, equine—Australia (28): (Queensland, New South Wales). ProMed Mail Archive Number: 20111013.3061

Arankalle VA, Bandyopadhyay BT, Ramdasi AY, Jadi R, Patil DR, Rahman M, Majumdar M, Banerjee PS, Hati AK, Goswami RP, Neogi DK, Mishra AC (2011) Genomic characterization of Nipah virus, West Bengal, India. Emerg Infect Dis 17:907–909

Berhane Y, Weingartl HM, Lopez J, Neufeld J, Czub S, Embury-Hyatt C, Goolia M, Copps J, Czub M (2008) Bacterial infections in pigs experimentally infected with Nipah virus. Transbound Emerg Dis 55:165–174

Bishop KA, Broder CC (2008) Hendra and Nipah: lethal zoonotic paramyxoviruses. In: Scheld WM, Hammer SM, Hughes JM (eds) Emerging infections, vol 8. American Society for Microbiology, Washington, pp 155–187

Bossart KN, Bingham J, Middleton D (2007) Targeted strategies for Henipavirus therapeutics. Open Virol J 1:14–25

Bossart KN, Zhu Z, Middleton D, Klippel J, Crameri G, Bingham J, McEachern JA, Green D, Hancock TJ, Chan YP, Hickey AC, Dimitrov DS, Wang LF, Broder CC (2009) A neutralizing human monoclonal antibody protects against lethal disease in a new ferret model of acute nipah virus infection. PLoS Pathog 5:e1000642

Bossart KN, Geisbert TW, Feldmann H, Zhu Z, Feldmann F, Geisbert JB, Yan L, Feng YR, Brining D, Scott D, Wang Y, Dimitrov AS, Callison J, Chan YP, Hickey AC, Dimitrov DS, Broder CC, Rockx B (2011) A neutralizing human monoclonal antibody protects african green monkeys from hendra virus challenge. Sci Transl Med 3: 105ra103

Chadha MS, Comer JA, Lowe L, Rota PA, Rollin PE, Bellini WJ, Ksiazek TG, Mishra A (2006) Nipah virus-associated encephalitis outbreak, Siliguri, India. Emerg Infect Dis 12:235–240

Chong HT, Tan CT (2003) Relapsed and late-onset Nipah encephalitis, a report of three cases. Neurol J Southeast Asia 8:109–112

Chua KB (2003) Nipah virus outbreak in Malaysia. J Clin Virol 26:265–275

Chua KB, Goh KJ, Wong KT, Kamarulzaman A, Tan PS, Ksiazek TG, Zaki SR, Paul G, Lam SK, Tan CT (1999) Fatal encephalitis due to Nipah virus among pig-farmers in Malaysia. Lancet 354:1257–1259

Chua KB, Bellini WJ, Rota PA, Harcourt BH, Tamin A, Lam SK, Ksiazek TG, Rollin PE, Zaki SR, Shieh W, Goldsmith CS, Gubler DJ, Roehrig JT, Eaton B, Gould AR, Olson J, Field H, Daniels P, Ling AE, Peters CJ, Anderson LJ, Mahy BW (2000) Nipah virus: a recently emergent deadly paramyxovirus. Science 288:1432–1435

Chua KB, Koh CL, Hooi PS, Wee KF, Khong JH, Chua BH, Chan YP, Lim ME, Lam SK (2002) Isolation of Nipah virus from Malaysian Island flying-foxes. Microbes Infect 4:145–151

Daniels P, Ksiazek T, Eaton BT (2001) Laboratory diagnosis of Nipah and Hendra virus infections. Microbes Infect 3:289–295

de Wit E, Bushmaker T, Scott D, Feldmann H, Munster V (2012) Nipah virus transmission in a hamster model. PLoS Negl Trop Dis 5:e1432

Eaton BT, Broder CC, Middleton D, Wang LF (2006) Hendra and Nipah viruses: different and dangerous. Nat Rev Microbiol 4:23–35

Field H, Young P, Yob JM, Mills J, Hall L, Mackenzie J (2001) The natural history of Hendra and Nipah viruses. Microbes Infect 3:307–314

Freiberg AN, Worthy MN, Lee B, Holbrook MR (2010) Combined chloroquine and ribavirin treatment does not prevent death in a hamster model of Nipah and Hendra virus infection. J Gen Virol 91:765–772

Geisbert TW, Daddario-DiCaprio KM, Hickey AC, Smith MA, Chan YP, Wang LF, Mattapallil JJ, Geisbert JB, Bossart KN, Broder CC (2010) Development of an acute and highly pathogenic nonhuman primate model of Nipah virus infection. PLoS ONE 5:e10690

Georges-Courbot MC, Contamin H, Faure C, Loth P, Baize S, Leyssen P, Neyts J, Deubel V (2006) Poly(I)-poly(C12U) but not ribavirin prevents death in a hamster model of Nipah virus infection. Antimicrob Agents Chemother 50:1768–1772

Goh KJ, Tan CT, Chew NK, Tan PS, Kamarulzaman A, Sarji SA, Wong KT, Abdullah BJ, Chua KB, Lam SK (2000) Clinical features of Nipah virus encephalitis among pig farmers in Malaysia. N Engl J Med 342:1229–1235

Guillaume V, Contamin H, Loth P, Georges-Courbot MC, Lefeuvre A, Marianneau P, Chua KB, Lam SK, Buckland R, Deubel V, Wild TF (2004) Nipah virus: vaccination and passive protection studies in a hamster model. J Virol 78:834–840

Guillaume V, Contamin H, Loth P, Grosjean I, Courbot MC, Deubel V, Buckland R, Wild TF (2006) Antibody prophylaxis and therapy against Nipah virus infection in hamsters. J Virol 80:1972–1978

Guillaume V, Wong KT, Looi RY, Georges-Courbot MC, Barrot L, Buckland R, Wild TF, Horvat B (2009) Acute Hendra virus infection: analysis of the pathogenesis and passive antibody protection in the hamster model. Virology 387:459–465

Gurley ES, Montgomery JM, Hossain MJ, Bell M, Azad AK, Islam MR, Molla MA, Carroll DS, Ksiazek TG, Rota PA, Lowe L, Comer JA, Rollin P, Czub M, Grolla A, Feldmann H, Luby SP, Woodward JL, Breiman RF (2007) Person-to-person transmission of Nipah virus in a Bangladeshi community. Emerg Infect Dis 13:1031–1037

Halpin K, Young PL, Field HE, Mackenzie JS (2000) Isolation of Hendra virus from pteropid bats: a natural reservoir of Hendra virus. J Gen Virol 81:1927–1932

Halpin K, Hyatt AD, Fogarty R, Middleton D, Bingham J, Epstein JH, Rahman SA, Hughes T, Smith C, Field HE, Daszak P (2011) Pteropid bats are confirmed as the reservoir hosts of henipaviruses: a comprehensive experimental study of virus transmission. Am J Trop Med Hyg 85:946–951

Hanna JN, McBride WJ, Brookes DL, Shield J, Taylor CT, Smith IL, Craig SB, Smith GA (2006) Hendra virus infection in a veterinarian. Med J Aust 185:562–564

Harit AK, Ichhpujani RL, Gupta S, Gill KS, Lal S, Ganguly NK, Agarwal SP (2006) Nipah/Hendra virus outbreak in Siliguri, West Bengal, India in 2001. Indian J Med Res 123:553–560

Hayman DT, Suu-Ire R, Breed AC, McEachern JA, Wang L, Wood JL, Cunningham AA (2008) Evidence of henipavirus infection in West African fruit bats. PLoS ONE 3:e2739

Homaira N, Rahman M, Hossain MJ, Epstein JH, Sultana R, Khan MS, Podder G, Nahar K, Ahmed B, Gurley ES, Daszak P, Lipkin WI, Rollin PE, Comer JA, Ksiazek TG, Luby SP (2010) Nipah virus outbreak with person-to-person transmission in a district of Bangladesh, 2007. Epidemiol Infect 138:1630–1636

Hooper PT, Gould AR, Russell GM, Kattenbelt JA, Mitchell G (1996) The retrospective diagnosis of a second outbreak of equine morbillivirus infection. Aust Vet J 74:244–245

Hooper PT, Ketterer PJ, Hyatt AD, Russell GM (1997a) Lesions of experimental equine morbillivirus pneumonia in horses. Vet Pathol 34:312–322

Hooper PT, Westbury HA, Russell GM (1997b) The lesions of experimental equine morbillivirus disease in cats and guinea pigs. Vet Pathol 34:323–329

Hooper P, Zaki S, Daniels P, Middleton D (2001) Comparative pathology of the diseases caused by Hendra and Nipah viruses. Microbes Infect 3:315–322

Hsu VP, Hossain MJ, Parashar UD, Ali MM, Ksiazek TG, Kuzmin I, Niezgoda M, Rupprecht C, Bresee J, Breiman RF (2004) Nipah virus encephalitis reemergence, Bangladesh. Emerg Infect Dis 10:2082–2087

Li Y, Wang J, Hickey AC, Zhang Y, Wu Y, Zhang H, Yuan J, Han Z, McEachern J, Broder CC, Wang LF, Shi Z (2008) Antibodies to Nipah or Nipah-like viruses in bats, China. Emerg Infect Dis 14:1974–1976

Li M, Embury-Hyatt C, Weingartl HM (2010) Experimental inoculation study indicates swine as a potential host for Hendra virus. Vet Res 41:33

Luby SP, Rahman M, Hossain MJ, Blum LS, Husain MM, Gurley E, Khan R, Ahmed BN, Rahman S, Nahar N, Kenah E, Comer JA, Ksiazek TG (2006) Foodborne transmission of Nipah virus, Bangladesh. Emerg Infect Dis 12:1888–1894

Luby SP, Gurley ES, Hossain MJ (2009a) Transmission of human infection with Nipah virus. Clin Infect Dis 49:1743–1748

Luby SP, Hossain MJ, Gurley ES, Ahmed BN, Banu S, Khan SU, Homaira N, Rota PA, Rollin PE, Comer JA, Kenah E, Ksiazek TG, Rahman M (2009b) Recurrent zoonotic transmission of Nipah virus into humans, Bangladesh, 2001–2007. Emerg Infect Dis 15:1229–1235

Marianneau P, Guillaume V, Wong T, Badmanathan M, Looi RY, Murri S, Loth P, Tordo N, Wild F, Horvat B, Contamin H (2010) Experimental infection of squirrel monkeys with Nipah virus. Emerg Infect Dis 16:507–510

Marsh GA, Haining J, Hancock TJ, Robinson R, Foord A, Barr JA, Riddell S, Heine HG, White JR, Crameri G, Field HE, Wang LF, Middleton D (2011) Experimental infection of horses with Hendra virus/Australia/Horse/2008/Redlands. Emerg Infect Dis 17:2232–2238

McEachern JA, Bingham J, Crameri G, Green DJ, Hancock TJ, Middleton D, Feng YR, Broder CC, Wang LF, Bossart KN (2008) A recombinant subunit vaccine formulation protects against lethal Nipah virus challenge in cats. Vaccine 26:3842–3852

Middleton DJ, Westbury HA, Morrissy CJ, van der Heide BM, Russell GM, Braun MA, Hyatt AD (2002) Experimental Nipah virus infection in pigs and cats. J Comp Pathol 126:124–136

Middleton DJ, Morrissy CJ, van der Heide BM, Russell GM, Braun MA, Westbury HA, Halpin K, Daniels PW (2007) Experimental Nipah virus infection in pteropid bats (*Pteropus poliocephalus*). J Comp Pathol 136:266–272

Mills JN, Alim AN, Bunning ML, Lee OB, Wagoner KD, Amman BR, Stockton PC, Ksiazek TG (2009) Nipah virus infection in dogs, Malaysia, 1999. Emerg Infect Dis 15:950–952

Mohd Nor MN, Gan CH, Ong BL (2000) Nipah virus infection of pigs in peninsular Malaysia. Rev Sci Tech 19:160–165

Mungall BA, Middleton D, Crameri G, Bingham J, Halpin K, Russell G, Green D, McEachern J, Pritchard LI, Eaton BT, Wang LF, Bossart KN, Broder CC (2006) Feline model of acute nipah virus infection and protection with a soluble glycoprotein-based subunit vaccine. J Virol 80:12293–12302

Mungall BA, Middleton D, Crameri G, Halpin K, Bingham J, Eaton BT, Broder CC (2007) Vertical transmission and fetal replication of Nipah virus in an experimentally infected cat. J Infect Dis 196:812–816

Murray K, Rogers R, Selvey L, Selleck P, Hyatt A, Gould A, Gleeson L, Hooper P, Westbury H (1995a) A novel morbillivirus pneumonia of horses and its transmission to humans. Emerg Infect Dis 1:31–33

Murray K, Selleck P, Hooper P, Hyatt A, Gould A, Gleeson L, Westbury H, Hiley L, Selvey L, Rodwell B et al (1995b) A morbillivirus that caused fatal disease in horses and humans. Science 268:94–97

O'Sullivan JD, Allworth AM, Paterson DL, Snow TM, Boots R, Gleeson LJ, Gould AR, Hyatt AD, Bradfield J (1997) Fatal encephalitis due to novel paramyxovirus transmitted from horses. Lancet 349:93–95

Pallister J, Middleton D, Crameri G, Yamada M, Klein R, Hancock TJ, Foord A, Shiell B, Michalski W, Broder CC, Wang LF (2009) Chloroquine administration does not prevent Nipah virus infection and disease in ferrets. J Virol 83:11979–11982

Pallister J, Middleton D, Wang LF, Klein R, Haining J, Robinson R, Yamada M, White J, Payne J, Feng YR, Chan YP, Broder CC (2011) A recombinant Hendra virus G glycoprotein-based subunit vaccine protects ferrets from lethal Hendra virus challenge. Vaccine 29:5623–5630

Paton NI, Leo YS, Zaki SR, Auchus AP, Lee KE, Ling AE, Chew SK, Ang B, Rollin PE, Umapathi T, Sng I, Lee CC, Lim E, Ksiazek TG (1999) Outbreak of Nipah-virus infection among abattoir workers in Singapore. Lancet 354:1253–1256

Playford EG, McCall B, Smith G, Slinko V, Allen G, Smith I, Moore F, Taylor C, Kung YH, Field H (2010) Human Hendra virus encephalitis associated with equine outbreak, Australia, 2008. Emerg Infect Dis 16:219–223

Rahman SA, Hassan SS, Olival KJ, Mohamed M, Chang LY, Hassan L, Saad NM, Shohaimi SA, Mamat ZC, Naim MS, Epstein JH, Suri AS, Field HE, Daszak P (2010) Characterization of Nipah virus from naturally infected *Pteropus vampyrus* bats, Malaysia. Emerg Infect Dis 16:1990–1993

Rahman MA, Hossain MJ, Sultana S, Homaira N, Khan SU, Rahman M, Gurley ES, Rollin PE, Lo MK, Comer JA, Lowe L, Rota PA, Ksiazek TG, Kenah E, Sharker Y, Luby SP (2011) Date palm sap linked to Nipah virus outbreak in Bangladesh, 2008. Vector Borne Zoonotic Dis

Reynes JM, Counor D, Ong S, Faure C, Seng V, Molia S, Walston J, Georges-Courbot MC, Deubel V, Sarthou JL (2005) Nipah virus in Lyle's flying foxes, Cambodia. Emerg Infect Dis 11:1042–1047

Rockx B, Bossart KN, Feldmann F, Geisbert JB, Hickey AC, Brining D, Callison J, Safronetz D, Marzi A, Kercher L, Long D, Broder CC, Feldmann H, Geisbert TW (2010) A novel model of lethal Hendra virus infection in African green monkeys and the effectiveness of ribavirin treatment. J Virol 84:9831–9839

Rockx B, Brining D, Kramer J, Callison J, Ebihara H, Mansfield K, Feldmann H (2011) Clinical outcome of henipavirus infection in hamsters is determined by the route and dose of infection. J Virol 85:7658–7671

Rogers RJ, Douglas IC, Baldock FC, Glanville RJ, Seppanen KT, Gleeson LJ, Selleck PN, Dunn KJ (1996) Investigation of a second focus of equine morbillivirus infection in coastal Queensland. Aust Vet J 74:243–244

Selvey LA, Wells RM, McCormack JG, Ansford AJ, Murray K, Rogers RJ, Lavercombe PS, Selleck P, Sheridan JW (1995) Infection of humans and horses by a newly described morbillivirus. Med J Aust 162:642–645

Smith I, Broos A, de Jong C, Zeddeman A, Smith C, Smith G, Moore F, Barr J, Crameri G, Marsh G, Tachedjian M, Yu M, Kung YH, Wang LF, Field H (2011) Identifying Hendra virus diversity in pteropid bats. PLoS ONE 6:e25275

Sohayati AR, Hassan L, Sharifah SH, Lazarus K, Zaini CM, Epstein JH, Shamsyul Naim N, Field HE, Arshad SS, Abdul Aziz J, Daszak P (2011) Evidence for Nipah virus recrudescence and serological patterns of captive *Pteropus vampyrus*. Epidemiol Infect 139:1570–1579

Tan CT, Wong KT (2003) Nipah encephalitis outbreak in Malaysia. Ann Acad Med Singapore 32:112–117

Tan CT, Goh KJ, Wong KT, Sarji SA, Chua KB, Chew NK, Murugasu P, Loh YL, Chong HT, Tan KS, Thayaparan T, Kumar S, Jusoh MR (2002) Relapsed and late-onset Nipah encephalitis. Ann Neurol 51:703–708

Torres-Velez FJ, Shieh WJ, Rollin PE, Morken T, Brown C, Ksiazek TG, Zaki SR (2008) Histopathologic and immunohistochemical characterization of Nipah virus infection in the guinea pig. Vet Pathol 45:576–585

Weingartl H, Czub S, Copps J, Berhane Y, Middleton D, Marszal P, Gren J, Smith G, Ganske S, Manning L, Czub M (2005) Invasion of the central nervous system in a porcine host by Nipah virus. J Virol 79:7528–7534

Weingartl HM, Berhane Y, Caswell JL, Loosmore S, Audonnet JC, Roth JA, Czub M (2006) Recombinant Nipah virus vaccines protect pigs against challenge. J Virol 80:7929–7938

Westbury HA, Hooper PT, Selleck PW, Murray PK (1995) Equine morbillivirus pneumonia: susceptibility of laboratory animals to the virus. Aust Vet J 72:278–279

Westbury HA, Hooper PT, Brouwer SL, Selleck PW (1996) Susceptibility of cats to equine morbillivirus. Aust Vet J 74:132–134

Williamson MM, Hooper PT, Selleck PW, Gleeson LJ, Daniels PW, Westbury HA, Murray PK (1998) Transmission studies of Hendra virus (equine morbillivirus) in fruit bats, horses and cats. Aust Vet J 76:813–818

Williamson MM, Hooper PT, Selleck PW, Westbury HA, Slocombe RF (2000) Experimental hendra virus infectionin pregnant guinea-pigs and fruit bats (*Pteropus poliocephalus*). J Comp Pathol 122:201–207

Williamson MM, Hooper PT, Selleck PW, Westbury HA, Slocombe RF (2001) A guinea-pig model of Hendra virus encephalitis. J Comp Pathol 124:273–279

Wong KT, Ong KC (2011) Pathology of acute henipavirus infection in humans and animals. Patholog Res Int 2011:567248

Wong SC, Ooi MH, Wong MN, Tio PH, Solomon T, Cardosa MJ (2001) Late presentation of Nipah virus encephalitis and kinetics of the humoral immune response. J Neurol Neurosurg Psychiatry 71:552–554

Wong KT, Shieh WJ, Kumar S, Norain K, Abdullah W, Guarner J, Goldsmith CS, Chua KB, Lam SK, Tan CT, Goh KJ, Chong HT, Jusoh R, Rollin PE, Ksiazek TG, Zaki SR (2002) Nipah virus infection: pathology and pathogenesis of an emerging paramyxoviral zoonosis. Am J Pathol 161:2153–2167

Wong KT, Grosjean I, Brisson C, Blanquier B, Fevre-Montange M, Bernard A, Loth P, Georges-Courbot MC, Chevallier M, Akaoka H, Marianneau P, Lam SK, Wild TF, Deubel V (2003) A golden hamster model for human acute Nipah virus infection. Am J Pathol 163:2127–2137

Wong KT, Robertson T, Ong BB, Chong JW, Yaiw KC, Wang LF, Ansford AJ, Tannenberg A (2009) Human Hendra virus infection causes acute and relapsing encephalitis. Neuropathol Appl Neurobiol 35:296–305

Yob JM, Field H, Rashdi AM, Morrissy C, van der Heide B, Rota P, bin Adzhar A, White J, Daniels P, Jamaluddin A, Ksiazek T (2001) Nipah virus infection in bats (order Chiroptera) in peninsular Malaysia. Emerg Infect Dis 7: 439–441

Young PL, Halpin K, Selleck PW, Field H, Gravel JL, Kelly MA, Mackenzie JS (1996) Serologic evidence for the presence in Pteropus bats of a paramyxovirus related to equine morbillivirus. Emerg Infect Dis 2:239–240

Diagnosis of Henipavirus Infection: Current Capabilities and Future Directions

Lin-Fa Wang and Peter Daniels

Abstract Since the last major review on diagnosis of henipavirus infection about a decade ago, significant progress has been made in many different areas of test development, especially in the development of molecular tests using real-time PCR and many novel serological test platforms. In addition to provide an updated review of the current test capabilities, this review also identifies key future challenges in henipavirus diagnosis.

Contents

L.-F. Wang (✉) · P. Daniels
CSRIO Livestock Industries, Australian Animal Health Laboratory,
Geelong, VIC, Australia
e-mail: Linfa.wang@csiro.au

Current Topics in Microbiology and Immunology (2012) 359: 179–196
DOI: 10.1007/82_2012_215

Published Online: 6 April 2012

1 Introduction

Hendra virus (HeV) and Nipah virus (NiV), classified in the genus *Henipavirus* in the family *Paramyxoviridae* (Eaton et al. 2007), are internationally recognized as dangerous zoonotic pathogens for which there are currently no available vaccines or effective, registered treatment regimes (Murray et al. 1995; Chua et al. 2000; Pallister et al. 2011a). Hence they are usually worked with in laboratories at the highest level of biorisk management; using containment approaches commonly designated biosecurity level (BSL) 4 or physical containment (PC) 4. The risks to biosafety and laboratory biosecurity pose special requirements for laboratory diagnosis in many countries that require detailed attention to physical infrastructure and operating procedures for both laboratory facility and scientific operations.

The natural history of these agents, the history of their becoming known to science and their subsequent association with documented disease outbreaks have been reviewed elsewhere in this volume (Field et al. 2001, 2007). Both HeV and NiV have been shown to have a natural reservoir in fruit bats of the genus *Pteropus* (Yob et al. 2001; Young et al. 1996; Field et al. 2001). However, the discovery of henipavirus infections in other species of bats in West Africa and China (Hayman et al. 2008; Li et al. 2008) reinforces the need for diagnostic capability beyond the range of *Pteropus spp.*

HeV has been identified only in Australia where it has been diagnosed in disease in horses in many of the years since its initial discovery in 1994 (Pallister et al. 2011a; Field et al. 2007) under circumstances where transmission from pteropid bats seemed plausible. Seven people have become infected resulting in four fatalities where they have been closely exposed to excretions and body fluids of diseased horses (Playford et al. 2010; Pallister et al. 2011a; Field et al. 2007). Experimental infections of horses with HeV have shown that the virus can be present in respiratory infections before temperature rise and the development of clinical signs, with increasing amounts of virus in respiratory excretions and other body fluids during the clinical phase of infection (Marsh et al. 2011), a situation that affects risk management approaches for attending equine veterinarians and horse owners in localities where the proximity of pteropid bats may lead to transmission to horses.

NiV disease has emerged in two forms. It was identified first in Malaysia in 1999 where there was hypothesized transmission from the wildlife reservoir to farmed pigs, and amplification among intensively farmed pigs. This in turn resulted in infection in other domestic animals in close proximity on affected farms, as well as in people engaged in occupations that required close contact with the excretions of infected pigs or the body fluids of diseased pigs. Human cases numbered 265, with 105 deaths at the time, a case fatality ratio of 38% (Pallister et al. 2011a). During the outbreak infections were diagnosed in dogs, horses, and cats (Chua et al. 1999; Hooper et al. 2001). Infection was spread geographically in Malaysia by the movement of infected pigs and internationally by the transport of asymptomatically infected pigs for slaughter (Chua 2003; Epstein et al. 2006).

NiV disease has also become frequently reported in certain parts of the Ganges delta, particularly in northern and western Bangladesh, where the farming system of collecting sweet palm sap for human consumption can bring people into direct contact with pteropid bat excretions. Here human illness and death result from consumption of contaminated palm juice without the involvement of a domestic animal amplifier host (Hsu et al. 2004; Luby et al. 2009a, b). Importantly, epidemiological analyses here indicate that human infections are being transmitted person to person. This mode of transmission has not been seen with human HeV infections or human NiV disease in Southeast Asia, and is a development warranting close diagnostic monitoring and public health support. Until early 2011, there had been 278 human cases of NiV disease in this region with 194 deaths, a case fatality ratio of 70% (Pallister et al. 2011a). Fortunately, the infection has not been highly contagious among people to the present time.

Clearly there is a need for diagnostic tests that can be deployed under a range of circumstances. These will include facilitation of studies of the natural history of henipavirus infections, diagnosis of infection or disease in a range of domestic animal species, in special circumstances facilitation of international trade in known susceptible species and the diagnosis of human infection and disease. Since NiV disease with human to human transmission occurs in remote locations without ready access to sophisticated laboratory facilities, there is a need for diagnostic technologies appropriate to the circumstances. Similarly, strategies are required to manage the risk of infection from horses at the primary point of care, which may include a role for diagnostic testing.

In all these circumstances accurate tests that are diagnostically sensitive and specific are needed. For such a dangerous zoonosis the consequences of false negative or false positive test results are difficult to manage, since the pressures to adopt a zero risk approach to incident management are considerable. Laboratories offering henipavirus testing pay detailed attention to quality assurance issues, but reproducible diagnostic test accuracy can be more challenging to deliver at the point of care.

This chapter will describe the diagnostic approaches to HeV and NiV testing in terms of their strengths and shortcomings for the range of purposes identified, and explore the diagnostic opportunities that are currently being investigated. Similar topics are covered to certain degree in various other review and book chapters, which may provide additional information not covered here (Daniels et al. 2001; Eaton et al. 2007; Rota et al. 2010; Rollin et al. 2012).

2 Clinical Diagnosis

The clinical signs of HeV and NiV infections in domestic animals may be suggestive of these diseases where epidemiologically the infections may feasibly occur, but are not specific to these diseases. Similarly the pathological changes are not pathognomonic. Importantly, the infections in important amplifier hosts such as horses and pigs may be asymptomatic. NiV and HeV infections in domestic animals have been reviewed (Hooper et al. 2001; Daniels et al. 2004).

3 Laboratory Diagnosis

In this intention of this chapter to be consistent with established standards such as the World Organization for Animal Health (OIE) Manual of Diagnostic Tests and Vaccines for Terrestrial Animals (Anonymous 2011) and also with the World Health Organization (WHO) International Health Regulations 2005 (IHR) (Anonymous 2007), noting that such standards are regularly updated to reflect scientific advances. The approaches to disease diagnosis and the supporting laboratory activities are based on both the rapid and accurate detection of the viral agent in samples using a variety of technologies and the detection of the bodily responses to infection, particularly through serology.

As in any diagnostic process, the collection of the appropriate specimen for the test that is to be performed requires a comprehensive understanding of the pathogenesis of the infection in the species concerned. This will allow the targeting of the appropriate tissue or excretion for sampling at the appropriate time post infection. Both the appropriateness of the specimen and the timing of its collection are critical in having a meaningful and interpretable test result.

To this end, the aspects of pathogenesis of henipavirus infection that lead to transmission, and hence to the identification of specimens in which virus will be present, have been comprehensively discussed (Daniels et al. 2007). Specimens for agent detection may be collected during the course of infection or at post mortem examination. Reliable detection of a serological response varies with the test and species of animal, but is usually late in the course of infection, 10–14 days after infection. There are currently no diagnostic tests to detect early animal responses to infection in use. Further elucidation of the excretion of HeV during the course of HeV infections in horses (Marsh et al. 2011) has shown that HeV can be detected in nasal excretions even in the preclinical phase, while early studies of NiV infections in pteropid bats (Middleton et al. 2007) showed that detection of virus even in known infected animals could be problematic. These examples serve to emphasize the importance of the knowledge of viral pathogenesis for effective diagnostic testing. Specimens appropriate for each test will be discussed in association with the description of each test.

3.1 Molecular Diagnosis

Due to their high sensitivity and the speed with which results can be available molecular tests have an increasingly important role in diagnosis of viral infection of both human and animal diseases. From the first application of conventional PCR to the latest investigation of outbreak virus strains in 2011 using next-generation sequencing (NGS), molecular tests are contributing to every aspect of henipavirus diagnosis, from early detection of spillover events to confirmation of human infections.

3.1.1 Conventional PCR

Until the wide application of real-time PCR (see Sect. 3.1.2), gel-based conventional PCR (also known as end point PCR) tests played a key role in henipavirus diagnosis for almost a decade. Two major tests were developed by the CSIRO Australian Animal Health Laboratory (AAHL) and the US Center for Disease Control (CDC), respectively. The AAHL test was based on the relatively conserved matrix protein gene or the M gene of the HeV (Hooper et al. 2000) while the CDC test targeted the more conserved nucleocapsid protein gene, the N gene of the NiV (Chua et al. 2000; Lo et al. 2012). These tests can be used for detection of viral genetic materials in different types of specimens, including fixed, or fresh tissue, various swabs, cerebrospinal fluid or urine samples. Sequencing of the PCR product also proved useful in rapid characterization of viral isolates from tissue culture supernatant.

Other than in a few laboratories in developing countries, tests based on conventional PCR are now considered suboptimal in most diagnostic laboratories and no longer in use for diagnostic purposes.

3.1.2 Real-Time PCR

Two major forms of real-time PCR or quantitative PCR (qPCR) have been used for the development of highly sensitive and specific molecular tests for henipaviruses. The most commonly used form is the TaqMan® test which utilizes a specific 3′-minor groove binder-DNA probe targeting a short sequence between the two priming sites (Kutyavin et al. 2000). The SYBR green qPCR does not require a specific probe and the test is less expensive, but can be less specific than the TagMan test (Zipper et al. 2004).

The first HeV TaqMan® assay was developed by Smith et al. in 2001 which was a single tube one-step PCR based on a probe targeting the M gene. In the initial study, the authors demonstrated that the TaqMan® assay was 1,000 times more sensitive than a conventional RT-PCR. The assay was shown to specifically detect HeV, but not other paramyxoviruses, including J paramyxovirus, measles virus, or Mossman virus (Smith et al. 2001).

In 2004, Guillaume et al. (2004b) developed a TaqMan® assay for NiV based on the N gene sequence. In this study, it was shown that linearity of the standard curve allowed accurate quantification of 10^3–10^9 RNA NiV N gene transcripts. The sensitivity of the test was close to 1 pfu. In an application designed to mimic the testing of field samples in an outbreak, it was demonstrated that the TaqMan® assay was able to detect NiV RNA in blood specimens from hamsters infected with NiV. The assay was NiV-specific since it failed to detect HeV RNA (Guillaume et al. 2004b).

Chang et al. (2006) developed a SYBR Green-based qPCR for NiV. Although this assay was also based on the N gene, its primers targeted a different region from that used by Guillaume et al. As a result, this SYBR qPCR was able to detect HeV RNA as well (Chang et al. 2006). This assay has a detection limit of approximately 100 pfu, which is much less sensitive than the one developed by Guillaume et al. However, the SYBR assay was developed for quantitative estimation of NiV replication kinetics in different cell types in vitro rather than for application as a robust diagnostic test for field specimens, and was considered fit for its intended purpose.

To fill the gap for the lack of a qPCR which is able to detect different henipaviruses, Feldman et al. (2009) developed and evaluated several consensus qPCRs for the detection of both HeV and NiV. In this study, the N, P, M and L genes were selected as candidates for TaqMan® assay design and conserved regions were identified across the available sequences. Degeneracies and multiple components were designed to accommodate for sequence discrepancies. Evaluation of the assay characteristics on both biological and in vitro transcribed RNA samples demonstrated the different performances that can result from consensus design based on different genes or PCR formats. In this study, the N gene SYBR Green assay was the most useful assay for investigation of potentially unknown henipaviruses and the P TaqMan® assay was the most useful for studies where high sensitivity was required to detect infection by know strains of NiV or HeV (Feldman et al. 2009).

3.1.3 Sequencing

Sanger sequencing played a key role in the initial diagnosis/discovery of both HeV and NiV (Murray et al. 1995; Chua et al. 2000; Daniels et al. 2001). Although this technique remains important in determining potential point mutations in targeted regions of the genome, the much more powerful NGS platforms are the method of choice for rapid whole genome analysis of any new henipavirus isolate. First such application was published recently by Smith et al. (2011) in which the HeV diversity in pteropid bats was studied by whole genome sequencing of four isolates using the 454 sequencing platform. During July to October of 2011, there were a total of 18 independent HeV cases detected in Queensland and New South Wales (Promed 2011c). This was totally unexpected and unprecedented considering that only a total of 14 events was recorded from 1994 to 2011, with a maximal of two events per year recorded from 2006 to 2009 (Pallister et al. 2011a). To determine where the sudden surge of Hendra outbreaks was due to the emergence of a new strain which is more transmissible from bats to horses, the whole genome sequencing approach was applied to both horse and bat isolates obtained from the 2011 outbreaks. From the genome sequences of five isolates, it was clear that the multiple spillover events were not a result of a single strain (Wang, unpublished results), instead they were caused by multiple strains with sequence variations at a similar level to those observed from previous outbreaks (Smith et al. 2011; Marsh et al. 2010). From these studies, it was clear that a number of HeV variants circulated at the one time at multiple locations, while at times the same variant was found circulating at disparate locations. The data collectively indicated that spillover events did not correlate with particular HeV isolates, suggesting that host and/or environmental factors are the primary determinants of bat-horse spillover (Smith et al. 2011). Direct sequencing of different NiV isolates revealed that while the NiV strains isolated from different hosts (human, pig and bat) in Malaysia were very similar at genome sequence level the NiV variants present in Bangladesh and India demonstrated a greater genetic diversity [see chapter by Lo and Rota in this volume, (Chadha et al. 2006; Lo et al. 2012)].

3.2 Virus Isolation

Virus isolation remains an important primary diagnostic approach for both HeV and NiV infections (Daniels et al. 2001). For confirmation of new cases or outbreaks or identification of new host(s) for henipaviruses virus isolation is highly desirable. However, due to their classification as dangerous zoonotic agents it is absolutely essential that isolation of virus from suspected henipavirus infections should only be conducted where laboratory biosafety can be guaranteed. As soon as the identity of henipavirus is confirmed, it is advised that all work dealing with live virus be conducted in a BSL4 laboratory. For many countries, this may mean partnering with an international reference laboratory. Development of relationships of trust and collaboration with such internationally designated partners will decrease the need for duplication of expensive BSL4 facilities.

Both HeV and NiV grow well in Vero cells, which is the cell line of choice for henipavirus isolation in most laboratories. At AAHL, different pteropid bat cell lines have recently been developed (Crameri et al. 2009), which have proven to be a useful additional tool for virus isolation during the investigation of the 2011 HeV outbreaks, especially for isolation of viruses from bat urine samples (Wang, unpublished results). As documented in a previous publication (Daniels et al. 2001), a range of specimens that have been successfully used for virus isolation in either natural or experimental cases has been summarized. Brain, lung, kidney and spleen should always be submitted. Tissues are handled under sterile conditions for preparation of 10% suspensions in cell culture media. These are clarified by centrifugation and the supernatant used for inoculation of cell cultures. Daily monitoring of cytopathic effect (CPE) is required for early detection of virus replication. A CPE usually develops within 3 days, but two 5-day passages are recommended before judging the attempt unsuccessful. The CPE is usually manifested by the formation of syncytia that may contain up to 20 or more nuclei. Subsequently syncytia lift from the substrate, leaving punctate holes in the cell monolayer. The syncytia formed by NiV in Vero cell monolayers are significantly larger than those created by HeV in the same time period. Interestingly, the distribution of nuclei differs between HeV- and NiV-induced syncytia and can be used to differentiate the two viruses (Hyatt et al. 2001).

It should be noted that due to the strain variation and cell line used, it is possible that different henipaviruses may present CPE morphology slightly different from those described above and in current literature. Final confirmation of henipavirus isolation is required. This can be done using a variety of techniques, with immunostaining of fixed, infected cells being the more traditional approach (Daniels et al. 2001; Bonaparte et al. 2005). PCR and sequencing can provide more definitive characterization of the virus isolate, especially if the new isolate is substantially different from previously known strains.

3.3 Immunohistochemistry

Immunohistochemistry is one of the most useful tests for HeV and NiV detection, especially where formalin-fixed tissues are presented as the main form of specimen for diagnosis. In the past, it has been used for retrospective diagnosis of a fetal HeV infection based on archival material (Hooper et al. 1996). In HeV and NiV infections there is a wide range of tissues in which viral antigen can be detected (Daniels et al. 2001), since the primary pathology occurs in the vascular endothelium (Hooper et al. 2001). Although lung was the main target tissue for viral antigen staining in the original HeV diagnosis, other tissues should also be submitted for diagnosis since it has been suspected that viral antigens may be cleared from lung tissue somewhat early in the course of infection. Spleen is another organ of importance for HeV diagnosis since HeV has been isolated from the spleen of a horse 21 days post infection (Williamson et al. 1998). Ideally a submission for immunohistochemistry would include samples of the brain at various levels, lung, mediastinal lymph nodes, spleen, and kidney. In pregnant animals the uterus, placenta, and foetal tissues should be included. The initial immunohistochemical investigation of HeV-infected tissue used a convalescent human serum, but a range of polyclonal and monoclonal antisera is now available (Imada et al. 2004; Tanimura et al. 2004a, b; White et al. 2005; Wang et al. 1998; Chiang et al. 2010; Zhu et al. 2008). A comparative study using various sera at AAHL found that hyperimmune rabbit sera raised against recombinant N proteins of NiV and HeV were most effective in immunohistochemical staining of different tissues infected by henipaviruses (Bingham and Wang, unpublished results).

3.4 Electron Microscopy

Electron microscopy (EM) played a major role in the initial discovery of HeV (Murray et al. 1995), and remains an important technique for henipavirus diagnosis (Daniels et al. 2001). Both HeV and NiV grow in cultured cells to high titers (up to

10^8 $TCID_{50}$ or PFU/ml), which makes feasible for direct visualization of viruses in the medium of infected cells by negative contrast electron microscopy and detection of virus-antibody interactions by immunoelectron microscopy. These techniques can rapidly provide valuable information on virus structure and antigenic reactivity, even during primary isolation of the virus. Other ultrastructural techniques such as grid cell culture, in which cells are grown, infected and visualized on electron microscope grids, and identification of replicating viruses and inclusion bodies in thin sections of fixed, embedded cell cultures and infected tissues complement the diagnostic effort (Hyatt et al. 2001).

3.5 Serology

Due to the very acute and highly lethal nature of henipavirus infections in humans, detection of serum IgG antibodies is considered to be not a very useful method of diagnosis. However, serological tests are highly important for investigation of infection status in the natural reservoir, bats, and in the identification of new animal hosts which may or may not present acute symptoms following henipavirus infection. This is best illustrated by the recent diagnosis of a sero-positive dog in a property where lethal infection of horses by HeV was confirmed as part of the 2011 HeV outbreak investigations jointly conducted by Biosecurity Queensland and AAHL (Promed 2011b). Serology has been used to identify both infected animals and infected premises after outbreaks, as part of surveillance and control measures, for both NiV infections in pigs in Malaysia and for HeV infections of horses in Australia. Some countries require serological testing of animals prior to international movements, although such testing is usually not recommended as being necessary.

3.5.1 Serum Neutralization Test

For serology the serum neutralization test (SNT) is still considered the "gold" standard, and usual laboratory biosafety risk management considerations lead to the recommendation that it be performed within a BSL4 facility (Daniels et al. 2001). In the test, sera are incubated with virus in the wells of 96-cell microtiter plates prior to the addition of Vero cells. Sera are screened at a 1:2 dilution, although this occasionally leads to problems with serum-induced cytotoxicity, especially common with bat sera for unknown reasons. Where sample quality is poor or sample volumes small, an initial dilution of 1:5 may be used. Cultures are read at 3 days, and those sera that completely block development of CPE are designated as positive.

A modified SNT was developed at AAHL based on a rapid immune plaque assay (Crameri et al. 2002). In this test, viruses are titrated on Vero cell monolayers in 96-well plates and after 18–24 h, foci of infection are detected immunologically in methanol-fixed cells using a hyperimmune rabbit serum to recombinant HeV P protein fragment. Briefly, serially diluted test sera were incubated with a predetermined number of plaque-forming units prior to adsorption to Vero cell monolayers. Unadsorbed virus is removed and virus-induced syncytia detected 24 h later by fixing the plate and conducting the immunostaining outside the BSL4 facility. Such a plaque-reduction neutralization test has merit if cytotoxicity is a problem, because virus-serum mixtures are removed after an adsorption period, and if the volumes of sera available are low. This test is also more quantitative since the number of stained plaques can be accurately determined outside of the BSL4 environment.

3.5.2 ELISA

Due to its simplicity and safe operation without the need for a BSL4 facility, ELISA remains the most affordable serological test for diagnosis of henipavirus infection. It is especially useful in epidemiological studies and for ongoing surveillance.

The most used indirect ELISA for detection of IgG antibodies to HeV is based on inactivated HeV-infected Vero cell lysate (Daniels et al. 2001). Problems with nonspecific reactions encountered in the original ELISA test was partially reduced by using a control antigen from uninfected Vero cells prepared in a manner identical to that used for virus infected cell lysates. Although this ELISA has proven to be highly sensitive, it still suffers specificity issue when dealing with field sera, often generating false positive results. Recently, highly purified recombinant soluble G proteins of HeV and NiV (see Sect. 3.5.3) have been trialed in a new ELISA test being developed at AAHL. Preliminary data so far has shown some promise in greatly increasing the test specificity (Lunt and Wang, unpublished results).

Two ELISA tests were developed at CDC in response to the NiV outbreaks in Malaysia (Daniels et al. 2001). These include an indirect ELISA for detection of IgG antibodies and a capture ELISA for detection of IgM antibodies. The CDC ELISAs played a pivotal role in confirming the diagnosis in both human and porcine populations, the identification of cases and in establishing the serological profile of infected pig herds as a prelude to designing a national surveillance program (Chua et al. 2000).

Various ELISA tests based on various recombinant proteins have been developed by different groups, including NiV N protein (Chen et al. 2006; Yu et al. 2006), NiV glycoprotein (Eshaghi et al. 2004, 2005a, b), and truncated P proteins of HeV and NiV (Chen et al. 2007). Due to the very limited number of positive sera tested in each of these ELISAs, their usefulness in field diagnosis of henipavirus infection is yet to be proved.

3.5.3 Luminex-Based Binding and Inhibition Tests

In 2007, a novel test platform was developed specifically for the detection of the henipavirus antibodies (Bossart et al. 2007). The success of this platform was dependent on three important progresses at the time: (1) the commercial availability of a bead-based liquid protein array system, commonly known as the Luminex system (Vignali 2000), which allows highly multiplex and sensitive detection of molecular interactions; (2) the development of highly purified soluble G proteins of HeV and NiV which exhibited total functionality as the intact G protein on virion surface (Bossart et al. 2005); and (3) the discovery of ephrin-B2 as the major functional cell entry receptor for henipaviruses (Bonaparte et al. 2005; Negrete et al. 2005) and the commercial availability of the soluble ephrin-B2 molecule. Two different test formats were developed, a binding assay and a blocking or inhibition assay.

For the binding assay, HeV and NiV sG proteins were coated on two beads with distinctive fluorescent spectra. A predetermined number of beads were mixed in a single well, followed by incubation with test antiserum at different dilutions. The bound antibodies were then detected using biotinylated Protein-A: G (1:1 mix), followed by streptavidin-phycoerythrin. The final readings were determined and analyzed using a Bio-Plex Protein Array System (BioRad, USA).

For the blocking assay, the same sG-coated beads were used. The biotinylated Protein-A:G mix was replaced by a biotinylated soluble ephrin-B2 molecule. In the absence of henipavirus antibodies, the biotin-ephrin-B2 molecule will bind to the sG proteins on the beads, which in turn will support the binding of the streptavidin-phycoerythrin molecule, resulting in a positive binding signal. However, in the presence of henipavirus antibodies which can block the G-receptor interaction (i.e., some of the classical neutralizing antibodies), the binding of biotin-ephrin-B2 will be completely or partially blocked, resulting a decrease in the fluorescent signal.

These assays have proved to be a huge success in both disease investigation, epidemiological, and surveillance studies. They have been used in confirming HeV infection in human cases by Queensland Health (Playford et al. 2010), in detecting henipavirus antibodies in bat populations in Africa (Hayman et al. 2008) and in the confirmation of HeV infection in a dog in the 2011 HeV outbreak investigation (Barr and Wang, unpublished results).

It is worth to note that the Luminex blocking assay is the first surrogate SNT developed for any virus, which does not require any live cell culture and can be performed in a standard laboratory without any biocontainment need.

3.5.4 Surrogate Neutralization Tests Based on Pseudotyped Viruses

A pseudotyped virus is an enveloped virus particle assembled with one or more foreign viral envelope proteins. The most used virus "backbone" systems are derived from vesicular stomatitis virus (VSV) and lentivirus (HIV). For henipaviruses, the pseudotyped virus system has one obvious attractiveness, i.e., the SNT can be conducted in a BSL2 laboratory instead of a BSL4 facility that only a handful of groups have access around the world. To date, a total of four different pseudotyped virus systems have been developed for Nipah virus.

The first published Nipah pseudotyped virus was developed by Tamin et al. at CDC in 2009. They utilized the recombinant VSV system with the luciferase as the reporter molecule. Using a panel of human and pig sera, the authors demonstrated that the assay based on pVSV-NiV-F/G has a high specificity (94–100%), but the sensitivity was not as good as the classical plaque reduction neutralization test (PRNT) for some serum samples (Tamin et al. 2009).

Almost at the same time, another VSV-based Nipah pseudotyped virus was developed by Kaku et al. in 2009 as collaboration between National Institute of Infectious Diseases (NIID) in Japan and AAHL in Australia. In this system, GFP was used as the reporter molecule and quantification was achieved either by counting GFP-expressing cells or by measuring fluorescence. Initial comparative studies between this new assay and conventional SNT was conducted using serum samples from human, bat, horse, cat, rabbit, and pig. Preliminary results indicated a greater sensitivity than classical SNT if a 50% reduction in VSV-NiV-GFP infected cells is used as a cut off (Kaku et al. 2009). It also demonstrated that the same system is equally effective in detecting HeV-specific antibodies, which was not unexpected considering the high level of cross-neutralization observed previously (Bossart et al. 2007).

A third NiV pseudotyped virus was developed by Khetawat and Broder in 2010 using a lentivirus system. Both GFP and luciferase were used as a reporter molecule. The main application of this system was to probe the virus-host interaction. There was no data reported in the paper to assess its suitability for detection of virus neutralizing antibodies.

Very recently, a second generation of pseudotyped virus was reported by Kaku et al. (2012), in which the reporter gene GFP was replaced with a secreted alkaline phosphatase (SEAP). The main advantage of this system is its ability to measure SEAP activity in culture supernatant, hence converting a SNT to an assay only requiring the common ELISA plate reader. Using a similar panel of sera as for the 2009 study (Kaku et al. 2009), the VSV-NiV-SEAP virus produced higher neutralizing antibody titer in comparison to conventional SNT, but requiring almost 10-fold less serum or 2 μl. Additionally, when 66 fruit bat sera were screened at a single dilution, the VSV-NiV-SEAP assay produced identical results to those obtained using live NiV virus (Kaku et al. 2012).

Currently, the standard first-line serological test for henipaviruses is ELISA, followed by SNT as a confirmatory test (Daniels et al. 2001). The various pseudotyped virus systems described above can be used either as a frontline or confirmatory test. If further validation with more serum samples can corroborate the high sensitivity and specificity observed so far, it is not impossible to consider the use of the pseudotyped assay platform to replace both ELISA and SNT, hence combining the frontline and confirmatory tests into a single assay.

4 Future Challenges

Since the first discovery of Hendra virus in 1994, significant progress has been made in the development of henipavirus diagnostics. However, many challenges still remain and some of them are discussed below.

4.1 Rapid Field Test

From an outbreak investigation/control point of view, it is most useful in some situations to have a rapid field test or point-of-care (POC) test for both human and animal applications. For example, current diagnosis of HeV infection in horses takes at least 24–48 h using the laboratory-based tests, with most of that time being taken in transporting specimens to the laboratory. In terms of controlling or preventing horse-to-human transmission of HeV infection, any time that could be saved with a rapid field test will be highly significant. However, any misdiagnosis, especially those with a false negative diagnosis, could have disastrous consequences. The most commonly used technology platforms for POC tests are largely based on lateral flow chromatography (Lewandrowski et al. 2011), and such devices are unlikely to deliver the sensitivity and specificity required for a POC test for diagnosing infection of henipaviruses. The development of a reliable POC test for henipavirus, therefore, remains a challenge awaiting more advanced technological development.

4.2 DIVA Test

Although development of a henipavirus vaccine for human use is considered to be nonpractical considering the limited number of human infections to date, the possibility of vaccinating horses to prevent transmission to humans has gained momentum in the last few years in Australia. It is expected that there is a high chance of success in marketing a commercial HeV vaccine for horses in the next few years (Promed 2011a).

To date, the most promising vaccine candidates are based on recombinant envelope proteins G or F (Weingartl et al. 2006; Guillaume et al. 2004a; McEachern et al. 2008; Mungall et al. 2006; Pallister et al. 2011b). Several studies indicated that recombinant G alone is able to provide protection against lethal challenges by either HeV or NiV (Pallister et al. 2011a). Recombinant G proteins delivered either as a purified protein with adjuvant or by the canary pox viral vector were proven to be equally effective against either NiV or HeV infection (Weingartl et al. 2006; McEachern et al. 2008; Pallister et al. 2011b). It can therefore be concluded that any future commercial vaccine will most likely to be based on the G protein.

If the vaccine is to be widely used in the Australian horse population in the near future, there is an urgent need to develop a serological test that can differentiate vaccinated animals from infected, for example, the DIVA test.

4.3 Capacity to Detect Infection by Henipa-Like Viruses

Recent serological and molecular surveillance studies indicated the presence of a diverse group of henipavirus or henipa-like viruses in bats at different geographic locations around the world, from Australia, Southeast Asia to China, Madagascar, and West Africa (Hayman et al. 2008; Drexler et al. 2009; Li et al. 2008; Lehle et al. 2007; Field et al. 2007). It is highly possible that any of these bat viruses have the potential to spillover into other mammalian hosts, including human, and cause diseases. It has already been demonstrated that these viruses may produce cross reactive, but not cross neutralizing, antibodies against known henipaviruses (Li et al. 2008; Hayman et al. 2008). It can also be predicted that the current molecular tests, including the recently developed consensus henipavirus PCRs (Feldman et al. 2009), may not be able to detect new viruses from the expanded henipavirus group. Ongoing genome sequencing and development of improved serological tests are required to meet the needs of future diagnostic challenges.

4.4 Diagnostic Capability in Developing Nations

As for most infection diseases, the hot spots of henipavirus outbreaks are largely located in regions where first class equipments for advanced diagnostic platforms are lacking. It is therefore absolutely essential for close collaborations between developing nations where most outbreaks occur and developed nations where latest diagnostic technologies are developed. In addition, it is also desirable to transfer well established, relatively low-cost methods to national and regional diagnostic laboratories in developing nations so that a faster response can be mounted to reduce the impact of future outbreaks.

References

Anonymous (2007) International Health Regulations (2005) 2nd edn. WHO, Geneva, p 82 http://whqlibdoc.who.int/publications/2008/9789241580410_eng.pdf. Accessed 18 Oct 2011

Anonymous (2011) Manual of diagnostic tests and vaccines for terrestrial animals (2011). OIE, Paris. http://www.oie.int/en/international-standard-setting/terrestrial-manual/access-online/. Accessed 18 Oct 2011

Bonaparte MI, Dimitrov AS, Bossart KN, Crameri G, Mungall BA, Bishop KA, Choudhry V, Dimitrov DS, Wang LF, Eaton BT, Broder CC (2005) Ephrin-B2 ligand is a functional receptor for Hendra virus and Nipah virus. Proc Natl Acad Sci U S A 102:10652–10657

Bossart KN, Crameri G, Dimitrov AS, Mungall BA, Feng YR, Patch JR, Choudhary A, Wang LF, Eaton BT, Broder CC (2005) Receptor binding, fusion inhibition and induction of cross-reactive neutralizing antibodies by a soluble G glycoprotein of Hendra virus. J Virol 79:6690–6702

Bossart KN, McEachern JA, Hickey AC, Choudhry V, Dimitrov DS, Eaton BT, Wang LF (2007) Neutralization assays for differential henipavirus serology using Bio-Plex protein array systems. J Virol Methods 142:29–40

Chadha MS, Comer JA, Lowe L, Rota PA, Rollin PE, Bellini WJ, Ksiazek TG, Mishra A (2006) Nipah virus-associated encephalitis outbreak, Siliguri, India. Emerg Infect Dis 12:235–240

Chang LY, Ali AR, Hassan SS, AbuBakar S (2006) Quantitative estimation of Nipah virus replication kinetics in vitro. Virol J 3:47

Chen JM, Yaiw KC, Yu M, Wang LF, Wang QH, Crameri G, Wang ZL (2007) Expression of truncated phosphoproteins of Nipah virus and Hendra virus in *Escherichia coli* for the differentiation of henipavirus infections. Biotechnol Lett 29:871–875

Chen JM, Yu M, Morrissy C, Zhao YG, Meehan G, Sun YX, Wang QH, Zhang W, Wang LF, Wang ZL (2006) A comparative indirect ELISA for the detection of henipavirus antibodies based on a recombinant nucleocapsid protein expressed in *Escherichia coli*. J Virol Methods 136:273–276

Chiang CF, Lo MK, Rota PA, Spiropoulou CF, Rollin PE (2010) Use of monoclonal antibodies against Hendra and Nipah viruses in an antigen capture ELISA. Virol J 7:115

Chua KB (2003) Nipah virus outbreak in Malaysia. J Clin Virol 26:265–275

Chua KB, Bellini WJ, Rota PA, Harcourt BH, Tamin A, Lam SK, Ksiazek TG, Rollin PE, Zaki SR, Shieh W, Goldsmith CS, Gubler DJ, Roehrig JT, Eaton B, Gould AR, Olson J, Field H, Daniels P, Ling AE, Peters CJ, Anderson LJ, Mahy BW (2000) Nipah virus: a recently emergent deadly paramyxovirus. Science 288:1432–1435

Chua KB, Goh KJ, Wong KT, Kamarulzaman A, Tan PS, Ksiazek TG, Zaki SR, Paul G, Lam SK, Tan CT (1999) Fatal encephalitis due to Nipah virus among pig-farmers in Malaysia. Lancet 354:1257–1259

Crameri G, Todd S, Grimley S, McEachern JA, Marsh GA, Smith C, Tachedjian M, De Jong C, Virtue ER, Yu M, Bulach D, Liu JP, Michalski WP, Middleton D, Field HE, Wang LF (2009) Establishment, immortalisation and characterisation of pteropid bat cell lines. PLoS ONE 4:e8266

Crameri G, Wang LF, Morrissy C, White J, Eaton BT (2002) A rapid immune plaque assay for the detection of Hendra and Nipah viruses and anti-virus antibodies. J Virol Methods 99: 41–51

Daniels P, Ksiazek T, Eaton BT (2001) Laboratory diagnosis of Nipah and Hendra virus infections. Microbes Infect 3:289–295

Daniels PW, Halpin K, Hyatt A, Middleton D (2007) Infection and disease in reservoir and spillover hosts: determinants of pathogen emergence. Curr Top Microbiol Immunol 315: 113–131

Daniels PW, Shahirudin S, Aziz J, Ong BL (2004) Nipah virus disease. In: Coetzer JAW, Tustin RC (eds) Infectious diseases of livestock, 2nd edn. Oxford University Press, Oxford, pp 692–697

Drexler JF, Corman VM, Gloza-Rausch F, Seebens A, Annan A, Ipsen A, Kruppa T, Muller MA, Kalko EK, Adu-Sarkodie Y, Oppong S, Drosten C (2009) Henipavirus RNA in African bats. PLoS ONE 4:e6367

Eaton BT, Mackenzie JS, Wang L-F (2007) Henipaviruses. In: Knipe DM, Griffin DE, Lamb RA et al (eds) Fields virology. Lippincott Williams & Wilkins, Philadelphia, pp 1587–1600

Epstein JH, Field HE, Luby S, Pulliam JR, Daszak P (2006) Nipah virus: impact, origins, and causes of emergence. Curr Infect Dis Rep 8:59–65

Eshaghi M, Tan WS, Chin WK, Yusoff K (2005a) Purification of the extra-cellular domain of Nipah virus glycoprotein produced in *Escherichia coli* and possible application in diagnosis. J Biotechnol 116:221–226
Eshaghi M, Tan WS, Mohidin TB, Yusoff K (2004) Nipah virus glycoprotein: production in baculovirus and application in diagnosis. Virus Res 106:71–76
Eshaghi M, Tan WS, Ong ST, Yusoff K (2005b) Purification and characterization of Nipah virus nucleocapsid protein produced in insect cells. J Clin Microbiol 43:3172–3177
Feldman KS, Foord A, Heine HG, Smith IL, Boyd V, Marsh GA, Wood JL, Cunningham AA, Wang LF (2009) Design and evaluation of consensus PCR assays for henipaviruses. J Virol Methods 161:52–57
Field H, Young P, Yob JM, Mills J, Hall L, Mackenzie J (2001) The natural history of Hendra and Nipah viruses. Microbes Infect 3:307–314
Field HE, Mackenzie JS, Daszak P (2007) Henipaviruses: emerging paramyxoviruses associated with fruit bats. Curr Top Microbiol Immunol 315:133–159
Guillaume V, Contamin H, Loth P, Georges-Courbot MC, Lefeuvre A, Marianneau P, Chua KB, Lam SK, Buckland R, Deubel V, Wild TF (2004a) Nipah virus: vaccination and passive protection studies in a hamster model. J Virol 78:834–840
Guillaume V, Lefeuvre A, Faure C, Marianneau P, Buckland R, Lam SK, Wild TF, Deubel V (2004b) Specific detection of Nipah virus using real-time RT-PCR (TaqMan). J Virol Methods 120:229–237
Hayman DT, Suu-Ire R, Breed AC, McEachern JA, Wang L, Wood JL, Cunningham AA (2008) Evidence of henipavirus infection in West African fruit bats. PLoS ONE 3:e2739
Hooper P, Zaki S, Daniels P, Middleton D (2001) Comparative pathology of the diseases caused by Hendra and Nipah viruses. Microbes Infect 3:315–322
Hooper PT, Gould AR, Hyatt AD, Braun MA, Kattenbelt JA, Hengstberger SG, Westbury HA (2000) Identification and molecular characterization of Hendra virus in a horse in Queensland. Aust Vet J 78:281–282
Hooper PT, Gould AR, Russell GM, Kattenbelt JA, Mitchell G (1996) The retrospective diagnosis of a second outbreak of equine morbillivirus infection. Aust Vet J 74:244–245
Hsu VP, Hossain MJ, Parashar UD, Ali MM, Ksiazek TG, Kuzmin I, Niezgoda M, Rupprecht C, Bresee J, Breiman RF (2004) Nipah virus encephalitis reemergence, Bangladesh. Emerg Infect Dis 10:2082–2087
Hyatt AD, Zaki SR, Goldsmith CS, Wise TG, Hengstberger SG (2001) Ultrastructure of Hendra virus and Nipah virus within cultured cells and host animals. Microbes Infect 3:297–306
Imada T, Abdul Rahman MA, Kashiwazaki Y, Tanimura N, Syed Hassan S, Jamaluddin A (2004) Production and characterization of monoclonal antibodies against formalin-inactivated Nipah virus isolated from the lungs of a pig. J Vet Med Sci 66:81–83
Kaku Y, Noguchi A, Marsh GA, Barr JA, Okutani A, Hotta K, Bazartseren B, Fukushi S, Broder CC, Yamada A, Inoue S, Wang LF (2012) Second generation of pseudotype-based serum neutralization assay for Nipah virus antibodies: sensitive and high-throughput analysis utilizing secreted alkaline phosphatase. J Virol Methods 179:226–232
Kaku Y, Noguchi A, Marsh GA, McEachern JA, Okutani A, Hotta K, Bazartseren B, Fukushi S, Broder CC, Yamada A, Inoue S, Wang LF (2009) A neutralization test for specific detection of Nipah virus antibodies using pseudotyped vesicular stomatitis virus expressing green fluorescent protein. J Virol Methods 160:7–13
Khetawat D, Broder CC (2010) A functional henipavirus envelope glycoprotein pseudotyped lentivirus assay system. Virol J 7:312
Kutyavin IV, Afonina IA, Mills A, Gorn VV, Lukhtanov EA, Belousov ES, Singer MJ, Walburger DK, Lokhov SG, Gall AA, Dempcy R, Reed MW, Meyer RB, Hedgpeth J (2000) 3′-minor groove binder-DNA probes increase sequence specificity at PCR extension temperatures. Nucleic Acids Res 28:655–661
Lehle C, Razafitrimo G, Razainirina J, Andriaholinirina N, Goodman SM, Faure C, Georges-Courbot MC, Rousset D, Reynes JM (2007) Henipavirus and Tioman virus antibodies in pteropodid bats, Madagascar. Emerg Infect Dis 13:159–161

Lewandrowski K, Gregory K, Macmillan D (2011) Assuring quality in point-of-care testing: evolution of technologies, informatics, and program management. Arch Pathol Lab Med 135:1405–1414
Li Y, Wang J, Hickey AC, Zhang Y, Li Y, Wu Y, Zhang H, Yuan J, Han Z, McEachern J, Broder CC, Wang LF, Shi Z (2008) Antibodies to Nipah or Nipah-like viruses in bats, China. Emerg Infect Dis 14:1974–1976
Lo MK, Lowe L, Hummel KB, Sazzad HMS, Gurley ES, Hossain MJ, Luby SP, Miller DB, Comer JA, Rollin PE, Bellini WJ, Rota PA (2012) Characterization of Nipah virus from outbreaks in Bangladesh, 2008–2010. Emerg Infect Dis. doi:10.3201/eid1802.111492
Luby SP, Gurley ES, Hossain MJ (2009a) Transmission of human infection with Nipah virus. Clin Infect Dis 49:1743–1748
Luby SP, Hossain MJ, Gurley ES, Ahmed BN, Banu S, Khan SU, Homaira N, Rota PA, Rollin PE, Comer JA, Kenah E, Ksiazek TG, Rahman M (2009b) Recurrent zoonotic transmission of Nipah virus into humans, Bangladesh, 2001–2007. Emerg Infect Dis 15:1229–1235
Marsh GA, Haining J, Hancock TJ, Robinson R, Foord AJ, Barr JA, Riddell S, Heine HG, White JR, Crameri G, Field HE, Wang LF, Middleton D (2011) Experimental infection of horses with hendra virus/australia/horse/2008/redlands. Emerg Infect Dis 17:2232–2238
Marsh GA, Todd S, Foord A, Hansson E, Davies K, Wright L, Morrissy C, Halpin K, Middleton D, Field HE, Daniels P, Wang LF (2010) Genome sequence conservation of Hendra virus isolates during spillover to horses, Australia. Emerg Infect Dis 16:1767–1769
McEachern JA, Bingham J, Crameri G, Green DJ, Hancock TJ, Middleton D, Feng YR, Broder CC, Wang LF, Bossart KN (2008) A recombinant subunit vaccine formulation protects against lethal Nipah virus challenge in cats. Vaccine 26:3842–3852
Middleton DJ, Morrissy CJ, van der Heide BM, Russell GM, Braun MA, Westbury HA, Halpin K, Daniels PW (2007) Experimental Nipah virus infection in pteropid bats (*Pteropus poliocephalus*). J Comp Pathol 136:266–272
Mungall BA, Middleton D, Crameri G, Bingham J, Halpin K, Russell G, Green D, McEachern J, Pritchard LI, Eaton BT, Wang LF, Bossart KN, Broder CC (2006) Feline model of acute Nipah virus infection and protection with a soluble glycoprotein-based subunit vaccine. J Virol 80:12293–12302
Murray K, Selleck P, Hooper P, Hyatt A, Gould A, Gleeson L, Westbury H, Hiley L, Selvey L, Rodwell B et al (1995) A morbillivirus that caused fatal disease in horses and humans. Science 268:94–97
Negrete OA, Levroney EL, Aguilar HC, Bertolotti-Ciarlet A, Nazarian R, Tajyar S, Lee B (2005) EphrinB2 is the entry receptor for Nipah virus, an emergent deadly paramyxovirus. Nature 436:401–405
Pallister J, Middleton D, Broder CC, Wang L-F (2011a) Henipavirus vaccine development. J Bioterr Biodef S1:005
Pallister J, Middleton D, Wang LF, Klein R, Haining J, Robinson R, Yamada M, White J, Payne J, Feng YR, Chan YP, Broder CC (2011b) A recombinant Hendra virus G glycoprotein-based subunit vaccine protects ferrets from lethal Hendra virus challenge. Vaccine 29:5623–5630
Playford EG, McCall B, Smith G, Slinko V, Allen G, Smith I, Moore F, Taylor C, Kung YH, Field H (2010) Human Hendra virus encephalitis associated with equine outbreak, Australia, 2008. Emerg Infect Dis 16:219–223
Promed (2011a) 20110525.1589 Hendra virus—Australia (06), Vaccine
Promed (2011b) 20110727.2257 Hendra virus, equine—Australia (18), Queensland, Canine
Promed (2011c) 20111013.3061 Hendra virus, equine—Australia (28), Queensland, NS
Rollin PE, Rota PA, Zaki S, Ksiazek TG (2012) Chapter 92. Hendra and Nipah viruses. In: Versalovic J (ed) Manual of clinical microbiology, 10th edn, vol 2. ASM Press, pp 1479–1487
Rota PA, Mungall B, Halpin K (2010) Ch 49. Nipah virus. Molecular detection of human viral pathogens. In: Liu D (ed) Molecular detection of human viral pathogens. CRC Press, pp 543–555
Smith I, Broos A, De Jong C, Zeddeman A, Smith C, Smith G, Moore F, Barr J, Crameri G, Marsh GA, Tachedjian M, Yu M, Wang L-F, Field H (2011) Identifying Hendra virus diversity in Pteropid bats. Emerg Infect Dis 6(9):e25275

Smith IL, Halpin K, Warrilow D, Smith GA (2001) Development of a fluorogenic RT-PCR assay (TaqMan) for the detection of Hendra virus. J Virol Methods 98:33–40

Tamin A, Harcourt BH, Lo MK, Roth JA, Wolf MC, Lee B, Weingartl H, Audonnet JC, Bellini WJ, Rota PA (2009) Development of a neutralization assay for Nipah virus using pseudotype particles. J Virol Methods 160:1–6

Tanimura N, Imada T, Kashiwazaki Y, Shahirudin S, Sharifah SH, Aziz AJ (2004a) Monoclonal antibody-based immunohistochemical diagnosis of Malaysian Nipah virus infection in pigs. J Comp Pathol 131:199–206

Tanimura N, Imada T, Kashiwazaki Y, Shamusudin S, Syed Hassan S, Jamaluddin A, Russell G, White J (2004b) Reactivity of anti-Nipah virus monoclonal antibodies to formalin-fixed, paraffin-embedded lung tissues from experimental Nipah and Hendra virus infections. J Vet Med Sci 66:1263–1266

Vignali DA (2000) Multiplexed particle-based flow cytometric assays. J Immunol Methods 243: 243–255

Wang LF, Michalski WP, Yu M, Pritchard LI, Crameri G, Shiell B, Eaton BT (1998) A novel P/V/C gene in a new member of the Paramyxoviridae family, which causes lethal infection in humans, horses, and other animals. J Virol 72:1482–1490

Weingartl HM, Berhane Y, Caswell JL, Loosmore S, Audonnet JC, Roth JA, Czub M (2006) Recombinant Nipah virus vaccines protect pigs against challenge. J Virol 80:7929–7938

White JR, Boyd V, Crameri GS, Duch CJ, van Laar RK, Wang LF, Eaton BT (2005) Location of, immunogenicity of and relationships between neutralization epitopes on the attachment protein (G) of Hendra virus. J Gen Virol 86:2839–2848

Williamson MM, Hooper PT, Selleck PW, Gleeson LJ, Daniels PW, Westbury HA, Murray PK (1998) Transmission studies of Hendra virus (equine morbillivirus) in fruit bats, horses and cats. Aust Vet J 76:813–818

Yob JM, Field H, Rashdi AM, Morrissy C, van der Heide B, Rota P, bin Adzhar A, White J, Daniels P, Jamaluddin A, Ksiazek T (2001) Nipah virus infection in bats (order Chiroptera) in peninsular Malaysia. Emerg Infect Dis 7:439–441

Young PL, Halpin K, Selleck PW, Field H, Gravel JL, Kelly MA, Mackenzie JS (1996) Serologic evidence for the presence in Pteropus bats of a paramyxovirus related to equine morbillivirus. Emerg Infect Dis 2:239–240

Yu F, Khairullah NS, Inoue S, Balasubramaniam V, Berendam SJ, Teh LK, Ibrahim NS, Abdul Rahman S, Hassan SS, Hasebe F, Sinniah M, Morita K (2006) Serodiagnosis using recombinant nipah virus nucleocapsid protein expressed in *Escherichia coli*. J Clin Microbiol 44:3134–3138

Zhu Z, Bossart KN, Bishop KA, Crameri G, Dimitrov AS, McEachern JA, Feng Y, Middleton D, Wang LF, Broder CC, Dimitrov DS (2008) Exceptionally potent cross-reactive neutralization of Nipah and Hendra viruses by a human monoclonal antibody. J Infect Dis 197:846–853

Zipper H, Brunner H, Bernhagen J, Vitzthum F (2004) Investigations on DNA intercalation and surface binding by SYBR Green I, its structure determination and methodological implications. Nucleic Acids Res 32:e103

Immunization Strategies Against Henipaviruses

Christopher C. Broder, Thomas W. Geisbert, Kai Xu, Dimitar B. Nikolov, Lin-Fa Wang, Deborah Middleton, Jackie Pallister and Katharine N. Bossart

C. C. Broder (✉)
Department of Microbiology and Immunology, Uniformed Services University, Bethesda, MD 20814, USA
e-mail: cbroder@usuhs.mil

T. W. Geisbert
Galveston National Laboratory and Department of Microbiology and Immunology, University of Texas Medical Branch, Galveston, TX 77550, USA
e-mail: tom.geisbert@utmb.edu

K. Xu · D. B. Nikolov
Structural Biology Program, Memorial Sloan-Kettering Cancer Center, New York, NY 10065, USA
e-mail: xuk@mskcc.org

D. B. Nikolov
e-mail: nikolovd@mskcc.org

L.-F. Wang · D. Middleton · J. Pallister
CSIRO Livestock Industries, Australian Animal Health Laboratory, 5 Portarlington Road, Geelong, VIC 3220, Australia
e-mail: Linfa.Wang@csiro.au

D. Middleton
e-mail: Deborah.Middleton@csiro.au

J. Pallister
e-mail: Jackie.Pallister@csiro.au

K. N. Bossart
Department of Microbiology, Boston University School of Medicine, Boston, MA 02118, USA
e-mail: kbossart@bu.edu

K. N. Bossart
National Emerging Infectious Diseases Laboratories Institute, Boston University School of Medicine, Boston, MA 02118, USA

Current Topics in Microbiology and Immunology (2012) 359: 197–223
DOI: 10.1007/82_2012_213

Published Online: 6 April 2012

Abstract Hendra virus and Nipah virus are recently discovered and closely related emerging viruses that now comprise the genus henipavirus within the subfamily *Paramyxoviridae* and are distinguished by their broad species tropism and ability to cause fatal disease in a wide variety of mammalian hosts including humans. The high mortality associated with human and animal henipavirus infections has highlighted the importance and necessity of developing effective immunization strategies. The development of suitable animal models of henipavirus infection and pathogenesis has been critical for testing the efficacy of potential therapeutic approaches. Several henipavirus challenge models have been used and recent successes in both active and passive immunization strategies against henipaviruses have been reported which have all targeted the viral envelope glycoproteins.

Contents

1 Introduction

Hendra virus (HeV) and Nipah virus (NiV) are recently identified members of the family *Paramyxoviridae* (Eaton et al. 2007). The henipaviruses are distinguished from all other paramyxoviruses particularly by their broad species tropism and in addition to bats can infect and cause fatal disease in multiple vertebrate hosts including humans, monkeys, pigs, horses, cats, dogs, ferrets, hamsters, and guinea pigs, spanning six mammalian Orders (Bossart et al. 2009; Geisbert et al. 2010; Guillaume et al. 2009; Hooper et al. 1997b, 2001; Li et al. 2010; Marianneau et al. 2010; Middleton et al. 2007; Mungall et al. 2006; Rockx et al. 2010, 2011; Weingartl et al. 2005; Westbury et al. 1995, 1996; Wong et al. 2003). HeV appeared first in eastern Australia in 1994 and was transmitted to humans from infected horses (reviewed in Murray et al. 1998); NiV later emerged in 1998–1999 in peninsular Malaysia and was primarily transmitted to humans from infected pigs, but several other animal species also became infected (reviewed in Bishop and Broder 2008; Eaton et al. 2006). Thus, both viruses may be amplified and cause disease in animals and may in turn be transmitted to humans, where infection is manifested as a severe respiratory illness and/or febrile encephalitis with associated high case fatality rates (Selvey et al. 1995; Tan and Wong 2003; Wong et al. 2002).

Since their recognition in the mid to late 1990s, both HeV and NiV have continued to re-emerge. Occasional outbreaks of HeV occurred in the years immediately following its appearance in 1994, but in 2006 HeV began to cause spillover events on an annual basis with all occurring in horses in Australia and a total of seven human cases of which four have been fatal (Anonymous 2009; Playford et al. 2010). In 2011, however, (June to October) the dynamics of HeV spillover events changed considerably, and an unprecedented 18 independent outbreaks of HeV among horses in Australia were recorded, leading to the death or euthanasia of 23 horses, one dog and the monitoring of more than 60 people for possible HeV infection (Anonymous 2011; Smith et al. 2011). There has also been a somewhat surprising early appearance of HeV infection in a horse reported in the first week of January, 2012 (Anonymous 2012a). There have now been a total of 33 separate occurrences of HeV spillover and infection of horses since 1994 in Queensland and New South Wales. Similarly, nearly annual outbreaks of NiV infection, primarily in Bangladesh but also including India, have occurred since 2001 (13 total) since NiV was first recognized from the Malaysian outbreak in 1998. These events have been associated with significantly higher case fatality rates (ranging from 10 to 100%) among the people that have been infected since 2001 following the first outbreak in 1998. To date, there have been a total of 570 reported cases of NiV infection in people of which 305 have been fatal (reviewed in Luby et al. 2009; Pallister et al. 2011a; Anonymous 2012b).

The natural hosts of HeV and NiV have been identified as several species of fruit bats (flying foxes) in the genus *Pteropus* (Chua et al. 2002; Field et al. 2007; Halpin et al. 2000). Although the spillovers and outbreaks of HeV and NiV have all been limited to Australia and Malaysia, Bangladesh, and India; respectively, accumulating serological and limited nucleic acid evidence among a variety of different species of bats suggests that at least antigenically related henipaviruses are circulating in other regions including Thailand, Indonesia, China, Madagascar, and West Africa (Drexler et al. 2009; Hayman et al. 2008; Iehle et al. 2007; Li et al. 2008; Sendow et al. 2006, 2010; Wacharapluesadee et al. 2005).

In addition, serological evidence (cross-reactive antibodies to NiV glycoproteins) has also suggested the apparent transmission of some antigenically related henipaviruses to domestic pigs in West Africa is possible (Hayman et al. 2011). The routes of transmission to humans are also notably different for the henipaviruses, with HeV being transmitted from bats to horses and then to humans, whereas NiV transmission has included transmissions from bats to pigs and then to humans, from bats to humans and from humans to human (Bishop and Broder 2008; Field et al. 2010; Gurley et al. 2007; Homaira et al. 2010; Luby et al. 2009).

The unusually broad species tropism and highly pathogenic capacity of HeV and NiV, together with their uniquely large genome size led to their classification into the new genus henipavirus in the family *Paramyxoviridae* (Lamb et al. 2005). Given the high morbidity and mortality rates associated with henipavirus infection in both humans and livestock, their recognized natural reservoirs in nature and ease of propagation, and a lack of any licensed vaccines or therapeutics, HeV and NiV pose significant biosecurity threats and are classified as biosafety level 4 (BSL-4) pathogens. There are presently no approved or commercially available active or passive therapeutic measures available for preventing or treating henipavirus infection as a result of a natural outbreak, laboratory accident, or deliberate misuse. However, significant developments in effective active and passive immunization strategies against HeV and NiV infection applicable for both human and livestock protection have been reported in recent years. Here, we will summarize these countermeasure developments and the immunization and virus challenge models that have been used to test their efficacy.

2 Henipavirus Tropism and Pathogenesis

As with most paramyxoviruses, henipavirus infection of a susceptible host cell is mediated by an attachment glycoprotein and a fusion (F) glycoprotein, and HeV and NiV possess an attachment glycoprotein (G) (Eaton et al. 2006; Lamb and Parks 2007). The F glycoprotein is a type I transmembrane glycoprotein with an extracellular NH_2-terminus and biologically active F consists of two disulfide-linked subunits, F_1 and F_2, that are generated by the proteolytic cleavage of an F_0 precursor. Biologically active F mediates fusion between the viral and host cell membranes via a class I fusion mechanism involving two α-helical domains known as heptad repeats that mediate the formation of a six-helix bundle during or concomitant with membrane merger (reviewed in Lee and Ataman 2011). The attachment G glycoprotein is a type II transmembrane glycoprotein with its NH_2-terminus oriented towards the cytoplasm and an extracellular COOH-terminus consisting of a stalk and globular head, and both the NiV and HeV head domain structures alone and in complex with ephrin receptors have recently been determined (Bowden et al. 2008, 2010; Colgrave et al. 2012; Xu et al. 2008, 2012). Henipaviruses require both G and F to mediate membrane fusion in a cooperative manner (reviewed in Dutch 2010).

Both HeV and NiV utilize the host cellular membrane proteins ephrin-B2 and ephrin-B3 as entry receptors via their attachment G glycoproteins (Bishop et al. 2007; Bonaparte et al. 2005; Negrete et al. 2005, 2006). Ephrin-B2 and ephrin-B3 are members of a large family of cell surface expressed ligands that bind to Eph receptor tyrosine kinases mediating important bidirectional cell–cell signaling events within the vascular, nervous, and skeletal systems playing critical roles in governing cell migration, attachment and repulsion (Lackmann and Boyd 2008; Pasquale 2010). Ephrin-B2 and ephrin-B3 are highly conserved proteins across vertebrate species (95–96% and 95–98% amino acid identity, respectively), including hosts known to be susceptible to henipavirus infection such as human, horse, pig, cat, dog, and flying foxes (Bossart et al. 2008). Ephrin-B2 is found in arteries, arterioles, capillaries in multiple organs, and tissues but appears absent from the venous components of the vasculature, whereas ephrin-B3 is more prominently found in the nervous system as well as the vasculature (Gale et al. 2001; Pasquale 2008; Su et al. 2004). The recognition of ephrin-B2 and ephrin-B3 as the virus entry receptors for HeV and NiV provided important insight into both the broad species tropisms of the viruses, because of the high sequence conservation of the molecules, as well as the distribution of viral antigen in arterial endothelial cells, smooth muscle, neurons, and some epithelial cells from infected hosts (reviewed in Hooper et al. 2001; Wong 2010).

2.1 Human Pathology

In people, acute henipavirus infection and pathogenesis results from a systemic infection that likely occurs via hematogenous spread of the virus from an undefined primary site of replication (Wong et al. 2002). The key findings of infection are a wide-spread vasculitis and endothelial cell tropism resulting in multinucleated syncytial cells which is considered to be diagnostic of henipavirus infection. There is also prominent parenchymal cell infection and pathogenesis of many, if not most, major organs with the brain, lung, heart, kidney, and spleen severely involved (Chua et al. 1999; Wong et al. 2002, 2009).

Clinically, severe henipavirus infection in humans will present as a severe respiratory disease, encephalitis or a combination of both. In humans, henipavirus infections can also result in a clinically quiescent period following an apparent recovery from an acute infection, which can later recrudesce as encephalitis. This was first observed in the second fatal human HeV case which occurred in an individual who experienced relapsed encephalitis 13 months after infection (O'Sullivan et al. 1997). Among the many more cases of human NiV infection during the initial Malaysian outbreak, it was noted that neurological disease could frequently present later (more than 10 weeks) after a recovery from an acute encephalitic or even an asymptomatic infection (Tan and Wong 2003). Relapsed encephalitis presented from several months to as late as 4 years after infection (Chong and Tan 2003; Tan et al. 2002). An analysis of the first two fatal human cases of HeV infection; one presenting as an acute respiratory disease and the other as relapsed encephalitis, revealed that HeV was neurotropic in each situation causing either acute encephalitis without apparent clinical symptoms or a relapsed encephalitis that resembled those cases of relapsed NiV encephalitis (Wong et al. 2009). These episodes of elapsed encephalitis are believed to be caused by the recrudescence of virus replication that is apparently restricted to central nervous system (CNS) (Tan et al. 2002; Wong et al. 2009). A disseminated endothelial cell infection, vasculitis, thrombosis, and CNS parenchymal cell infection all appear to play essential roles in the fatal outcome of human henipavirus infection (Eaton et al. 2007; Hooper et al. 2001; Wong et al. 2002).

2.2 Animal Pathology and Virus Challenge Models

The development of animal modeling systems of henipavirus infection has been essential for understanding their pathogenic processes and for the evaluation of potential antiviral approaches. A detailed review of all natural and experimental infections of various mammalian species, including their bat reservoir hosts, with either NiV or HeV or both is reviewed in a separate chapter within this series (Geisbert et al.). Here, we will only briefly summarize the salient features of henipavirus in vivo pathogenesis in those well established animal models that have been used in the evaluation of promising antiviral therapeutic strategies.

A major challenge faced by researchers in animal experimentation with the henipaviruses has been the restriction of live virus use to BSL-4 containment; nevertheless, remarkable progress has been made by a number of research teams over the past decade in evaluating the outcomes of experimentally infecting animals with the henipaviruses. It has also been essential that consistent outcomes from experimental animal infection experiments be possible to establish a successful modeling approach. For the henipaviruses, these challenges have been exacerbated by the early observations that neither HeV nor NiV could cause a productive infection with disease in several typical small animal model systems; including mice, rats, and rabbits (Westbury et al. 1995; Wong et al. 2003). Further, for agents such as HeV and NiV more than one accepted animal modeling platform would be required to move forward a potential vaccine or antiviral pharmaceutical for possible human use, falling under the "animal rule" requirement set forth by US food and drug administration (FDA) in 2002 as an alternative licensing pathway for therapeutics against highly pathogenic agents when human efficacy studies are not feasible or ethical (Crawford 2002) (recently reviewed in Aebersold 2012). Several animal model platforms of henipavirus infection have now emerged that essentially mirror the pathogenesis seen in either naturally infected humans or in economically important livestock (horses and pigs).

2.2.1 Pathogenic Natural and Experimental Henipavirus Infections

All occurrences of natural HeV infection in Australia have been in horses, whereas NiV was associated with pigs in Malaysia, although dogs, cats, and horses were also involved (reviewed in Eaton et al. 2006). The pathology caused by either virus in horses (natural or experimental infection with HeV or natural infection with NiV) is more severe than that caused by either virus in pigs (reviewed in Weingartl et al. 2009). The exact mode of transmission from bats to animal hosts is not known, but the most likely scenarios are that virus-contaminated residual fruit pulp spat out by flying foxes are ingested by horses or pigs (Yob et al. 2001) or that urine and or fetal tissues and fluids from infected and/or virus shedding bats contaminates pastures or other food sources for livestock (Halpin et al. 2000).

Naturally acquired HeV infection in horses is often associated with severe disease, and experimental infections are essentially uniformly fatal. The incubation period in naturally infected horses is between 8 and 11 days, and the animals initially become anorexic and depressed with general uneasiness and ataxia, with a developing fever with sweating. Respiration becomes rapid, shallow and labored with pulmonary edema and congestion with nasal discharge being a common terminal feature 1–3 days following the onset of clinical signs. Neurologic disease is also present but less frequent and noted in both terminally ill horses and in those that recovered from respiratory infection (Rogers et al. 1996; Williamson et al. 1998). Infection is wide spread with an endothelial cell tropism with syncytia (Hooper et al. 2001) and virus can be readily recovered from a number of internal organs (Hooper et al. 1997a; Marsh et al. 2011; Murray et al. 1995; Williamson et al. 1998). Experimental infection of horses with NiV has not been carried out, but the brain and spinal cord of one naturally infected horse was confirmed and revealed nonsuppurative meningitis (Hooper et al. 2001).

Experimental NiV infection of pigs has revealed the respiratory system as a primary target organ of virus replication and pathology (Hooper et al. 2001; Middleton et al. 2002), but virus was also present in the kidneys (Middleton et al. 2002) which was less common in naturally infected pigs (Hooper et al. 2001). The involvement of the CNS in pigs was less prevalent and meningitis or meningo-encephalitis more common than encephalitis (Middleton et al. 2002). Other experimental NiV infections (landrace piglets) resulted in a mild clinical disease with fever and respiratory signs, but could cause neurological disease depending on the challenge route of infection (Weingartl et al. 2005). Recoverable virus was obtained from the respiratory, lymphatic and nervous systems, with virus shedding in nasal, pharyngeal and ocular fluids. Experimental HeV infection of pigs has also confirmed their susceptibility to disease, and again infection manifested primarily as a respiratory disease in both Landrace piglets and older Gottingen minipigs, with possible CNS involvement observed in minipigs, and virus shedding was noted in nasal, oral, rectal, and ocular fluids (Li et al. 2010). Both horses and pigs now serve as models for the testing of livestock vaccines against henipaviruses, discussed below.

2.2.2 Pathogenic Experimental Henipavirus Infections

Cats were recognized as a naturally susceptible host for NiV during the 1998–1999 Malaysian outbreaks (Hooper et al. 2001). Experimental infections of cats revealed they are highly susceptible to productive infection by both HeV and NiV and disease is severe. HeV infected cats develop fever and elevated respiratory rates, and there is rapid progression to severe illness and death within 24 h of the onset of clinical signs (Westbury et al. 1995, 1996). HeV disease in cats is quite similar to that seen in horses, with widespread vasculitis and parenchymal lesions in a wide range of organ systems particularly the lungs (Hooper et al. 1997b, 2001). Experimental NiV infection in the cat is essentially identical in outcome as compared to HeV infection and closely resembles most of the pathogenic processes seen in cases of henipavirus infection of people (Hooper et al. 2001; Middleton et al. 2002; Mungall et al. 2006, 2007). The cat model has been used in the testing of the first henipavirus subunit vaccine, discussed below.

The only successful small animal model for henipavirus infection and pathogenesis studies described is the golden hamster, carried out first with NiV challenge experiments (Wong et al. 2003) and more recently with HeV (Guillaume et al. 2009). Hamsters infected by an intranasal or intraperitoneal route with NiV revealed neurological disease signs that progressed rapidly followed by death in 5–8 days. Intranasally challenged animals succumbed to infection 9–15 days later. NiV infection is systemic and endothelial tropic and viral pathology is particularly prevalent in CNS and somewhat to a lesser extent in the lung (Wong et al. 2003). HeV infection of golden hamsters also resembled the pathology seen in acute human NiV cases, including both respiratory and brain pathology with widespread endothelial infection and vasculitis especially in the CNS (Guillaume et al. 2009; Rockx et al. 2011). The golden hamster model has been extensively used to evaluate vaccines and antivirals against henipaviruses, discussed below.

Ferrets have more recently emerged as a highly suitable model for both NiV and HeV infection and pathogenesis (Bossart et al. 2009; Pallister et al. 2009, 2011b). Ferrets challenged with NiV develop severe respiratory and neurological disease within 6–10 days, with generalized and widespread vasculitis and parenchymal lesions following oral nasal challenge with low doses of virus (Bossart et al. 2009; Pallister et al. 2009). Parenchymal lesions and virus antigen were found in the CNS including neurons and virus could be isolated from brain and a wide variety of organs. HeV challenged ferrets, by the oral nasal route results in essentially identical outcomes as reported for NiV challenge (Pallister et al. 2011b), and both models reproduce all the hallmarks seen in henipavirus-infected people. The ferret model has been used in the first human monoclonal antibody passive immunization strategy and also the subunit vaccination approach against henipaviruses, discussed below.

The first nonhuman primate models of NiV and HeV infection have also recently been reported in challenge studies using the African green monkey (AGM) (Geisbert et al. 2010; Rockx et al. 2010). Both NiV and HeV infection of AGMs result in a uniformly lethal disease with low dose challenge by intratracheal virus inoculation within 7–10 days postinfection. Subjects develop severe respiratory disease with congestion and hemorrhage along with fibrosis (Geisbert et al. 2010). There is widespread vasculitis and endothelial and arterial smooth muscle cell virus infection in most organs and tissues. The development of respiratory disease was seen within 7 days postinfection and the progression of lung pathology was observed by radiological examination following an intratracheal inoculation (Geisbert et al. 2010; Rockx et al. 2010). Neurological disease, along with vascular and parenchymal lesions in the brain including infection of neurons and the brainstem is significant in both NiV and HeV infected animals (Geisbert et al. 2010; Rockx et al. 2010). The severe respiratory and neurological disease along with the generalized vasculitis observed in henipavirus-infected AGMs provides the most accurate reflection of what has been reported in henipavirus-infected people. Indeed, for the purposes of moving potential therapeutics forward to human application, the development of the AGM model has been a significant milestone.

3 Active Immunization

The use of safe and efficacious vaccines for several important viral pathogens has been the mainstay of prevention strategies in humans. As with all paramyxoviruses the attachment and fusion glycoproteins are the principal antigens to which virtually all neutralizing antibodies are directed (Lamb and Parks 2007), and neutralizing antibodies are the key vaccine-induced protective mechanisms in the case of some well known paramyxovirus diseases of humans (mumps and measles viruses) (Griffin 1995; Pantaleo and Koup 2004). It is therefore likely that a successful henipavirus vaccine will be one that can elicit neutralizing antibodies.

Paramyxovirus F glycoproteins are homotrimers and the attachment glycoproteins including the henipavirus G glycoproteins are tetramers composed of disulfide-linked dimers (Lee and Ataman 2011), and the native oligomeric structures of viral glycoproteins can influence their antigenicity and immunogenicity (Broder et al. 1994; Wiley and Skehel 1987). For the henipaviruses, the development of vaccines and therapeutics has focused on targeting virus attachment and entry, processes facilitated by their oligomeric viral envelope glycoprotein spikes. Here, we will summarize the successful active and passive immunization and henipavirus challenge studies that have been reported to date (Table 1).

3.1 Live Recombinant Vaccines

The first vaccination and challenge experiments were carried out with NiV in the hamster model using recombinant vaccinia viruses (Guillaume et al. 2004; Table 1). Here, NiV F and G encoding recombinant vaccinia viruses were examined individually and in combination by immunizing hamsters twice with a 1 month interval, using 10^7 PFU of either the F or G encoding recombinants or 5×10^6 of each recombinant together. Antibody titers measured by ELISA and virus neutralization following the second immunization were modest with the sera raised against the NiV G recombinant eliciting the strongest response ($\sim$1:4,000 and <1:50, respectively). All animals were completely protected following an intraperitoneal challenge of 1,000 PFU of NiV, regardless of whether they were

Table 1 Immunization strategies against Nipah virus and Hendra virus trialed in one or more animal models

Immunization strategy	Viral antigen used	Animal challenge model
Active immunization		
Recombinant vaccinia virus[a]	Nipah F and/or G glycoprotein	Golden Hamster
Recombinant canarypox virus[b]	Nipah F and/or G glycoprotein	Pig
Glycoprotein subunit	Hendra or Nipah soluble G glycoprotein	Cat[c]
	Hendra soluble G glycoprotein	Ferret[d]
		African Green Monkey[e]
		Horse[f]
Passive immunization		
Polyclonal antibody[g]	Nipah F and/or G glycoproteins	Golden Hamster
Monoclonal antibody[h]	Nipah F and/or G glycoproteins	Golden Hamster
	Nipah/Hendra G glycoprotein	Ferret[i]
		African green monkey[j]

[a] Nipah F and/or G glycoprotein encoding recombinant vaccinia viruses used to immunize hamsters protects against intraperitoneal NiV challenge (Guillaume et al. 2004)

[b] Nipah F and/or G encoding recombinant canarypox viruses used to immunize pigs protects against intranasal NiV challenge (Weingartl et al. 2006)

[c] Recombinant HeV-sG subunit, in CSIRO triple adjuvant (Montanide, Quil A, and DEAE-dextran) used to immunize cats (three doses of 100 μg, at three week intervals) protects against subcutaneous NiV challenge (Mungall et al. 2006); HeV-sG in CpG (ODN 2007) and Allhydrogel™ used to immunize cats (two doses of 50, 25 μg or 5 μg, day 0 and 21) protects against oronasal NiV challenge (McEachern et al. 2008)

[d] Recombinant HeV-sG in CpG (ODN 2007) and Allhydrogel™ used to immunize ferrets (two doses of 100, 20 μg or 4 μg; day 0 and 20) protects against oronasal HeV challenge (Pallister et al. 2011b); or NiV challenge (Pallister, Middleton and Broder, unpublished)

[e] Recombinant HeV-sG in CpG (ODN 2006) and Allhydrogel™ used to immunize African green monkeys can protect against intratracheal NiV challenge (Hickey et al. 2011); or HeV challenge (Geisbert and Broder, unpublished)

[f] Recombinant HeV-sG used to immunize horses (two dose regime) protects against high dose oronasal HeV challenge and prevents virus replication and shedding (Middleton and Broder, unpublished) (Balzer 2011)

[g] Polyclonal hamster serum against NiV F or G glycoprotein administered by intraperitoneal injection can protect against intraperitoneal NiV challenge (Guillaume et al. 2004)

[h] Mouse mAbs against NiV F or G glycoprotein administered by intraperitoneal injection can protect against intraperitoneal NiV challenge (Guillaume et al. 2006), and mouse mAbs to NiV F administered by intraperitoneal injection pre and postexposure can protect against intraperitoneal HeV challenge (Guillaume et al. 2009)

[i] A cross-reactive neutralizing human mAb against henipavirus G glycoprotein (m102.4) provides postexposure protection in ferrets by intravenous infusion following high dose oronasal NiV challenge (Bossart et al. 2009)

[j] African green monkeys are protected against lethal intratracheal HeV challenge by postexposure intravenous infusion of human mAb m102.4 at 10, 24 or 72 h (Bossart et al. 2011); similar results were obtained against NiV challenge (Geisbert and Broder, unpublished)

immunized with the G or F or both vaccinia virus recombinants (Guillaume et al. 2004). However, both ELISA and neutralizing antibody titers against NiV were considerably elevated following virus challenge indicating virus replication and a humoral anamnestic response. Although the contribution of cell-mediated immunity to protection in this experiment cannot be excluded, passive transfer experiments with antisera, prepared using these recombinant vaccinia viruses, demonstrated protection (discussed below), and together these studies suggested a major role of specific neutralizing antibody in protection. However, even though the highly attenuated vaccinia virus strain, NYVAC, was employed, such a vaccine platform for use in humans is unlikely although there potential for use in livestock is possible although protection against HeV challenge in hamsters has not been reported.

The development of livestock vaccines for HeV and NiV has been a desirable goal because of the association of pigs as an amplifying host for NiV and the fact that all HeV outbreaks in Australia have occurred in horses. The first vaccine that was explored and tested was a recombinant canarypox-based vaccine candidate for swine (Weingartl et al. 2006; Table 1). Similar to the vaccinia virus constructs described above, the NiV F and G glycoprotein genes were used to produce recombinant canarypox virus (ALVAC) vaccine vectors. These ALVAC NiV F and G recombinants were used to immunize 4-week-old pigs twice with a 2 week interval. Similar to the study carried out in hamsters, each ALVAC recombinant was tested alone and in combination, and 10^8 PFU of either the F or G recombinant were employed or a 10^8 PFU dose of each. Piglets were challenged intranasally with a 2.5×10^5 PFU dose of NiV, which was divided between each nostril at 28 days post vaccination. Even though NiV disease in pigs is often less severe than in other susceptible mammalian hosts, the virus does replicate and disseminate to a variety of organ systems with significant levels of recoverable virus present in the respiratory system associated with virus shedding. This vaccination study aimed to prevent disease and also to impede or prevent virus shedding. The study demonstrated protection from NiV-mediated disease in all vaccinated animals by the G, F or combination of ALVAC recombinant vectors. Further, only low levels of viral RNA were detectable in only a few tissue samples and importantly no isolatable virus was observed in the vaccine recipients. In contrast, both high levels of viral RNA as well as isolatable virus were consistently obtained in the control challenged pigs, notably in the throat and nose. A more detailed description of the clinical and pathological findings of NiV infection in the pig model can be found in a separate review in this series (Geisbert, Feldmann and Broder). Neutralizing antibody was elicited by both vectors, with ALVAC-G yielding the strongest response approximately five-fold higher ($\sim$1:1,280). The combined ALVAC-F/G vaccination appeared to be only marginally better than that of G alone, all together the data indicated that either formulation could serve as a protective vaccine against NiV for swine (Weingartl et al. 2006). The ALVAC henipavirus vaccine use in protection against HeV challenge in pigs has not been reported.

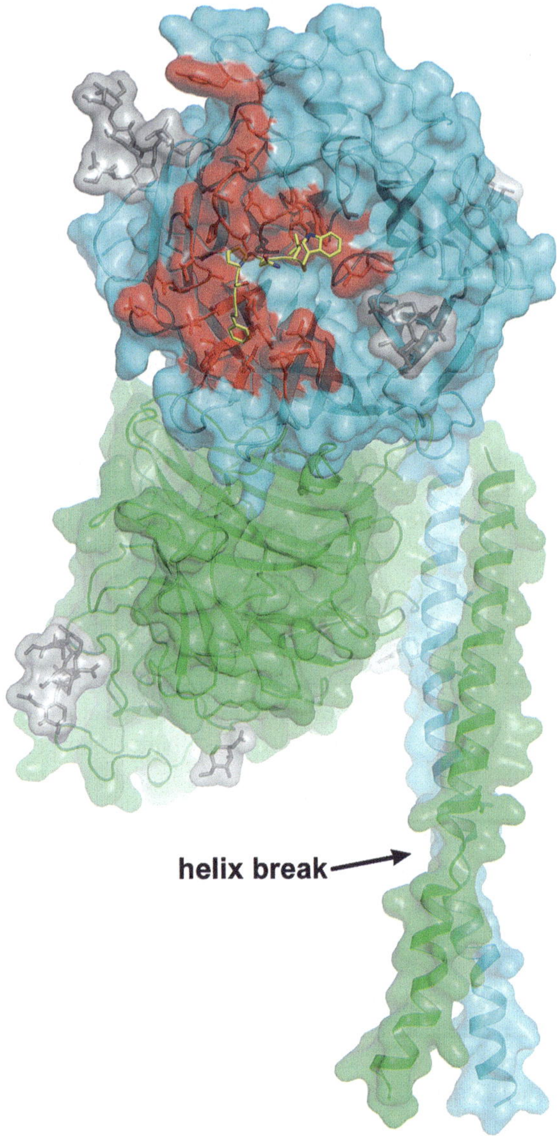

3.2 Subunit Vaccines

In contrast to live or replication competent recombinant viral vaccines, subunit immunogen based vaccines represent a viable option for vaccine development, especially in the case for henipaviruses. These types of vaccines can be relatively quickly implemented and tested, are simple, and can be administered with no risk of infection. Soluble and secreted, oligomeric versions of the G glycoprotein from

◀**Fig. 1** Hendra virus soluble G glycoprotein subunit vaccine. The recombinant HeV-sG glycoprotein subunit vaccine candidate is the entire predicted ectodomain (residues 76–604). Here, HeV-sG is shown as the dimer, with secondary structure and surface exposed elements modeled. HeV-sG dimer has been the purified form of HeV-sG used in all vaccine studies to date. One monomer in the dimer is colored *cyan* and the other is *green*. The secondary structure elements of the two globular head domains (*cyan* and *green*) are derived from the crystal structure of the HeV G head domain, which also revealed that all five predicted N-linked glycosylation sites (N306, N378, N417, N481 and N529) were occupied by carbohydrate moieties (Xu et al. 2012). The N-linked carbohydrate modifications are illustrated as *gray* sticks, but N378 was not modeled in the figure due to weak electron density. The G glycoprotein head domain folds as a six-bladed β-propeller. The structure of the entire HeV G stalk domain (residues 71–173) has not been determined, but here the stalk regions (residues 77–136) of each monomer are modeled (Kelley and Sternberg 2009) and are not a continuous helix (labeled helix break). There are two helical ranges, Thr-77 to Lys-95 and Thr-98 to Ser-135. The hydrophobic residues distribution (most hydrophobic side chains point to the same direction) suggests a bundling tendency. The HeV-sG stalk residues 98–135 appear equivalent to the HN glycoprotein stalk helix domain of the recently reported NDV structure (Yuan et al. 2011). Here, the position of HeV sG head dimer and stalks are oriented based on the alignment to the NDV structure. The ephrin receptor binding face of the *cyan* monomer is facing out and that of the *green* monomer is facing *left*. The ephrin receptor binding region is colored *red* in the *cyan* globular head, and an overlay of the ephrin-B2 G-H loop is shown in *yellow*

both HeV and NiV (sG) were generated in the course of analyzing henipavirus host cell interaction and virus entry mechanisms, and these reagents represented potential subunit vaccines that could be tested in vivo (Bossart et al. 2005). The Hendra virus sG glycoprotein subunit vaccine (Fig. 1) is an engineered secreted version of the molecule in which the transmembrane and cytoplasmic tail domains have been deleted from the N-terminal coding sequence. On expression, the molecule is produced and released from cells because it lacks its transmembrane anchor sequence, and sG has been shown to retain many native characteristics, such as its oligomerization into dimers and tetramers (Bossart et al. 2005), and ability to bind ephrin receptor (Bonaparte et al. 2005). HeV sG has typically been produced in mammalian cell culture expression systems, and is N-linked glycosylated at its predicted sites; recently shown to be variably occupied (between four and seven) depending on the recombinant expression platform utilized (Colgrave et al. 2012; Xu et al. 2012).

HeV sG was first shown to elicit a potent cross-reactive neutralizing (HeV and NiV) antibody response in rabbits (Bossart et al. 2005), suggesting that its application as a subunit vaccine would be promising. Rabbit anti-HeV G antibodies could neutralize both HeV and NiV in cell culture, displaying a slightly higher titer against the homologous virus; however, follow-up studies have indicated that Hendra virus sG appears to present more cross-reactive epitopes (anti-NiV G antibodies) as compared to sG from NiV (Bossart et al. 2007). Four neutralizing epitopes have been mapped on the globular head of the Hendra virus G protein. Two are located on the base of the head and two on the top, in locations resembling those identified as neutralizing sites in other paramyxoviruses (White et al. 2005). Further studies have indicated that there are at least three more independent neutralizing epitopes on HeV G (seven in total) and only one of the seven maps to the ephrin-B2/B3 receptor binding site (Hickey and Broder, unpublished).

Immunization and challenge studies using the HeV-sG subunit immunogen (produced by recombinant vaccinia virus in human cell culture) were first carried out in the cat model and a side-by-side comparison was made using a similarly designed and expressed NiV-sG. Both HeV-sG and NiV-sG could elicit a completely protective immune response against a lethal subcutaneous NiV challenge (NiV-Malaysia isolate) (Mungall et al. 2006; Table 1). Both HeV-sG and NiV-sG were formulated in CSIRO triple adjuvant (Montanide, Quil A, and DEAE-dextran) and three doses of 100 μg each were given at 3 week intervals. This protocol elicited a very strong neutralizing antibody response, and homologous serum neutralizing titers were greater than 1:20,000 and heterologous titers were greater than 1:20,000 to 16-fold lower. Immunized animals and two additional Naïve controls were challenged subcutaneously with 500 $TCID_{50}$ of NiV. Naïve animals developed clinical disease (fever and respiratory distress) and succumbed to infection 6–13 days postnfection, while none of the immunized animals showed any sign of fever or disease. Taqman PCR analysis of samples from Naïve animals revealed high levels of NiV genome in a wide range of tissues, while genome was evident in only two animals and in only four samples with signals below the limit of accurate detection (Mungall et al. 2006). These data suggested that a single vaccine (HeV-sG) could be effective against both HeV and NiV. Further, an analysis of the antibody responses in sera from naturally infected or other immunized sources has also shown that HeV elicited high levels of NiV G cross-reactive antibodies; whereas NiV gave rise to limited cross-reactive antibodies to HeV G. Together, these data suggested that the HeV G stimulates a more cross-reactive immune response (Bossart et al. 2007).

A follow-up study in the cat model explored the protection levels of prechallenge virus neutralizing titers in combination with a high oronasal challenge, the likely natural route of zoonotic exposure (McEachern et al. 2008). Here, two doses of 50, 25 μg or 5 μg of HeV-sG formulated with CpG and Alhydrogel™ were administered at day 0 and 21, and neutralizing antibody titers present at day 35 ranged from 1:32 (low vaccine dose) to 1:512 (high vaccine dose). Animals were challenged on day 42, and the study revealed that even low vaccine dose and just a two-dose immunization protocol could fully protect animals from a high dose oronasal challenge (50,000$TCID_{50}$ of NiV-Malaysia) with prechallenge neutralizing titers as low as 1:32 offering full protection.

During the vaccination and challenge studies carried out in the cat model, the ferret was explored as an alternative and more suitable henipavirus animal model and recently, a HeV-sG immunization study was carried out in the ferret with a HeV challenge (Pallister et al. 2011b). Here, ferrets (two in each group) were vaccinated twice (day 0 and 20) with either a 100, 20 μg or 4 μg dose of HeV-sG formulated in CpG and Alhydrogel™ using similar parameters as the second cat vaccination and challenge study. In this case, recombinant HeV-sG was produced by stable human 293-F cell culture (and all subsequent vaccination studies have since used human cell line derived HeV-sG). Here again, the strong immunogenic

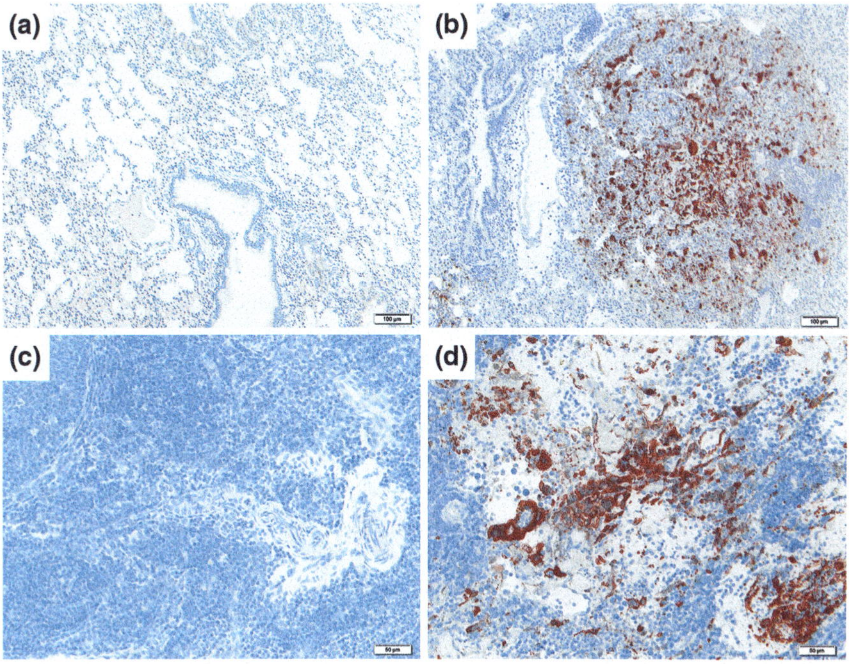

Fig. 2 HeV sG subunit vaccine protection against HeV challenge in the ferret. Immunohistochemical analysis using a NiV nucleoprotein (N) specific rabbit polyclonal antibody following lethal HeV challenge: lung tissue (**a**, **b**) and lymph node tissue (**c**, **d**) in a ferret immunized with recombinant HeV-sG glycoprotein prior to challenge (*left panel*) and a control ferret (*right panel*) (Pallister et al. 2011b). Scale bar (**a**, **b**) = 100 μm, scale bar (**c**, **d**) = 50 μm

characteristics of HeV-sG were noted and at the time of the booster vaccination at day 20, significant neutralizing titers in sera were seen which correlated to the dose of HeV-sG used. Specifically, titers ranged from 1:8,192 in the 100 μg vaccination groups, to 1:1,024 and 1:2,048 in the 20 μg vaccination group, and 1:64 and 1:128 for the 4 μg vaccination group. Prior to challenge, neutralizing titers in all six vaccinated animals were 1:8,192 or greater. All vaccinated ferrets remained completely free of any signs of fever or clinical disease from HeV infection following an oronasal challenge with 5,000 $TCID_{50}$ of a low passage isolate of HeV (Redlands 2008 isolate), whereas the control ferrets developed fever, depression, lack of grooming and tremors, and necropsy revealed HeV induced gross pathological lesions. Histological and immunohistological analysis of control animals revealed widespread systemic vasculitis and parenchymal lesions in many organs and tissues (Fig. 2). In HeV-sG vaccinated ferrets at postmortem, all were found to be gross pathologically and histopathologically normal (except one of four animals in the lose dose (4 μg) group. Further, there was no evidence of virus or viral genome in any tissues or body fluids in any animals in the 100 μg and 20 μg vaccination groups; whereas a low level of genome was detected in the nasal washes from one animal (of four) in the 4 μg vaccinated group; and no isolatable virus was recovered from any vaccinated ferrets. Together, these findings indicate that 100 or 20 μg doses of HeV-sG vaccine can completely prevent a productive HeV infection in the ferret, confirming in a second animal model of henipavirus challenge, that subunit protein vaccination strategy to prevent the infection and shedding of HeV is possible.

In addition to the cat and ferret HeV-sG vaccination studies, HeV-sG vaccination of nonhuman primates (AGMs) followed by intratracheal NiV challenge (Hickey et al. 2011) or intratracheal HeV challenge (Geisbert and Broder, unpublished) has been piloted and here, complete protection from henipavirus-induced disease was achievable. Follow-up experiments and further preclinical development of HeV-sG in vaccine formulations that could potentially be suitable for use in humans are in progress. Finally, because of these successes in HeV-sG vaccine-mediated protection in multiple animal challenge models, and the recent escalation and geographic spread of recurrent HeV outbreaks in Australia, the HeV-sG subunit was recently licensed by a multinational animal health company and is in commercial development as an equine vaccine for use in Australia. Preliminary horse HeV-sG vaccination and HeV challenge studies have recently been conducted in Australia at the high containment BSL-4 facilities of the Australian Animal health laboratories (AAHL) in Geelong. This research program has been a collaborative effort between the Uniformed Services University and Henry M. Jackson foundation, AAHL and a corporate partner. Findings of these preliminary studies were reported at Australian veterinary association, Annual conference in Adelaide, in May 2011. The HeV-sG was used to immunize horses (two dose regime, 3 week interval) and a high dose and low dose antigen formulation was used, and prechallenge HeV neutralizing titers ranged from 1:128 to 1:2,048. Following a high dose oronasal challenge with 2×10^6 $TCID_{50}$ of HeV, all vaccinated horses remained clinically disease-free, and there was no evidence of virus replication or virus shedding in any of the immunized horses (Balzer 2011). A target date for full registration of the equine HeV-sG subunit vaccine is mid2013.

4 Passive Immunization

Traditionally, the antibody response has been the immunologic measure of vaccine efficacy. While the neutralizing antibodies elicited by a vaccine can be highly effective, purified neutralizing antibodies used to passively immunize infected individuals can be equally efficacious. Passive antibody therapy is routinely used as an effective antiviral therapy or as a prophylactic measure for hepatitis B, CMV, varicella, Respiratory syncytial virus (RSV) and rabies virus (Casadevall et al. 2004). The development of mouse-human chimeric mAbs, the ability to "humanize" murine mAbs (Kang et al. 1991; Wright et al. 1992) and the advancement of phage-displayed human antibody libraries has made passive antibody therapy more feasible. Here we will summarize the passive immunization studies that have been carried out as antiviral strategies against the henipaviruses.

4.1 Polyclonal Antibody

The first evidence of passive protection against a NiV challenge was shown using the hamster model discussed above (Guillaume et al. 2004; Table 1). Here, monospecific polyclonal antiserums against NiV F and G were prepared in hamsters by using recombinant vaccinia viruses and immunization with cell lysates prepared from BHK21 hamster cells expressing NiV F or G in complete Freund's adjuvant. The protective efficacy of the various antisera to either NiV F or G or both mixed together, were tested in hamsters. Animals were given 200 μl of antisera followed 1 h later by challenge virus, and another 200 μl of antisera was administered at 24 h, all by intraperitoneal injection. Based on prior studies, the 1,000 PFU challenge of NiV was used and here all the antisera could provide protection (Guillaume et al. 2004) demonstrating the importance of the humoral response to the NiV envelope glycoproteins as a mechanism of protection.

4.2 Monoclonal Antibody

The polyclonal antisera protection studies in hamsters were followed-up using two murine mAbs against NiV F and NiV G as passive immunotherapies (Guillaume et al. 2006; Table 1). Here, the mAbs were examined as ascitic fluid preparations and delivered to hamsters by intraperitoneal administration. A series of experiments using various amounts of mAbs administered 24 h before NiV virus challenge and again at 1 h following challenge were conducted and hamsters were inoculated with 100 LD_{50} of NiV. Also, the efficacy of the anti-F or anti-G mAbs was examined as a postexposure therapy from 1 to 96 h after virus challenge. Hamsters that received mAb before and immediately following the intraperitoneal challenge of NiV were completely protected (Guillaume et al. 2006). However, only partial protection (50%) was observed with the anti-G mAbs when the animals were inoculated up to 24 h after challenge. Although F-specific mAb could protect, higher amounts were required in comparison to anti-G mAb. High levels of either anti-G or anti-F mAbs appeared to yield sterilizing immunity, when administered intraperitoneally with virus, while lower amounts of antibody could still protect against fatal infection but did result in measurable increases in anti-NiV antibodies following virus challenge. Similar studies using the hamster with a HeV challenge have also been conducted, and cross-reactive mouse mAbs specific for the NiV F glycoprotein could protect hamsters from HeV disease if given before virus challenge (Guillaume et al. 2009; Table 1). Together, these studies provided the proof-of-concept that passively immunotherapy against NiV infection by targeting the viral envelope glycoproteins is possible.

However, passive immunization using antibody administered systemically following virus infection would be the more likely scenario a person would encounter and this would require a more potent mAb therapeutic. Further the humanization of mouse mAbs is a considerable undertaking in order to formulate an acceptable therapeutic product suitable for human use, and also one without guaranteed success.

The development of molecular-based platforms for the identification and isolation of human-derived recombinant mAbs has accelerated the development of antibodies with the potential for human use. A major advance has been in the development of the phage display platform of combinatorial antibody libraries (Rader and Barbas 1997) that can encode human antibodies in the form of Fab fragments or single-chain variable region fragments (scFvs). Using recombinant antibody techniques, neutralizing human mAbs specific for the henipavirus G glycoprotein have been isolated and characterized (Zhu et al. 2006). These mAbs were generated by panning a large na human phage-display antibody library containing about 10^{10} different phage-displayed Fabs using recombinant HeV-sG (the subunit vaccine immunogen described above). In particular, two Fabs, m101 and m102, had significant neutralizing activities against live virus and m101 was converted to a full length human IgG1 antibody. IgG1 m101 was exceptionally potent in neutralizing infectious HeV; complete (100%) neutralization was achieved with 12.5 μg/ml and 98% neutralization with 1.6 μg/ml using a 96-well plate-based assay using 200 $TCID_{50}$ of virus with Vero cells. The other mAb, m102, exhibited the highest level of cross-reactive neutralization of both NiV and HeV, and m102 was affinity maturated by light-chain shuffling combined with random mutagenesis of its heavy-chain variable domain and clones were reisolated using HeV-sG. One of the selected antibody Fab clones, m102.4, had improved affinity of binding to HeV-sG and it was converted to IgG1 format and tested against infectious NiV and HeV. The mAb m102.4 exhibited exceptionally potent and cross-reactive inhibitory activity against NiV and HeV with 50% inhibitory concentrations below 0.04 and 0.6 μg/ml, respectively (Zhu et al. 2008). These were the first human mAbs identified against the henipaviruses and because of their potency in vitro, they have been piloted in vivo in NiV and HeV animal challenge studies.

The first in vivo efficacy study of human mAb m102.4 was carried out in the ferret model (Bossart et al. 2009; Table 1), and groups of three animals each were given a single 50 mg dose ($\sim$25 mg/kg) either 24 h before (pre-) or 10 h after (post-) NiV challenge. The mAb m102.4 was administered via intravenous catheter, but one animal was given m102.4 intraperitoneally because of a catheter block and control ferrets were given PBS. Ferrets were challenged by oronasal inoculation with a 5,000 $TCID_{50}$ dose of NiV (ten-fold the minimal infectious dose, 50%). Control ferrets became febrile with signs of clinical illness by day 6 following challenge, and by day 8 were severely depressed with subcutaneous edema of the head and cutaneous hemorrhages and were euthanized. By day 8 one animal in the pregroup and all ferrets in the postgroup were febrile with variable levels of depression and suppression of play activity, but by day 10, fever in some animals started to fall, and these animals had moderate (pregroup) or mild (postgroup) edema of the throat; while one animal in each group remained febrile with no other clinical signs. By day 13, however, two of three ferrets in the pregroup were depressed and inappetant with cutaneous ecchymoses, and one animal had marked hind limb paresis and generalized tremor and both animals were euthanized. However, at day 13 all other ferrets (3/3 in the postgroup and 1/3 pregroup) were well and free of any disease signs and remained so until the end of the study at day 20 postchallenge. Gross and microscopic pathology in control animals revealed severe systemic pathology; however, in the two pregroup animals that experienced a delayed disease course, the frequency of pinpoint hemorrhagic lesions observed in the pulmonary parenchyma was reduced and lesions were much smaller suggesting the disease progression in the respiratory tract had been dampened, consistent with their survival to 13 dpi. There were no significant pathological abnormalities found in any of the surviving ferrets which included 3/3 animals that received a signal dose of m102.4 10 h following NiV challenge (Bossart et al. 2009). Additional testing of m102.4 also confirmed its potent cross-reactive neutralization activity against the NiV-Malaysia (Chua et al. 1999), the original HeV-1994 (Murray et al. 1995), the recent HeV-Redlands (Anonymous 2008), and NiV-Bangladesh isolates (Harcourt et al. 2005). Although this 10 h postexposure treatment scenario is relevant to a known exposure, such as a laboratory accident, the therapeutic window of m102.4 needed further characterization. In addition, the average distribution and elimination halftimes of m102.4 in ferrets were calculated to be 1.48 and 3.53 days; respectively, and although distribution was good, clearance of the human mAb in ferrets was rapid suggesting that more dosing could have had a beneficial effect especially in the pretreated animals. Nevertheless, this study represented the first human mAb therapy successfully evaluated in vivo for prevention of lethal henipavirus infection.

To meet the requirements for potential development and approved use of m102.4 in people, additional in vivo studies are required as well as a second animal model showing efficacy. To accomplish these goals, m102.4 was examined in the recently developed, lethal henipavirus infection model in a nonhuman primate (the AGM) (Geisbert et al. 2010; Rockx et al. 2010; Table 1). This first study focused on the potential real life scenario of requiring mAb therapy as a postexposure treatment against virus infection and HeV was examined first, building on the prior results obtained with NiV challenge and m102.4 postexposure treatment. Fourteen AGMs were challenged intratracheally with a lethal dose of HeV and 12 animals were infused twice with a 100 mg dose ($\sim$20 mg/kg) of m102.4 beginning at either 10 and 24 h or 72 h p.i. and again approximately 48 h later. The presence of viral RNA, infectious virus and HeV-specific immune responses demonstrated that all subjects became infected following challenge; however, all 12 AGMs that received m102.4 survived infection; whereas the untreated control subjects succumbed to severe systemic disease on day 8 following virus challenge. It was noted that animals in the 72 h treatment group exhibited neurological signs of disease, but all these animals also recovered by day 16. In AGMs a distribution halftime of $\sim$1 day and the elimination halftime of $\sim$11 days of mAb m102.4 were calculated.

HeV-specific pathological changes were not observed in any of the m102.4 treated animals, whereas immunohistochemistry analysis revealed high levels of HeV antigen in the lung and brain of the control subjects along with high levels

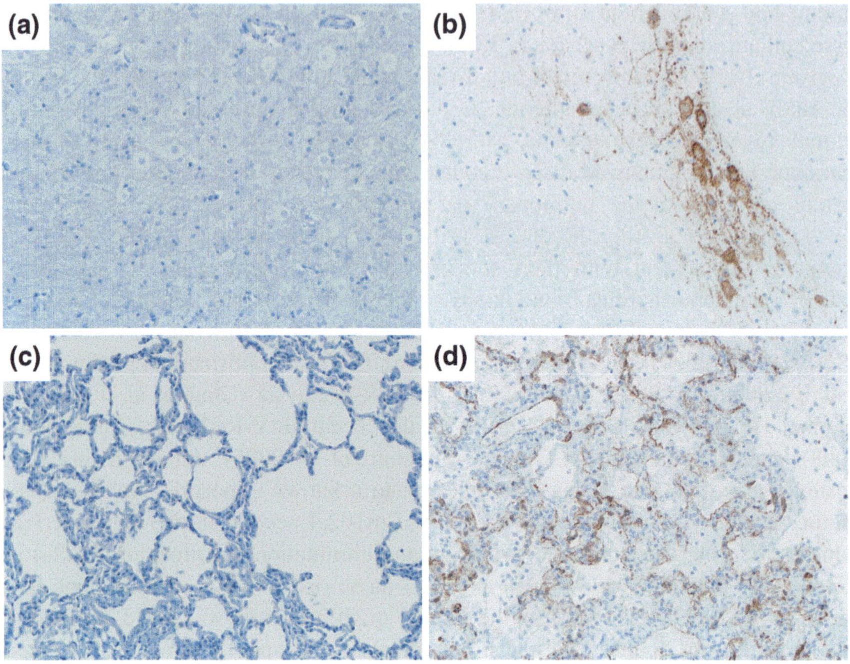

Fig. 3 Passive Immunotherapy against HeV challenge in the African green monkey. Localization of HeV antigen by immunohistochemical stain in the brain stem (**a**, **b**) and lower lung (**c**, **d**) of m102.4 treated subject (72 h/d5; *left panels*) or control subject (*right panels*). Sections were stained with a NiV nucleoprotein (N) specific rabbit polyclonal antibody and images were obtained at an original magnification of at 20X. Figure is modified and reproduced from original data previously published, with permission (Bossart et al. 2011)

of HeV RNA in the lung, spleen, lymph nodes, and brain. HeV was recovered from numerous tissues, highlighting the extensive dissemination of HeV within the body of control subjects. Blood samples collected over the course of infection were assayed for infectious HeV and HeV RNA, and consistent with previous findings, a gradual rise in viral RNA over time was evident in the control subjects and HeV was isolated from blood of both control subjects. Only very low levels of viral RNA could be detected on days 6 and 10 in some m102.4-treated subjects and all blood samples from all m102.4-treated subjects were negative for HeV isolation. Tissue samples collected from all m102.4-treated subjects on necropsy were assayed for the presence of HeV RNA and infectious virus, and only occasionally, were very low levels of viral RNA detected (spleen, lung, and brain) and predominantly only in the late treatment cohort (72 h/d5). Importantly, no infectious HeV could be recovered from any of the tissues from m102.4-treated subjects. Together, these data demonstrated that mAb m102.4 could prevent widespread HeV dissemination in the challenged subjects. Lung, brain, and spleen tissues from surviving subjects were also assayed for the presence of HeV antigen. All tissue architecture appeared normal and all survivor tissues examined were negative for HeV antigen (Fig. 3), whereas control subject tissues showed significant HeV antigen staining. These results were the first successful postexposure in vivo therapy against HeV and highlight the importance of further developing human mAbs, such as m102.4, to combat emerging viral pathogens like the henipaviruses.

5 Concluding Remarks

NiV and HeV are the only examples of zoonotic paramyxoviruses that are capable of infecting and causing disease with a remarkably broad host range. At present, there are no approved commercially available vaccines or passive immunization therapeutics for preventing or treating henipavirus infection in humans or animals. Because of the risk from natural infection, laboratory accident or the potential of deliberate misuse of HeV and NiV, and the high morbidity and mortality associated with infection, development of effective countermeasures has been a priority. Over the past several years, a focus on developing viable henipavirus animal challenge models has often gone hand-in-hand with targeted research strategies on preventing henipavirus entry by neutralizing antibodies; and together these efforts have led to the development of highly effective active and passive immunization strategies.

At present, there is a cross-reactive human mAb, m102.4, which is capable of neutralizing all known isolates of HeV and NiV. The m102.4 mAb has been demonstrated to be exceptionally efficacious as a postexposure therapy by protecting both ferrets and nonhuman primates from lethal henipavirus disease, even under conditions of high dose virus challenge by oronasal or intratracheal routes; and in monkeys it can afford protection as late as 3 days following infection (Bossart et al. 2009, 2011). This human mAb has already been administered to healthy individuals on a compassionate use basis and no adverse reactions were seen. The m102.4 mAb is presently in preclinical development in both the United States and in Australia. As an active vaccination strategy, a recombinant, subunit vaccine that consists of entire ectodomain of the G glycoprotein of HeV has shown remarkable efficacy and is cross protective against HeV and NiV. HeV-sG vaccination can protect against henipavirus challenge in cats, ferrets, monkeys, and horses. Immunization protocols of only two doses with several adjuvant systems, elicits a cross-reactive neutralizing antibody response that can completely protect from high dose virus challenge, with no evidence of virus replication, or anamnestic humoral response. Further, there is no evidence of virus shedding in vaccinated animals. The HeV-sG subunit vaccine immunogen has been licensed for commercial development as a livestock vaccine and currently being developed in Australia for use in horses. Finally, the HeV-sG subunit vaccine could be further developed for potential use in humans, and preliminary data in nonhuman primates using sG formulations in Alhydrogel™ appear promising.

Acknowledgments The views expressed in the manuscript are solely those of the authors, and they do not represent official views or opinions of the Department of Defense or The Uniformed Services University of the Health Science. Supported in part by the Department of Health and Human Services, National Institutes of Health, grants AI054715, AI077995, AI082121.

Conflict of Interest C.C.B is a United States federal employee; C.C.B is coinventor on patents relating to human monoclonal antibodies against Hendra and Nipah viruses and C.C.B and K.N.B are coinventors on patents relating to soluble forms of Hendra and Nipah envelope glycoproteins and vaccines; assignees are The United States of America as represented by the Department of Health and Human Services (Washington, DC), and the Henry M. Jackson Foundation for the Advancement of Military Medicine, Inc. (Bethesda, MD). All other authors declare no competing interests.

References

Aebersold P (2012) FDA experience with medical countermeasures under the animal rule. Adv Prev Med 2012:507–571

Anonymous (2008) Hendra virus, human, equine—Australia (04), Queensland. Pro-MED-mail, International Society for Infectious Diseases, 25 July, archive no. 20080725.2260. Available at www.promedmail.org

Anonymous (2009) Human, Equine—Australia (04), Queensland Fatal. Pro-MED-mail, International Society for Infectious Diseases, 3 Sep, archive no. 20090903.3098. Available at www.promedmail.org

Anonymous (2011) Hendra Virus, Equine—Australia (28), Queensland, New South Wales. Pro-MED-mail International Society for Infectious Diseases, 12 Oct, archive no. 20111013.3061. Available at www.promedmail.org

Anonymous (2012a) Hendra virus, equine—Australia, Queensland. Pro-MED-mail International Society for Infectious Diseases, 6 Jan, archive no. 20120106.1001359. Available at www.promedmail.org

Anonymous (2012b) Nipah encephalitis, human—Bangladesh, Jipurhat. Pro-MED-mail, International Society for Infectious Diseases, 7 Feb, archive no. 20120212.1040138. Available at www.promedmail.org

Balzer M (2011) Hendra vaccine success announced. Aust Vet J 89(7):N2–N3

Bishop KA, Broder CC (2008) Hendra and Nipah: lethal zoonotic paramyxoviruses. In: Scheld WM, Hammer SM, Hughes JM (eds) Emerging infections. American Society for Microbiology, Washington, pp 155–187

Bishop KA, Stantchev TS, Hickey AC, Khetawat D, Bossart KN, Krasnoperov V, Gill P, Feng YR, Wang L, Eaton BT, Wang LF, Broder CC (2007) Identification of hendra virus G glycoprotein residues that are critical for receptor binding. J Virol 81(11):5893–5901

Bonaparte MI, Dimitrov AS, Bossart KN, Crameri G, Mungall BA, Bishop KA, Choudhry V, Dimitrov DS, Wang LF, Eaton BT, Broder CC (2005) Ephrin-B2 ligand is a functional receptor for Hendra virus and Nipah virus. Proc Natl Acad Sci U S A 102(30):10652–10657

Bossart KN, Crameri G, Dimitrov AS, Mungall BA, Feng YR, Patch JR, Choudhary A, Wang LF, Eaton BT, Broder CC (2005) Receptor binding, fusion inhibition and induction of cross-reactive neutralizing antibodies by a soluble G glycoprotein of hendra virus. J Virol 79(11):6690–6702

Bossart KN, McEachern JA, Hickey AC, Choudhry V, Dimitrov DS, Eaton BT, Wang LF (2007) Neutralization assays for differential henipavirus serology using bio-plex protein array systems. J Virol Methods 142(1–2):29–40

Bossart KN, Tachedjian M, McEachern JA, Crameri G, Zhu Z, Dimitrov DS, Broder CC, Wang LF (2008) Functional studies of host-specific ephrin-B ligands as henipavirus receptors. Virology 372(2):357–371

Bossart KN, Zhu Z, Middleton D, Klippel J, Crameri G, Bingham J, McEachern JA, Green D, Hancock TJ, Chan YP, Hickey AC, Dimitrov DS, Wang LF, Broder CC (2009) A neutralizing human monoclonal antibody protects against lethal disease in a new ferret model of acute Nipah virus infection. PLoS Pathog 5(10):e1000642

Bossart KN, Geisbert TW, Feldmann H, Zhu Z, Feldmann F, Geisbert JB, Yan L, Feng YR, Brining D, Scott D, Wang Y, Dimitrov AS, Callison J, Chan YP, Hickey AC, Dimitrov DS, Broder CC, and Rockx B (2011) A neutralizing human monoclonal antibody protects african green monkeys from hendra virus challenge. Sci Transl Med 3(105):103–105

Bowden TA, Aricescu AR, Gilbert RJ, Grimes JM, Jones EY, Stuart DI (2008) Structural basis of Nipah and Hendra virus attachment to their cell-surface receptor ephrin-B2. Nat Struct Mol Biol 15(6):567–572

Bowden TA, Crispin M, Harvey DJ, Jones EY, Stuart DI (2010) Dimeric architecture of the hendra virus attachment glycoprotein: evidence for a conserved mode of assembly. J Virol 84(12):6208–6217

Broder CC, Earl PL, Long D, Abedon ST, Moss B, Doms RW (1994) Antigenic implications of human immunodeficiency virus type 1 envelope quaternary structure: oligomer-specific and sensitive monoclonal antibodies. Proc Natl Acad Sci U S A 91(24):11699–11703

Casadevall A, Dadachova E, Pirofski LA (2004) Passive antibody therapy for infectious diseases. Nat Rev Microbiol 2(9):695–703

Chong HT, Tan CT (2003) Relapsed and late-onset Nipah encephalitis, a report of three cases. Neurol J Southeast Asia 8:109–112

Chua KB, Goh KJ, Wong KT, Kamarulzaman A, Tan PS, Ksiazek TG, Zaki SR, Paul G, Lam SK, Tan CT (1999) Fatal encephalitis due to Nipah virus among pig-farmers in Malaysia. Lancet 354(9186):1257–1259

Chua KB, Lek Koh C, Hooi PS, Wee KF, Khong JH, Chua BH, Chan YP, Lim ME, Lam SK (2002) Isolation of Nipah virus from Malaysian island flying-foxes. Microbes Infect 4(2):145–151

Colgrave ML, Snelling HJ, Shiell BR, Feng YR, Chan YP, Bossart KN, Xu K, Nikolov DB, Broder CC, Michalski WP (2012) Site occupancy and glycan compositional analysis of two soluble recombinant forms of the attachment glycoprotein of hendra virus. Glycobiology 22(4):572–584

Crawford LM (2002) New drug and biological drug products; evidence needed to demonstrate effectiveness of new drugs when human efficacy studies are not ethical or feasible. Office of the Federal Register, National Archives and Records Administration (NARA), Washington, pp 37988–37998

Drexler JF, Corman VM, Gloza-Rausch F, Seebens A, Annan A, Ipsen A, Kruppa T, Muller MA, Kalko EK, Adu-Sarkodie Y, Oppong S, Drosten C (2009) Henipavirus RNA in African bats. PLoS One 4(7):e6367

Dutch RE (2010) Entry and fusion of emerging paramyxoviruses. PLoS Pathog 6(6):e1000881

Eaton BT, Broder CC, Middleton D, Wang LF (2006) Hendra and Nipah viruses: different and dangerous. Nat Rev Microbiol 4(1):23–35

Eaton BT, Mackenzie JS, Wang L-F (2007) Henipaviruses. In: Knipe DM, Howley PM (eds) Fields virology. Lippincott Williams & Wilkins, Philadelphia, pp 1587–1600

Field HE, Mackenzie JS, Daszak P (2007) Henipaviruses: emerging paramyxoviruses associated with fruit bats. Curr Top Microbiol Immunol 315:133–159

Field H, Schaaf K, Kung N, Simon C, Waltisbuhl D, Hobert H, Moore F, Middleton D, Crook A, Smith G, Daniels P, Glanville R, Lovell D (2010) Hendra virus outbreak with novel clinical features Australia. Emerg Infect Dis 16(2):338–340

Gale NW, Baluk P, Pan L, Kwan M, Holash J, DeChiara TM, McDonald DM, Yancopoulos GD (2001) Ephrin-B2 selectively marks arterial vessels and neovascularization sites in the adult, with expression in both endothelial and smooth-muscle cells. Dev Biol 230(2):151–160

Geisbert TW, Daddario-DiCaprio KM, Hickey AC, Smith MA, Chan YP, Wang LF, Mattapallil JJ, Geisbert JB, Bossart KN, Broder CC (2010) Development of an acute and highly pathogenic nonhuman primate model of Nipah virus infection. PLoS One 5(5):e10690

Griffin DE (1995) Immune responses during measles virus infection. Curr Top Microbiol Immunol 191:117–134

Guillaume V, Contamin H, Loth P, Georges-Courbot MC, Lefeuvre A, Marianneau P, Chua KB, Lam SK, Buckland R, Deubel V, Wild TF (2004) Nipah virus: vaccination and passive protection studies in a hamster model. J Virol 78(2):834–840

Guillaume V, Contamin H, Loth P, Grosjean I, Courbot MC, Deubel V, Buckland R, Wild TF (2006) Antibody prophylaxis and therapy against Nipah virus infection in hamsters. J Virol 80(4):1972–1978

Guillaume V, Wong KT, Looi RY, Georges-Courbot MC, Barrot L, Buckland R, Wild TF, Horvat B (2009) Acute hendra virus infection: analysis of the pathogenesis and passive antibody protection in the hamster model. Virology 387(2):459–465

Gurley ES, Montgomery JM, Hossain MJ, Bell M, Azad AK, Islam MR, Molla MA, Carroll DS, Ksiazek TG, Rota PA, Lowe L, Comer JA, Rollin P, Czub M, Grolla A, Feldmann H, Luby SP, Woodward JL, Breiman RF (2007) Person-to-person transmission of Nipah virus in a Bangladeshi community. Emerg Infect Dis 13(7):1031–1037

Halpin K, Young PL, Field HE, Mackenzie JS (2000) Isolation of hendra virus from pteropid bats: a natural reservoir of hendra virus. J Gen Virol 81(Pt 8):1927–1932

Harcourt BH, Lowe L, Tamin A, Liu X, Bankamp B, Bowden N, Rollin PE, Comer JA, Ksiazek TG, Hossain MJ, Gurley ES, Breiman RF, Bellini WJ, Rota PA (2005) Genetic characterization of Nipah virus, Bangladesh 2004. Emerg Infect Dis 11(10):1594–1597

Hayman DT, Suu-Ire R, Breed AC, McEachern JA, Wang L, Wood JL, Cunningham AA (2008) Evidence of henipavirus infection in west african fruit bats. PLoS ONE 3(7):e2739

Hayman DT, Wang LF, Barr J, Baker KS, Suu-Ire R, Broder CC, Cunningham AA, Wood JL (2011) Antibodies to henipavirus or henipa-like viruses in domestic pigs in Ghana, West Africa. PLoS One 6(9):e25256

Hickey AC, Bossart KN, Rockx B, Feldmann F, Geisbert JB, Yan L, Feng YR, Feldmann H, Geisbert TW, Broder CC (2011) Vaccination of nonhuman primates with a recombinant soluble henipavirus attachment G glycoprotein protects against lethal Nipah virus challenge. Am Soc Virol 40–2:195

Homaira N, Rahman M, Hossain MJ, Epstein JH, Sultana R, Khan MS, Podder G, Nahar K, Ahmed B, Gurley ES, Daszak P, Lipkin WI, Rollin PE, Comer JA, Ksiazek TG, Luby SP (2010) Nipah virus outbreak with person-to-person transmission in a district of Bangladesh 2007. Epidemiol Infect 138(11):1630–1636

Hooper PT, Ketterer PJ, Hyatt AD, Russell GM (1997a) Lesions of experimental equine morbillivirus pneumonia in horses. Vet Pathol 34(4):312–322

Hooper PT, Westbury HA, Russell GM (1997b) The lesions of experimental equine morbillivirus disease in cats and guinea pigs. Vet Pathol 34(4):323–329

Hooper P, Zaki S, Daniels P, Middleton D (2001) Comparative pathology of the diseases caused by hendra and Nipah viruses. Microbes Infect 3(4):315–322

Iehle C, Razafitrimo G, Razainirina J, Andriaholinirina N, Goodman SM, Faure C, Georges-Courbot MC, Rousset D, Reynes JM (2007) Henipavirus and tioman virus antibodies in pteropodid bats. Madagascar. Emerg Infect Dis 13(1):159–161

Kang AS, Jones TM, Burton DR (1991) Antibody redesign by chain shuffling from random combinatorial immunoglobulin libraries. Proc Natl Acad Sci U S A 88(24):11120–11123

Kelley LA, Sternberg MJ (2009) Protein structure prediction on the web: a case study using the phyre server. Nat Protoc 4(3):363–371

Lackmann M, Boyd AW (2008) Eph, a protein family coming of age: more confusion, insight, or complexity? Sci Signal 1(15): re2

Lamb RA, Parks GD (2007) *Paramyxoviridae*: the viruses and their replication. In: Knipe DM, Howley PM (eds) Fields virology. Lippincott Williams & Wilkins, Philadelphia, pp 1449–1496

Lamb RA, Collins PL, Kolakofsky D, Melero JA, Nagai Y, Oldstone MBA, Pringle CR, Rima BK (2005) Family Paramyxoviridae. In: Fauquet CM, Mayo MA, Maniloff J, Desselberger U, Ball LA (eds) Virus taxonomy: the classification and nomenclature of viruses. The eighth report of the international committee in taxonomy of viruses. Elsevier Academic Press, San Diego, pp 655–668

Lee B, Ataman ZA (2011) Modes of paramyxovirus fusion: a henipavirus perspective. Trends Microbiol 19(8):389–399

Li Y, Wang J, Hickey AC, Zhang Y, Wu Y, Zhang H, Yuan J, Han Z, McEachern J, Broder CC, Wang LF, Shi Z (2008) Antibodies to Nipah or Nipah-like viruses in bats. China. Emerg Infect Dis 14(12):1974–1976

Li M, Embury-Hyatt C, Weingartl HM (2010) Experimental inoculation study indicates swine as a potential host for hendra virus. Vet Res 41(3):33

Luby SP, Hossain MJ, Gurley ES, Ahmed BN, Banu S, Khan SU, Homaira N, Rota PA, Rollin PE, Comer JA, Kenah E, Ksiazek TG, Rahman M (2009) Recurrent zoonotic transmission of Nipah virus into humans, Bangladesh, 2001–2007. Emerg Infect Dis 15(8):1229–1235

Marianneau P, Guillaume V, Wong T, Badmanathan M, Looi RY, Murri S, Loth P, Tordo N, Wild F, Horvat B, Contamin H (2010) Experimental infection of squirrel monkeys with Nipah virus. Emerg Infect Dis 16(3):507–510

Marsh GA, Haining J, Hancock TJ, Robinson R, Foord A, Barr JA, Riddell S, Heine H, White JR, Crameri G, Field HE, Middleton D, Wang LF (2011) Experimental infection of horses with hendra virus/Australia/Horse/2008/Redlands; implications for control of transmission to people and horses. Emerg Infect Dis 17(12):2232–2238

McEachern JA, Bingham J, Crameri G, Green DJ, Hancock TJ, Middleton D, Feng YR, Broder CC, Wang LF, Bossart KN (2008) A recombinant subunit vaccine formulation protects against lethal Nipah virus challenge in cats. Vaccine 26(31):3842–3852

Middleton DJ, Westbury HA, Morrissy CJ, van der Heide BM, Russell GM, Braun MA, Hyatt AD (2002) Experimental Nipah virus infection in pigs and cats. J Comp Pathol 126(2–3):124–136

Middleton DJ, Morrissy CJ, van der Heide BM, Russell GM, Braun MA, Westbury HA, Halpin K, Daniels PW (2007) Experimental Nipah Virus Infection in Pteropid Bats (Pteropus poliocephalus). J Comp Pathol 136(4):266–272

Mungall BA, Middleton D, Crameri G, Bingham J, Halpin K, Russell G, Green D, McEachern J, Pritchard LI, Eaton BT, Wang LF, Bossart KN, Broder CC (2006) Feline model of acute Nipah virus infection and protection with a soluble glycoprotein-based subunit vaccine. J Virol 80(24):12293–12302

Mungall BA, Middleton D, Crameri G, Halpin K, Bingham J, Eaton BT, Broder CC (2007) Vertical transmission and fetal replication of Nipah virus in an experimentally infected cat. J Infect Dis 196(6):812–816

Murray K, Selleck P, Hooper P, Hyatt A, Gould A, Gleeson L, Westbury H, Hiley L, Selvey L, Rodwell B et al (1995) A morbillivirus that caused fatal disease in horses and humans. Science 268(5207):94–97

Murray K, Eaton B, Hooper P, Wang L, Williamson M, Young P (1998) Flying foxes, horses, and humans: a zoonosis caused be a new member of the *Paramyxoviridae*. In: Scheld WM, Armstrong D, Hughes JM (eds) Emerging infections. ASM Press, Washington, pp 43–58

Negrete OA, Levroney EL, Aguilar HC, Bertolotti-Ciarlet A, Nazarian R, Tajyar S, Lee B (2005) EphrinB2 is the entry receptor for Nipah virus, an emergent deadly paramyxovirus. Nature 436(7049):401–405

Negrete OA, Wolf MC, Aguilar HC, Enterlein S, Wang W, Muhlberger E, Su SV, Bertolotti-Ciarlet A, Flick R, Lee B (2006) Two key residues in Ephrinb3 are critical for its use as an alternative receptor for Nipah virus. PLoS Pathog 2(2):e7

O'Sullivan JD, Allworth AM, Paterson DL, Snow TM, Boots R, Gleeson LJ, Gould AR, Hyatt AD, Bradfield J (1997) Fatal encephalitis due to novel paramyxovirus transmitted from horses. Lancet 349(9045):93–95

Pallister J, Middleton D, Crameri G, Yamada M, Klein R, Hancock TJ, Foord A, Shiell B, Michalski W, Broder CC, Wang LF (2009) Chloroquine administration does not prevent Nipah virus infection and disease in ferrets. J Virol 83(22):11979–11982

Pallister J, Middleton D, Broder CC, Wang LF (2011a) Henipavirus vaccine development. J Bioterr Biodef S1:005. doi:10.4172/2157-2526.S1-005

Pallister J, Middleton D, Wang LF, Klein R, Haining J, Robinson R, Yamada M, White J, Payne J, Feng YR, Chan YP, Broder CC (2011b) A recombinant hendra virus G glycoprotein-based subunit vaccine protects ferrets from lethal hendra virus challenge. Vaccine 29(34):5623–5630

Pantaleo G, Koup RA (2004) Correlates of immune protection in HIV-1 infection: what we know, what we don't know, what we should know. Nat Med 10(8):806–810

Pasquale EB (2008) Eph-ephrin bidirectional signaling in physiology and disease. Cell 133(1):38–52

Pasquale EB (2010) Eph receptors and ephrins in cancer: bidirectional signalling and beyond. Nat Rev Cancer 10(3):165–180

Playford EG, McCall B, Smith G, Slinko V, Allen G, Smith I, Moore F, Taylor C, Kung YH, Field H (2010) Human hendra virus encephalitis associated with equine outbreak, Australia 2008. Emerg Infect Dis 16(2):219–223

Rader C, Barbas CF 3rd (1997) Phage display of combinatorial antibody libraries. Curr Opin Biotechnol 8(4):503–508

Rockx B, Bossart KN, Feldmann F, Geisbert JB, Hickey AC, Brining D, Callison J, Safronetz D, Marzi A, Kercher L, Long D, Broder CC, Feldmann H, Geisbert TW (2010) A novel model of lethal hendra virus infection in African green monkeys and the effectiveness of ribavirin treatment. J Virol 84(19):9831–9839

Rockx B, Brining D, Kramer J, Callison J, Ebihara H, Mansfield K, Feldmann H (2011) Clinical outcome of henipavirus infection in hamsters is determined by the route and dose of infection. J Virol 85(15):7658–7671

Rogers RJ, Douglas IC, Baldock FC, Glanville RJ, Seppanen KT, Gleeson LJ, Selleck PN, Dunn KJ (1996) Investigation of a second focus of equine morbillivirus infection in coastal Queensland. Aust Vet J 74(3):243–244

Selvey LA, Wells RM, McCormack JG, Ansford AJ, Murray K, Rogers RJ, Lavercombe PS, Selleck P, Sheridan JW (1995) Infection of humans and horses by a newly described morbillivirus. Med J Aust 162(12):642–645

Sendow I, Field HE, Curran J, Darminto, Morrissy C, Meehan G, Buick T, Daniels P (2006) Henipavirus in *Pteropus vampyrus* bats Indonesia. Emerg Infect Dis 12(4):711–712

Sendow I, Field HE, Curran J, Darminto, Morrissy C, Daniels P (2010) Screening for Nipah virus infection in west kalimantan province Indonesia. Zoonoses Public Health 57(7–8):499–503

Smith I, Broos A, de Jong C, Zeddeman A, Smith C, Smith G, Moore F, Barr J, Crameri G, Marsh G, Tachedjian M, Yu M, Kung YH, Wang LF, Field H (2011) Identifying hendra virus diversity in pteropid bats. PLoS One 6(9):e25275

Su AI, Wiltshire T, Batalov S, Lapp H, Ching KA, Block D, Zhang J, Soden R, Hayakawa M, Kreiman G, Cooke MP, Walker JR, Hogenesch JB (2004) A gene atlas of the mouse and human protein-encoding transcriptomes. Proc Natl Acad Sci U S A 101(16):6062–6067

Tan CT, Wong KT (2003) Nipah encephalitis outbreak in Malaysia. Ann Acad Med Singapore 32(1):112–117

Tan CT, Goh KJ, Wong KT, Sarji SA, Chua KB, Chew NK, Murugasu P, Loh YL, Chong HT, Tan KS, Thayaparan T, Kumar S, Jusoh MR (2002) Relapsed and late-onset Nipah encephalitis. Ann Neurol 51(6):703–708

Wacharapluesadee S, Lumlertdacha B, Boongird K, Wanghongsa S, Chanhome L, Rollin P, Stockton P, Rupprecht CE, Ksiazek TG, Hemachudha T (2005) Bat Nipah virus Thailand. Emerg Infect Dis 11(12):1949–1951

Weingartl H, Czub S, Copps J, Berhane Y, Middleton D, Marszal P, Gren J, Smith G, Ganske S, Manning L, Czub M (2005) Invasion of the central nervous system in a porcine host by Nipah virus. J Virol 79(12):7528–7534

Weingartl HM, Berhane Y, Caswell JL, Loosmore S, Audonnet JC, Roth JA, Czub M (2006) Recombinant Nipah virus vaccines protect pigs against challenge. J Virol 80(16):7929–7938

Weingartl HM, Berhane Y, Czub M (2009) Animal models of henipavirus infection: a review. Vet J 181(3):211–220

Westbury HA, Hooper PT, Selleck PW, Murray PK (1995) Equine morbillivirus pneumonia: susceptibility of laboratory animals to the virus. Aust Vet J 72(7):278–279

Westbury HA, Hooper PT, Brouwer SL, Selleck PW (1996) Susceptibility of cats to equine morbillivirus. Aust Vet J 74(2):132–134

White JR, Boyd V, Crameri GS, Duch CJ, van Laar RK, Wang LF, Eaton BT (2005) Location of, immunogenicity of and relationships between neutralization epitopes on the attachment protein (G) of hendra virus. J Gen Virol 86(Pt 10):2839–2848

Wiley DC, Skehel JJ (1987) The structure and function of the hemagglutinin membrane glycoprotein of influenza virus. Ann. Rev. Biochem. 56:365–394

Williamson MM, Hooper PT, Selleck PW, Gleeson LJ, Daniels PW, Westbury HA, Murray PK (1998) Transmission studies of hendra virus (equine morbillivirus) in fruit bats, horses and cats. Aust Vet J 76(12):813–818

Wong KT (2010) Emerging epidemic viral encephalitides with a special focus on henipaviruses. Acta Neuropathol 120(3):317–325

Wong KT, Shieh WJ, Kumar S, Norain K, Abdullah W, Guarner J, Goldsmith CS, Chua KB, Lam SK, Tan CT, Goh KJ, Chong HT, Jusoh R, Rollin PE, Ksiazek TG, Zaki SR (2002) Nipah virus infection: pathology and pathogenesis of an emerging paramyxoviral zoonosis. Am J Pathol 161(6):2153–2167

Wong KT, Grosjean I, Brisson C, Blanquier B, Fevre-Montange M, Bernard A, Loth P, Georges-Courbot MC, Chevallier M, Akaoka H, Marianneau P, Lam SK, Wild TF, Deubel V (2003) A golden hamster model for human acute Nipah virus infection. Am J Pathol 163(5):2127–2137

Wong KT, Robertson T, Ong BB, Chong JW, Yaiw KC, Wang LF, Ansford AJ, Tannenberg A (2009) Human hendra virus infection causes acute and relapsing encephalitis. Neuropathol Appl Neurobiol 35(3):296–305

Wright A, Shin SU, Morrison SL (1992) Genetically engineered antibodies: progress and prospects. Crit Rev Immunol 12(3–4):125–168

Xu K, Rajashankar KR, Chan YP, Himanen JP, Broder CC, Nikolov DB (2008) Host cell recognition by the henipaviruses: crystal structures of the Nipah G attachment glycoprotein and its complex with ephrin-B3. Proc Natl Acad Sci U S A 105(29):9953–9958

Xu K, Khetawat D, Rajashankar KR, Chan YP, Kolev MV, Broder CC, Nikolov DB (2012) Crystal structures of the hendra virus G glycoprotein and its complex with ephrin-B2 reveal new insights into the virus attachment and entry process submitted (in press)

Yob JM, Field H, Rashdi AM, Morrissy C, van der Heide B, Rota P, bin Adzhar A, White J, Daniels P, Jamaluddin A, Ksiazek T (2001) Nipah virus infection in bats (order Chiroptera) in peninsular Malaysia. Emerg Infect Dis 7(3):439–441

Yuan P, Swanson KA, Leser GP, Paterson RG, Lamb RA, Jardetzky TS (2011) Structure of the newcastle disease virus hemagglutinin-neuraminidase (HN) ectodomain reveals a four-helix bundle stalk. Proc Natl Acad Sci U S A 108(36):14920–14925

Zhu Z, Dimitrov AS, Bossart KN, Crameri G, Bishop KA, Choudhry V, Mungall BA, Feng YR, Choudhary A, Zhang MY, Feng Y, Wang LF, Xiao X, Eaton BT, Broder CC, Dimitrov DS (2006) Potent neutralization of Hendra and Nipah viruses by human monoclonal antibodies. J Virol 80(2):891–899

Zhu Z, Bossart KN, Bishop KA, Crameri G, Dimitrov AS, McEachern JA, Feng Y, Middleton D, Wang LF, Broder CC, Dimitrov DS (2008) Exceptionally potent cross-reactive neutralization of Nipah and Hendra viruses by a human monoclonal antibody. J Infect Dis 197(6):846–853

Index

Current Topics in Microbiology and Immunology (2012) 359: 225–226
DOI: 10.1007/978-3-642-29819-6

The manufacturer's authorised representative in the EU is Springer Nature Customer Service Centre GmbH, Europaplatz 3, 69115 Heidelberg, Germany. If you have any concerns regarding our products, please contact ProductSafety@springernature.com

Printed and bound by CPI Group (UK) Ltd, Croydon, CR0 4YY
30/09/2026
02225689-0008